Methods in Molecular Biology

Series Editor
John M. Walker
School of Life and Medical Sciences
University of Hertfordshire
Hatfield, Hertfordshire, AL10 9AB, UK

For further volumes:
http://www.springer.com/series/7651

Microarray Technology

Methods and Applications

Edited by

Paul C.H. Li

Department of Chemistry, Simon Fraser University, Burnaby, BC, Canada

Abootaleb Sedighi

Department of Chemistry, Simon Fraser University, Burnaby, BC, Canada

Lin Wang

Department of Biomedical Engineering, University of British Columbia, Vancouver, BC, Canada

Editors
Paul C.H. Li
Department of Chemistry
Simon Fraser University
Burnaby, BC, Canada

Abootaleb Sedighi
Department of Chemistry
Simon Fraser University
Burnaby, BC, Canada

Lin Wang
Department of Biomedical Engineering
University of British Columbia
Vancouver, BC, Canada

ISSN 1064-3745 ISSN 1940-6029 (electronic)
Methods in Molecular Biology
ISBN 978-1-4939-3135-4 ISBN 978-1-4939-3136-1 (eBook)
DOI 10.1007/978-1-4939-3136-1

Library of Congress Control Number: 2015952014

Springer New York Heidelberg Dordrecht London

Printed on acid-free paper

Humana Press is a brand of Springer
Springer Science+Business Media LLC New York is part of Springer Science+Business Media (www.springer.com)

Dedication

Dedicated to Cindy Woo, Paul Chan who have encouraged and supported my quest for analytical tools to fight cancer, in Australia and beyond.

Preface

The microarray method has been a widely applied technology. There are considerable amounts of technical and scientific literature published recently. The aim of this book is to provide updates of this established technology in topics such as methods and applications, as well as advances in applications of the microarray method to assay substances such as proteins, glycans, and whole cells. The 19 chapters, which are written by recognized experts in each topic, include reviews, detailed protocols and outlook of the nucleic acid microarrays and the cell/tissue microarrays.

Burnaby, BC, Canada — *Paul C.H. Li, Ph.D.*
Vancouver, BC, Canada — *Lin Wang, Ph.D.*
Burnaby, BC, Canada — *Abootaleb Sedighi, Ph.D.*

Contents

Contributors

TOMOHIRO ARITA • *Division of Digestive Surgery, Department of Surgery, Kyoto Prefectural University of Medicine, Kyoto, Japan*

MARÍA-JOSÉ BAÑULS • *Departamento de Química, Instituto interuniversitario de Reconocimiento Molecular y Desarrollo Tecnológico, Universitat Politècnica de València, Valencia, Spain*

JUAN CASADO-VELA • *Spanish National Research Council (CSIC)—Spanish National Biotechnology Centre (CNB), Madrid, Spain*

ZHIFANG CHAI • *CAS Key Laboratory for Biomedical Effects of Nanomaterials and Nanosafety, CAS Key Lab of Nuclear Radiation and Nuclear Energy Technology, Center for Multidisciplinary Research, Institute of High Energy Physics, Chinese Academy of Sciences (CAS), Beijing, China*

VERENA CHARWAT • *Institute of Applied Microbiology, University of Natural Resources and Life Sciences Vienna, Vienna, Austria*

JUN CHEN • *CAS Key Laboratory for Biomedical Effects of Nanomaterials and Nanosafety, CAS Key Lab of Nuclear Radiation and Nuclear Energy Technology, Center for Multidisciplinary Research, Institute of High Energy Physics, Chinese Academy of Sciences (CAS), Beijing, China*

SAU W. CHEUNG • *Department of Molecular and Human Genetics, Baylor College of Medicine—Medical Genetics Laboratories, Houston, TX, USA*

MARCIN K. CHMIELEWSKI • *Institute of Bioorganic Chemistry, Polish Academy of Sciences, Poznan, Poland*

FRÉDÉRIQUE DEISS • *Department of Chemistry, University of Alberta, Edmonton, AB, Canada*

JÉRÉMIE DENONFOUX • *Genomic Platform and R&D, Genoscreen, Campus de l'Institut Pasteur, Lille, France*

RATMIR DERDA • *Department of Chemistry, University of Alberta, Edmonton, AB, Canada*

C. DESMET • *Equipe Génie Enzymatique, Membranes Biomimétiques et Assemblages Supramoléculaires, Institut de Chimie et Biochimie Moléculaires et Supramoléculaires, Université Lyon, Lyon, France*

PAULA DÍEZ • *Proteomics Unit, Department of Medicine and General Cytometry Service-Nucleus, Cancer Research Center/IBMCC (USAL/CSIC)—IBSAL, University of Salamanca, Salamanca, Spain*

ERIC DUGAT-BONY • *Génie et Microbiologie des Procédés Alimentaires, Centre de Biotechnologies Agro-Industrielles, INRA, AgroParisTech, UMR 782, Thiverval-Grignon, France*

PETER ERTL • *BioSensor Technologies, AIT Austrian Institute of Technology GmbH, Vienna, Austria*

MARIA CHIARA FONTANA • *Department of Experimental, Diagnostic and Specialty Medicine, Institute of Hematology "L. and A. Seràgnoli", University of Bologna, Bologna, Italy*

EMILIA FRYDRYCH-TOMCZAK • *Poznan Science and Technology Park, Adam Mickiewicz University Foundation, Poznan, Poland*
MANUEL FUENTES • *Proteomics Unit, Department of Medicine and General Cytometry Service-Nucleus, Cancer Research Center/IBMCC (USAL/CSIC)—IBSAL, University of Salamanca, Salamanca, Spain*
TAKASHI FUNATSU • *Graduate School of Pharmaceutical Sciences, The University of Tokyo, Tokyo, Japan*
LIQIAN GAO • *Institute of Bioengineering and Nanotechnology, Singapore, Singapore; Department of Chemistry, National University of Singapore, Singapore, Singapore*
SEBASTIAN HENKEL • *Medizinisches Proteom-Center, Ruhr-Universität Bochum, Bochum, Germany*
JUN HIRABAYASHI • *Department of Functional Glycomics, Life Science Research Center, Kagawa University, Kagawa, Japan*
WEIHUA HU • *Institute for Clean Energy and Advanced Materials, Southwest University, Chongqing, People's Republic of China*
YI HU • *CAS Key Laboratory for Biomedical Effects of Nanomaterials and Nanosafety, CAS Key Lab of Nuclear Radiation and Nuclear Energy Technology, Center for Multidisciplinary Research, Institute of High Energy Physics, Chinese Academy of Sciences (CAS), Beijing, China*
ILARIA IACOBUCCI • *Department of Experimental, Diagnostic and Specialty Medicine, Institute of Hematology "L. and A. Seràgnoli", University of Bologna, Bologna, Italy*
DAISUKE ICHIKAWA • *Division of Digestive Surgery, Department of Surgery, Kyoto Prefectural University of Medicine, Kyoto, Japan*
RYO IIZUKA • *Graduate School of Pharmaceutical Sciences, The University of Tokyo, Tokyo, Japan*
HIROTAKA KONISHI • *Division of Digestive Surgery, Department of Surgery, Kyoto Prefectural University of Medicine, Kyoto, Japan*
SU SEONG LEE • *Institute of Bioengineering and Nanotechnology, Singapore, Singapore*
CHANGMING LI • *Institute for Clean Energy and Advanced Materials, Southwest University, Chongqing, People's Republic of China*
PAUL C.H. LI • *Department of Chemistry, Simon Fraser University, Burnaby, BC, Canada*
PEIWU LI • *Oil Crops Research Institute of the Chinese Academy of Agricultural Sciences, Wuhan, People's Republic of China*
ANNALISA LONETTI • *Department of Biomedical and Neuromotor Sciences, University of Bologna, Bologna, Italy*
ANTOINE MAHUL • *CRRI, Clermont Université, Aubière, France*
ÁNGEL MAQUIEIRA • *Departamento de Química, Instituto interuniversitario de Reconocimiento Molecular y Desarrollo Tecnológico, Universitat Politècnica de València, Valencia, Spain*
KATRIN MARCUS • *Medizinisches Proteom-Center, Ruhr-Universität Bochum, Bochum, Germany*
C.A. MARQUETTE • *Equipe Génie Enzymatique, Membranes Biomimétiques et Assemblages Supramoléculaires, Institut de Chimie et Biochimie Moléculaires et Supramoléculaires, Université Lyon, Lyon, France*

GIOVANNI MARTINELLI • *Department of Experimental, Diagnostic and Specialty Medicine, Institute of Hematology "L. and A. Seràgnoli", University of Bologna, Bologna, Italy*

MAHSA GHARIBI MARZANCOLA • *Department of Chemistry, Simon Fraser University, Burnaby, BC, Canada*

CAROLINE MAY • *Medizinisches Proteom-Center, Ruhr-Universität Bochum, Bochum, Germany*

SERGI B. MORAIS • *Departamento de Química, Instituto interuniversitario de Reconocimiento Molecular y Desarrollo Tecnológico, Universitat Politècnica de València, Valencia, Spain*

SHIN-ICHI NAKAKITA • *Department of Functional Glycomics, Life Science Research Center, Kagawa University, Kagawa, Japan*

EIGO OTSUJI • *Division of Digestive Surgery, Department of Surgery, Kyoto Prefectural University of Medicine, Kyoto, Japan*

NICOLAS PARISOT • *Université d'Auvergne, EA 4678, CIDAM, Clermont Université, Clermont-Ferrand, France*

ANKITA PATEL • *Department of Molecular and Human Genetics, Baylor College of Medicine—Medical Genetics Laboratories, Houston, TX, USA*

PIERRE PEYRET • *Université d'Auvergne, EA 4678, CIDAM, Clermont Université, Clermont-Ferrand, France*

ERIC PEYRETAILLADE • *Université d'Auvergne, EA 4678, CIDAM, Clermont Université, Clermont-Ferrand, France*

HOLGER PROKISCH • *Institute of Human Genetics, Helmholtz Zentrum München, Munich, Germany; German Research Center for Environmental Health, Neuherberg, Germany; Institute of Human Genetics, Technical University Munich, Munich, Germany*

TOMASZ RATAJCZAK • *Institute of Bioorganic Chemistry, Polish Academy of Sciences, Poznan, Poland*

MARIO ROTHBAUER • *BioSensor Technologies, AIT Austrian Institute of Technology GmbH, Vienna, Austria*

ARNE SCHILLERT • *Institut für Medizinische Biometrie und Statistik, Universitätsklinikum Schleswig-Holstein, Universität zu Lübeck, Lübeck, Germany; DZHK (German Centre for Cardiovascular Research), partner site Hamburg/Kiel/Lübeck, Lübeck, Germany*

KATHARINA SCHRAMM • *Institute of Human Genetics, Helmholtz Zentrum München, Munich, Germany; German Research Center for Environmental Health, Neuherberg, Germany; Institute of Human Genetics, Technical University Munich, Munich, Germany*

CLAUDIA SCHURMANN • *The Charles Bronfman Institute for Personalized Medicine, Icahn School of Medicine at Mount Sinai, New York, NY, USA; The Genetics of Obesity and Related Metabolic Traits Program, Icahn School of Medicine at Mount Sinai, New York, NY, USA; Interfaculty Institute for Genetics and Functional Genomics, University Medicine Greifswald, Greifswald, Germany*

ABOOTALEB SEDIGHI • *Department of Chemistry, Simon Fraser University, Burnaby, BC, Canada*

HONGYAN SUN • *Department of Biology and Chemistry, City University of Hong Kong, Kowloon, Hong Kong, China*

ALEXANDER TEUMER • *Institute for Community Medicine, University Medicine Greifswald, Greifswald, Germany; Interfaculty Institute for Genetics and Functional Genomics, University Medicine Greifswald, Greifswald, Germany*

LUIS A. TORTAJADA-GENARO • *Departamento de Química, Instituto interuniversitario de Reconocimiento Molecular y Desarrollo Tecnológico, Universitat Politècnica de València, Valencia, Spain*

TARO UENO • *Graduate School of Pharmaceutical Sciences, The University of Tokyo, Tokyo, Japan*

BARBARA USZCZYŃSKA • *Centre for Genomic Regulation (CGR), Barcelona, Catalonia, Spain; Universitat Pompeu Fabra (UPF), Barcelona, Spain*

ROBERT WELLHAUSEN • *Fraunhofer Institute for Cell Therapy and Immunology Branch Bioanalytics and Bioprocesses (IZI-BB), Potsdam-Golm, Germany*

DIRK WOITALLA • *St. Josef Hospital Bochum, Bochum, Germany; St. Josef Hospital Kupferdreh, Essen, Germany*

YANG YANG • *Department of Chemistry, University of Alberta, Edmonton, AB, Canada*

QI ZHANG • *Oil Crops Research Institute of the Chinese Academy of Agricultural Sciences, Wuhan, People's Republic of China*

ZHAOWEI ZHANG • *Oil Crops Research Institute of the Chinese Academy of Agricultural Sciences, Wuhan, People's Republic of China*

YULIANG ZHAO • *CAS Key Laboratory for Biomedical Effects of Nanomaterials and Nanosafety, CAS Key Lab of Nuclear Radiation and Nuclear Energy Technology, Center for Multidisciplinary Research, Institute of High Energy Physics, Chinese Academy of Sciences (CAS), Beijing, China*

ANDREAS ZIEGLER • *Institut für Medizinische Biometrie und Statistik, Universität zu Lübeck, Universitätsklinikum Schleswig-Holstein, Campus Lübeck, Lübeck, Germany; Center for Clinical Trials, University of Lübeck, Lübeck, Germany; DZHK (German Centre for Cardiovascular Research), partner site Hamburg/Kiel/Lübeck, Lübeck, Germany; School of Mathematics, Statistics and Computer Science, University of KwaZulu-Natal, Pietermaritzburg, South Africa*

Part I

Introduction

Chapter 1

Overview of Microarray Technology

Paul C.H. Li

Abstract

Microarray technology, with its high-throughput advantage, has been applied to analyze various biomaterials, such as nucleic acids, proteins, glycans, peptides, and cells.

Key words Microarray, Nucleic acids, Proteins, Glycans, Peptides, Cells

In view of next-generation sequencing, the microarray technology has faced strong competition in its applications for nucleic acid assays. Details on limitations of nucleic acid microarrays and on strategies to overcome the limitations are given in Chapter 12. Nevertheless, the microarray has the high-throughput capability that remains to be an advantage. Other than nucleic acid assays [1], microarrays can be applied to analyze proteins, glycans, peptides, and cells, as well as to screen products that bind to these substances arrayed on a solid substrate. Several review articles are in place to summarize such efforts in advancing protein microarrays [2], tissue microarrays [3], glycan microarrays [4], peptide microarrays [5], and small molecule microarrays [5].

This book is divided into three sections, namely methods (Chapters 2–6), applications of nucleic acid microarrays (Chapters 7–12), and applications to other microarray assays (Chapters 13–19). Recent advances in nucleic acid microarrays have also resulted in several products for biomedical applications, namely MammaPrint (for breast cancer), AmpliChip (for drug metabolism), and PailloCheck (for human papillomavirus).

In the method section, great progress has been made in probe immobilization and grafting chemistries, which are described in a review article in Chapter 2. In particular, the "click" chemistry, which is used to anchor nucleic acid probes on azide-functionalized glass chip surface, has been detailed in the first protocol in Chapter 3. The design of oligonucleotide probe sequences has become more systematic now than ever, thanks to the algorithms developed to

Paul C.H. Li et al. (eds.), *Microarray Technology: Methods and Applications*, Methods in Molecular Biology, vol. 1368, DOI 10.1007/978-1-4939-3136-1_1, © Springer Science+Business Media New York 2016

generate probes that are either gene-specific or sequence-specific, see Chapter 6 on probe design strategies. Chapter 4 describes the protocols of creating microarrays on surfaces of plastic especially polycarbonate. In terms of detection methods, Chapter 5 describes the detailed protocol of using ligase-assisted sandwich hybridization (LASH) to detect extracellular microRNAs.

In the section for nucleic acid applications, Chapter 7 describes the protocol for analysis of gene expression microarray data obtained from human whole blood cell samples. Chapter 8 uses the microarrays to detect plasma microRNA biomarkers pertaining to digestive tract cancers. Genomic copy number variations are detectable using the DNA microarray assays conducted on prenatal and postnatal samples, as illustrated in the protocols in Chapter 9. High-throughput DNA microarrays for KRAS mutation detection have been achieved using a centrifugal microfluidic device, as described in detail in Chapter 10. Analysis of single nucleotide polymorphisms (SNP) has been applied for pharmacogenomic studies, as depicted in the protocols in Chapter 11. A review article on DNA microarray-based diagnostics concludes this section on nucleic acid microarrays, see Chapter 12.

Microarrays can also be applied to detect biomaterials other than DNAs and RNAs. This section starts to include the first protocol (Chapter 13) to screen protein tyrosine phosphatases on the phosphopeptide array. Chapter 14 describes the nanotechnology employed to construct protein microarrays. The peptide microarray has been used for epitope mapping of autoantibodies, as detailed in Chapter 15. This is followed by Chapter 16 which describes the preparation of glycan microarrays used to detect the binding with lectins. Chapter 17 depicts the protocol for immunoassays based on an antigen microarray. Construction of the peptide microarray on paper has been described in the detailed protocol in Chapter 18, intended for ligand binding assays. This section is concluded with a review chapter on live-cell microarrays aiming for high-throughput biological assays.

References

1. Nsofor CA (2014) DNA microarrays and their applications in medical microbiology. Biotechnol Mol Biol Rev 9(1):1–11
2. Chandra H, Reddy PJ, Srivastava S (2011) Protein microarrays and novel detection platforms. Expert Rev Proteomics 8(1):61–79
3. La Spada A et al (2014) Application of tissue microarray technology to stem cell research. Microarrays 3(3):159–167
4. Arthur CM, Cummings RD, Stowell SR (2014) Using glycan microarrays to understand immunity. Curr Opin Chem Biol 18: 55–61
5. Foong YM et al (2012) Current advances in peptide and small molecule microarray technologies. Curr Opin Chem Biol 16(1): 234–242

Part II

Advances on Techniques in Microarray Technology

Chapter 2

Surface Functionalization for Immobilization of Probes on Microarrays

C. Desmet and C.A. Marquette

Abstract

The microarray technology has been a tremendous advance in molecular-based testing methods for biochemical and biomedical applications. As a result, the immobilization techniques and grafting chemistries of biochemical molecules have experienced great progress. The particularities of the grafting techniques adapted to the microarray development will be presented here.

Key words Biomolecules, Grafting, Immobilization, Microarray, Surface chemistry

1 Introduction

To manufacture biochips, the selection of suitable probes and their immobilization on a solid support are required. The immobilization will differ according to the selected probes (i.e., oligonucleotides, protein, cells, or small molecules) and the types of detection systems (i.e., optical, or electrochemical). This chapter presents an overview of the different approaches and techniques for each combination of probe/detection/solid support.

Different deposition techniques are currently available to create matrices of probes. All chip manufacturing techniques, largely robotized, target the same goal: (a) to deposit small amounts of probe molecules uniformly on the chip surface, (b) to minimize the consumption of costly probe solutions, and (c) to avoid any risk of contamination between the probes. Two deposition modes, which are "contact" and "non-contact" [1], are described in the following paragraphs.

Paul C.H. Li et al. (eds.), *Microarray Technology: Methods and Applications*, Methods in Molecular Biology, vol. 1368, DOI 10.1007/978-1-4939-3136-1_2, © Springer Science+Business Media New York 2016

2 Contact Printing

Contact methods are used to produce matrices through direct contact between the deposition tool and the substrate of the chip. These technologies are based on the use of tips (solid or hollow) or micro-stamps for depositing probe solutions in a row.

2.1 Printing with Tips

The sampling is performed by first dipping a printing tip or a set of tips in the probe solutions. When solid tips are used, it is necessary to reload them for each new deposition. On the other hand, hollow tips are used to perform multiple depositions with the same probe solutions which are loaded into the tips by capillary action, and then deposition of the probes is achieved by contacting the tips on the support. This operation can be repeated several times until the solutions are depleted. Then, the tips are cleaned and new probe solutions are loaded. The diameter of the spots obtained using this contact printing method is directly related to the properties of the surface, the probe solution viscosity, and the geometry of the tips. The main advantage of this technique is its simplicity. The main drawbacks are contamination and tip or chip surface damage.

2.1.1 Printing with Micro-stamps

In this method, a probe sample is first adsorbed on the surface of a micro-stamp and then transferred to the chip surface by physical contact. Elastomer materials, such as PDMS (polydimethylsiloxane), enable a good contact with the chip surface and are generally used for the micro-stamps. The fabrication of PDMS stamps is simple and low cost, is based on the principle of soft lithography developed by Whiteside et al. [2], and has been widely used for different applications. This stamping method has been applied for the contact printing of proteins on glass and PDMS substrates [3]. Thanks to the use of micro-stamps, the deposition of hundreds of spots in parallel on the microarrays can be achieved for high-throughput applications.

2.1.2 Printing with Nano-tip

The dip-pen nanolithography, described in 1999 by Piner et al., is another probe deposition method [4]. The atomic force microscope (AFM) tip is used, and the tip which contains the probe solution to deposit, is employed to draw nanopatterns on a support. In the study by Piner et al., thiolated probes were deposited and immobilized on a gold substrate in a matrix of 25 spots of 0.46 μm diameter and 0.54 μm apart. This method achieves a higher deposition resolution than the one obtained with photolithographical methods and a similar solution to more expensive techniques such as electron-beam lithography [5].

2.2 Non-contact Printing

2.2.1 Electrospray

Electrospray is a non-contact deposition technique. As soon as a strong electric field is applied to the probe solution located at the printing tip, micro-droplets are formed and the solution dries instantly. The electrospray method is believed to help preserve the structure of most proteins since the printed proteins are lyophilized on the surface. Avseenko et al. have used this technique to deposit a matrix of antigens onto an aluminum support using different surface modifications for the detection of antibodies by immunoassay [6].

2.2.2 Thermic Printing

This method can be applied for the deposition of sub-nanoliter size drops (typically 100–300 picoliters) on the surface of the chip without any contact with the printing tip. The ejection of the droplet is based on inkjet printing, in which the solution is heated to a temperature of around 200 °C, allowing the evaporation of a small volume of "ink" of the probes, which leads to the ejection of a drop from the tip. This technique is flexible and highly reproducible, and the printed spots have good uniformity. Contamination is prevented by the absence of contact between the tip and the chip surface, and tip damage on the chip surfaces which are not necessarily perfectly flat is also avoided.

2.2.3 Piezoelectric Printing

This method is similar to thermic printing except that the ejection of the drop is achieved by the vibration of a piezoelectric membrane at the printing tip. This second method, which is usually preferred over the thermic method, avoids overheating of the probe solution.

2.3 Microbead Array

Because of the large specific accessible surface on microbeads, they allow a high density of probes immobilized and provide a high sensitivity of the tests. The microspheres, after functionalization with the probe molecules, may be detected by different approaches.

2.3.1 Microbeads in Suspension

The microbeads used in suspension are optically encoded microspheres, which carry specific receptors to bind with probes that can capture targets in a liquid sample (Fig. 1). The distinct optical properties of the targets and the microspheres allow their multiplex identification by flow cytometry [7]. Technologies such as *x-MAP* from LUMINEX are operated on this principle. The nanobarcode particles [8] also exploit the principle of optical encoding, and the multiplex analysis is performed through microscopic observation. The microbead technology has the advantage of generating thousands of unique probes to react with targets in the same sample.

2.3.2 Microbeads Immobilized on Structured Surface

Another approach of the use of microspheres is to immobilize them on a structured matrix. Marquette et al. proposed the coupling of various types of probes (e.g., proteins or oligonucleotides) with sepharose or latex microbeads immobilized in a structured matrix of PDMS [9]. Walt et al. described a method that the

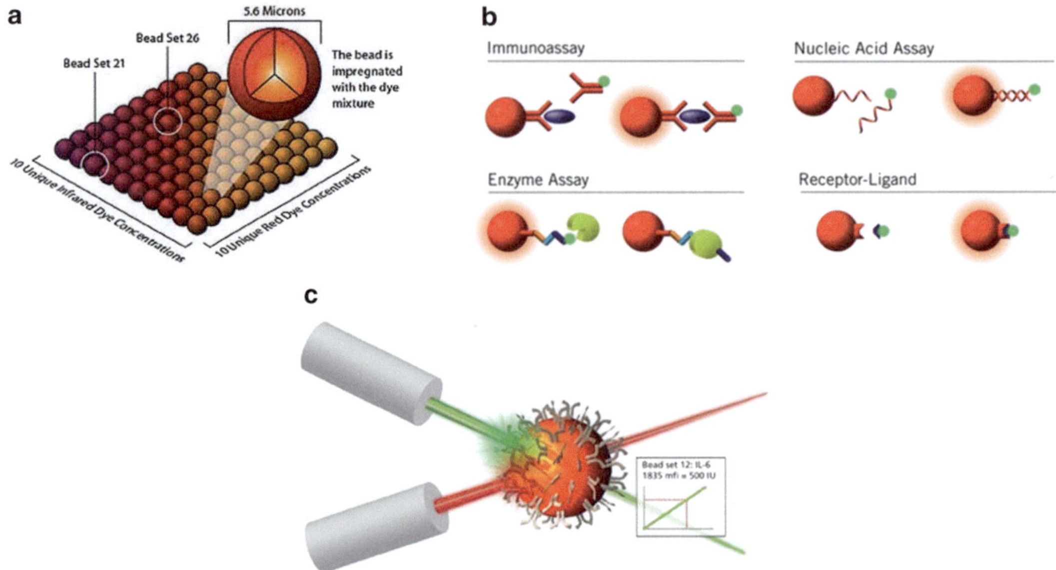

Fig. 1 Principle of the LUMINEX-type assay using optically encoded microspheres in suspension. (**a**) microspheres encoded by a mixture of red and infra-red dyes, (**b**) different types of binding assays on the beads, (**c**) detection of the microspheres and the binding events using 2-color flow cytometry for immunoassay of IL-6. *MFI* median fluorescence intensity, *IU* international unit (reproduced from http://www.ebioscience.com/)

microspheres functionalized with DNA strands are immobilized at the end surface of the optical fiber bundles. This method allowed multiplex target detection by fluorescence measurement [10, 11]. The company ILLUMINA commercializes DNA biochips based on this model called the *BeadArray*. These matrices are composed of silicon beads of 3 μm in diameter, self-assembled in microwells pre-etched on the surface of optical fibers or silica slides (Fig. 2).

2.3.3 Microbeads Immobilized Through Magnetic Control

Magnetic microspheres are also widely used, enabling the generation of arrays of probes assembled through magnetic control. Yu et al. developed this kind of assay with a magnetic field gradient to assemble magnetic microspheres on a nickel substrate for detection of cancer markers on a microfluidic chip [12]. This approach can reach a high detection sensitivity and specificity because any nonspecifically adsorbed contaminants can be eliminated, through the continuous liquid flow in the microfluidic chip, while the probes strongly bound by the magnetic field.

3 Immobilization of Probes

Probe immobilization on the chip is a critical step on which the quality of the test will depend, since it lies on the successful interactions between a target and its probe immobilized on the microarray.

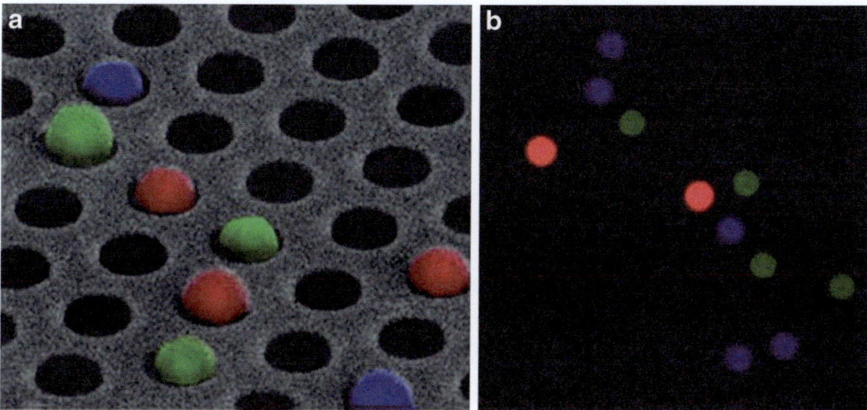

Fig. 2 Principle of the ILLUMINA-type assay using optically encoded microspheres self-assembled on a solid surface matrix. (**a**) false-color electron microscopic image, (**b**) fluorescent image of the chip showing the easy tracking of the different beads on the surface (reproduced from Illumina®)

For this purpose, different conditions have to be taken in account: (a) the probes have to be accessible to the targets and (b) the structure and function of the probes have to be preserved. Immobilization strategies specific to the solid support or to the immobilized probe will be described.

3.1 Adsorption

Probe immobilization by adsorption involves electrostatic, polar, and hydrophobic interactions, as well as Van der Waals interactions and hydrogen bonds, between the probe molecule and the substrate surface. The types of interactions actually involved depend on the surface chosen. Immobilization by adsorption is simple, fast, generally nondenaturing (except in the case of hydrophobic interaction), and it allows for noncovalent attachment of proteins and other small-molecule probes to the substrate. This immobilization strategy generates a heterogeneous layer of non-oriented probes because many contacts at different orientations can be made by the probe molecules with the substrate.

Random physisorption of probe molecules on the hydrophobic polystyrene surface of microplates is by far the most widely used method for immobilization of proteins, as typically implemented in ELISA. Adsorption strategies are also conventionally used with 3-dimensional substrate. For example, a polypropylene membrane modified with polyaniline, which permits protein adhesion due to electrostatic and hydrophobic interactions, has demonstrated high affinity for, and compatibility with, different proteins [13]. Nitrocellulose membranes are also widely used. The adsorption of proteins on this membrane surface is well known because of its use in Western blot and diagnostic devices. The interactions involved in nitrocellulose membrane are predominantly hydrophobic but also electrostatic [14].

The main disadvantage of the adsorption strategy is the risk of desorption due to changes of pH, temperature, and ionic strength, and this risk limits the storage and stability of microarrays. In addition, the risk of nonspecific adsorption can be high and it is necessary to treat the chip surface with blocking agents (usually a solution saturated with inactive proteins and nondenaturing surfactants) in order to obtain the best signal-to-background ratio.

3.2 Inclusion

The inclusion method consists of trapping the probe molecules in an insoluble gel such as polyacrylamide or agarose. This method preserves the conformation and activity of the protein probes. The probe is mechanically retained in the polymer matrix. The matrix should have an optimal mesh size that provides the retention of the probes, while allows the diffusion of the targets and reagents. However, low molecular weight proteins, and more generally small probe molecules, cannot be retained by loose meshes. A recent development of this method is the functionalization of the 3-dimensional matrices by chemical groups that can attach protein probes by creating covalent bonds [15]. Inclusion can also be a co-reticulation process of the probe in a matrix consisting of other molecules such as bovine serum albumin (BSA) or polyethylene glycol [16].

3.3 Electro-polymerization

In this method, a conductive polymer is formed electrochemically. The probe molecules are immobilized either by their mechanical trapping during the formation of the polymer or by co-polymerization of a monomer mixed with the probes. Shi et al. applied this method to prepare an array of antibodies immobilized within polypyrrole. They were able to achieve a sensitive and specific detection of different antigens in serum samples [17]. Arrays of screen-printed electrodes can also be used as active support for electropolymerization, as described for the immobilization of a capsid protein of HIV-1 virus [18]. This protein was pre-immobilized on a first polymer, then co-electropolymerized in polypyrrole in a localized way on the surface of microelectrodes. This protocol allowed for the detection of specific HIV antibodies in the sera of patients to determine whether they are HIV positive or HIV negative.

3.4 Affinity Immobilization

This probe immobilization technique involves specific affinity interactions that occur between the molecules previously fixed to a support and the probe molecules to be immobilized. This technique provides an oriented immobilization of the probe molecules. The most widely used methods of affinity immobilization will be described in this section.

3.4.1 Avidin/Biotin Interaction

This biotin/avidin interaction is widely used for affinity immobilization since the affinity of the couple is very high ($K_D = 10^{-15}$ M). Avidin is a soluble glycoprotein, stable in aqueous solutions over a

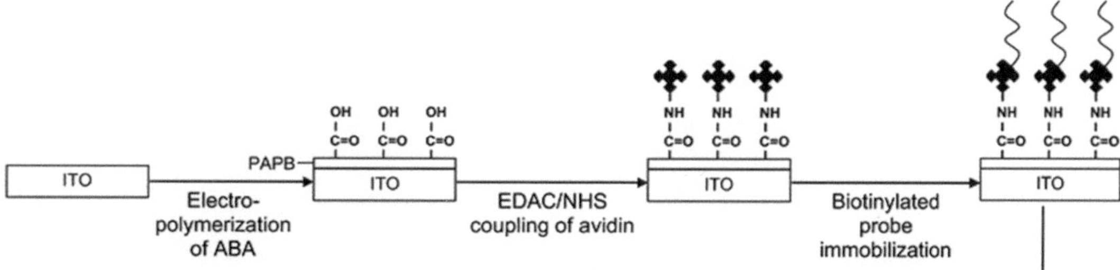

Fig. 3 Example of the use of avidin for the immobilization of probe on surface. *ITO* indium tin oxide, *ABA* 2-aminobenzoic acid, *PABA* poly(2-aminobenzoic acid), *EDAC* 1-ethyl-3-(3-dimethylaminopropyl)carbodiimide, *NHS N*-hydroxysuccinimide. Reproduced from Ref. 21 with permission of Wiley

wide range of pH values and temperatures, and streptavidin is a tetrameric protein very similar to avidin. The conjugation of a biotin molecule to a protein (biotinylation) allows for its immobilization on a support that has been functionalized with streptavidin or avidin (Fig. 3).

The carbodiimide chemistry in the presence of *N*-hydroxysuccinimide (NHS) is the most widely used method to conjugate the carboxyl group of the side chain of biotin to an amine group of the protein. Moreover, a hydrazide group can also be generated on biotin for it to bind to the carboxyl group of a protein. Since biotin, or vitamin H, is a small molecule (molecular weight: 244 g/mol), its coupling with macromolecules affects neither their conformation nor their function [19].

This biotin–avidin interaction may also be used for immobilization of larger particles, such as polystyrene microspheres of 1 μm in diameter, as has been described by Andersson in 2002. The beads, as the immobilization support for probe molecules, which are labeled with streptavidin, are immobilized on a surface functionalized with biotin [20].

3.4.2 Polyhistidine Tag

An alternative to immobilize a protein is to prepare a fusion protein that contains a polyhistidine tag (Fig. 4). This tag, with hexahistidine (6-His) being the most popular, is placed at a defined position on the protein, preferably away from the active site of recognition, so as to avoid any disruption of the affinity interaction with a ligand. The 6-His tag has several advantages such as (a) small size, (b) compatibility with organic solvents, (c) low immunogenicity, and (d) effectiveness of purification processes in native or denaturing conditions. Proteins having a 6-His tag at their N- or C-terminus can thus be immobilized through the formation of complexes with nickel ions, using surfaces functionalized by nitrilotriacetic acid (NTA) loaded with divalent metal ions, generally Ni^{2+}. The chelation between Ni-NTA and 6-His tag occurs through the involvement of the Ni^{2+} ion in an octahedral coordination, with

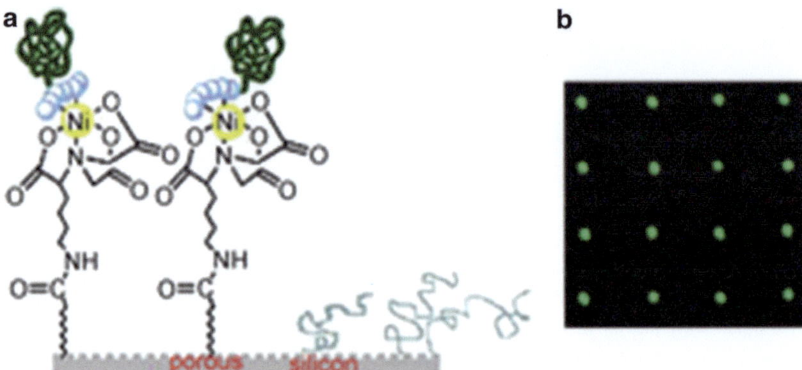

Fig. 4 Immobilization of the 6-His tagged protein on the Ni-NTA-functionalized silicon surface. (**a**) two 6-His tagged proteins grafted on the Ni-NTA surface, (**b**) fluorescent image of the FITC-labeled aprotinin grafted on a microarray surface (reproduced from Ref. 22 with permission from Springer)

two imidazole groups of the 6-His tag and three carboxylate group and one amine group of NTA, as illustrated in Fig. 6. The manufacturing of microarrays with functionalized NTA can be carried out by different strategies, most often by covalent grafting. The advantage of this polyhistidine method is its high specificity, and also its reversibility. Indeed, the release of 6-His tags may be achieved by the addition of a competitive ligand such as imidazole, or a complexing agent such as EDTA. This reversibility property is an important advantage, since the microarray chips can be reused and modified again. Moreover, being widely used, 6-His tag is commercially available for functionalization of proteins. However, this method has some drawbacks, such as nonspecific metal-dependent surface adsorption of proteins and dissociation of proteins due to a relatively low affinity of the 6-His tag to the Ni-NTA complex ($K_D = 10^{-6}$ M).

3.4.3 Protein–Oligonucleotide Conjugate

Hybridization of complementary oligonucleotides provides an immobilization solution for different types of proteins, thanks to the very high stability of oligonucleotides and the unique selectivity of base-pairing. The production of microarrays using this type of interaction is based on the established technology of DNA chips, in which oligonucleotide probes are arrayed on the solid support. The protein probe molecules are conjugated with the oligonucleotides complementary to those present on the surface, and incubated on the support for protein immobilization. This approach has recently been developed and the complications associated with the grafting of an oligonucleotide on the proteins have been raised. To resolve these complications, different strategies have been developed, such as the use of semi-synthetic conjugates of oligonucleotides and proteins [23].

3.4.4 Protein A and G

The immobilization of antibodies can be performed using the specific interaction between the Fc region of immunoglobulin G (IgG) and Protein A (or Protein G). Such immobilization ensures the full accessibility of the antibody recognition site, being located on the Fab region, for interaction with the antigens to be detected. An underlying difficulty is the need to orient Protein A in order to allow for its optimal interaction with IgG. To resolve this difficulty, different approaches have been presented, such as the use of Protein A synthesized with a 6-histidine tag at its C-terminus by Johnson et al., to allow for oriented immobilization of Protein A on a surface functionalized by Ni-NTA [24]. Various examples of using the Protein A affinity interaction for immobilization of immunoglobulins of different subtypes and origins are given in Table 1. It is also worth noting that the technique is not to be limited to any antibody.

4 Covalent Immobilization

This immobilization strategy is based on the formation of covalent bonds, which provide a highly stable binding of the probe molecules to the support. This method generally requires the functionalization of the solid support, generating reactive functional groups on the surface, enabling the formation of covalent bonds between the functional groups and the probes. The protein probe may be immobilized through a reactive group of the amino acids side chain. The reactive groups available depend on the types of amino acids present on the surface of the protein, and the covalent bond formation requires different functionalization of the substrate, as summarized in Table 2. Different covalent immobilization approaches will be described here.

4.1 Classical Covalent Bonding

4.1.1 Amine Chemistry

The most commonly used immobilization reaction involves the amine group of the lysine side chain, which represents about 10 % of the amino acid residues of a protein. Conventionally, this reaction uses substrates that contain carboxylic acid groups activated using carbodiimide in the presence of *N*-hydroxysuccinimide, and generates stable amide bonds. The extent of immobilization depends on pH, protein concentration, ionic strength, and reaction time, and these conditions should be optimized for each protein coupling reaction. However, the resulted immobilized proteins are non-oriented and heterogeneous. Immobilization via carbodiimide reaction on self-assembled monolayers (SAM) was also described. For instance, Patel et al. generated SAM of 3-mercaptopropionic acid and 11-mercaptoundecanoic acid on gold surfaces for the immobilization of an enzyme [26].

Table 1
Relative affinity of protein A for antibodies with various immunoglobulin subtypes in different species

	Protein A
Human IgG1	++++
Human IgG2	++++
Human IgG3	–
Human IgG4	++++
Human IgGA	++
Human IgGD	++
Human IgGE	++
Human IgGM	++
Mouse IgG1	+
Mouse IgG2a	++++
Mouse IgG2b	+++
Mouse IgG3	++
Mouse IgGM	+/–
Rat IgG1	–
Rat IgG2a	–
Rat IgG2b	–
Rat IgG2c	+
Rat IgGM	+/–
Rabbit Ig	++++
Hamster Ig	+
Guinea Pig Ig	++++
Bovine Ig	++
Sheep Ig	+/–
Goat Ig	+/–
Pig Ig	+++
Chicken Ig	–

(–) represents quasi-nul affinity while (++++) represent ultra-high affinity

Another strategy for the production of protein chips, reported for the first time by MacBeath and Schreiber and currently widely used, consists of the use of glass slides functionalized with aldehyde groups [27]. The aldehyde group reacts with the amine group of proteins via the formation of a Schiff base (imine).

Table 2
Functional groups of the proteins involved in the formation of covalent bond and the associated surface reactive groups

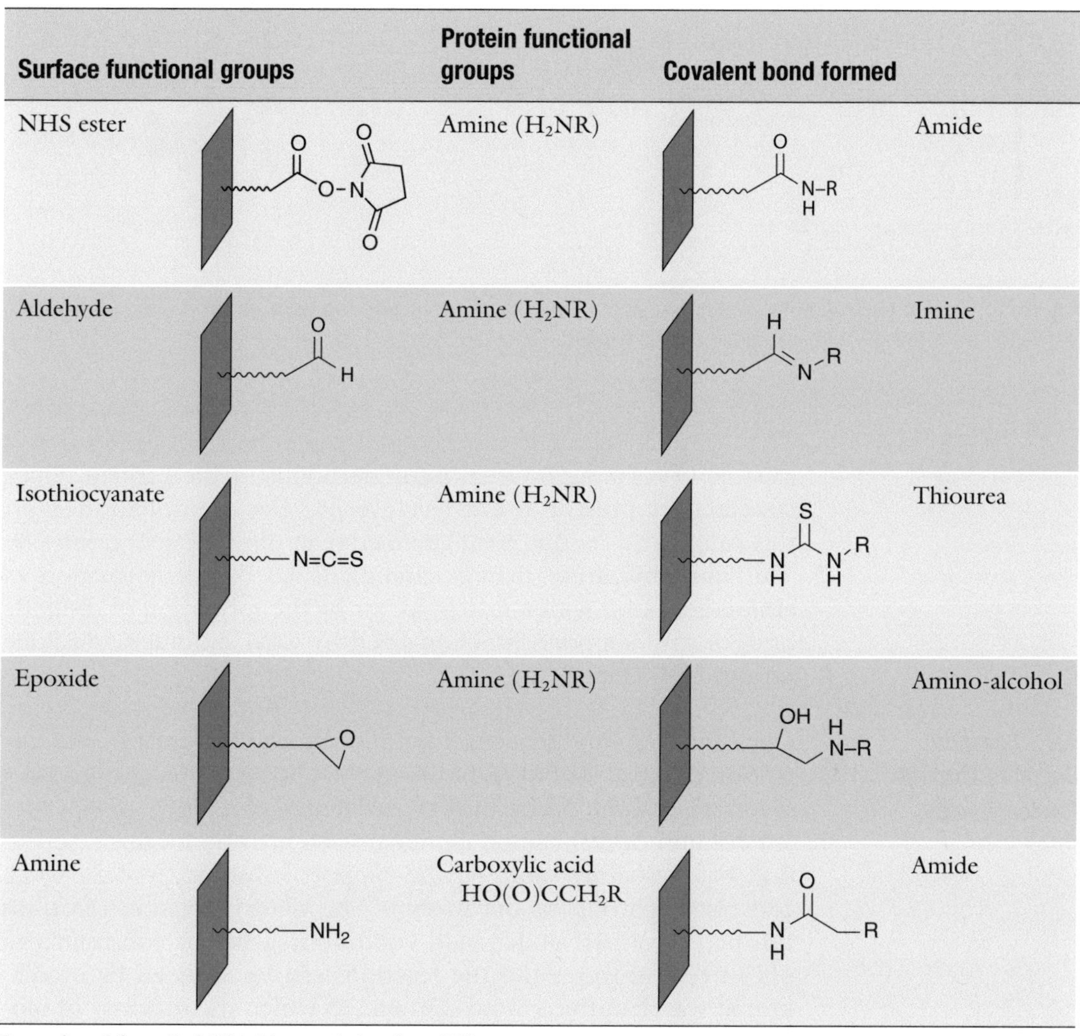

Surface functional groups	Protein functional groups	Covalent bond formed
NHS ester	Amine (H_2NR)	Amide
Aldehyde	Amine (H_2NR)	Imine
Isothiocyanate	Amine (H_2NR)	Thiourea
Epoxide	Amine (H_2NR)	Amino-alcohol
Amine	Carboxylic acid $HO(O)CCH_2R$	Amide

Reproduced from Ref. 25 with permission from Wiley

4.1.2 Carboxylic Acid Chemistry

An alternative to the amine group for protein immobilization is to use the carboxylic acid group provided by amino acids such as aspartic acid and glutamic acid, which are generally present on the surface of proteins. An effective coupling strategy relies on the formation of amide bonds with an amine-functionalized surface, after activation of the carboxylic acid groups of proteins using carbodiimide. Fernandez-lafuente et al., who used this method innovatively in 1993, prepared agarose gels containing primary amines, allowed for the adsorption of a protein in a low ionic strength medium, and then added carbodiimide to transform adsorption of the protein to its covalent immobilization [28].

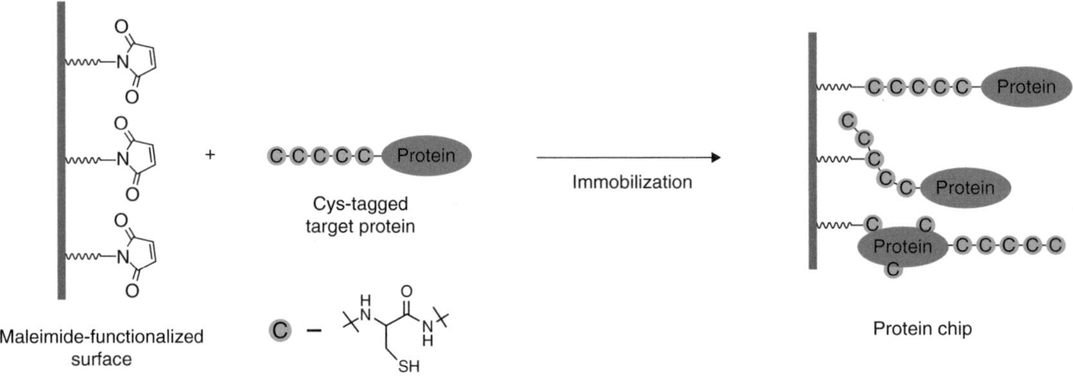

Fig. 5 Protein immobilization by cysteine residues and the oligocysteine tag on a target protein (e.g., EGFP) (extracted from Ref. 25 with permission from Wiley)

4.1.3 Thiol Chemistry

The cysteine residue in a protein contains the thiol group that is capable of forming internal disulfide bonds in the protein. Since cysteines are present at low levels in proteins, immobilization via this amino acid residue is well defined regarding the anchoring site. This immobilization may be accomplished by the formation of athioether bond with a maleimide or by the formation of disulfide bridges with supports functionalized by vinyl sulfone or disulfide groups [19] (Fig. 5).

4.2 Covalent Coupling Through Photoactivation

Covalent coupling generated by photoactivation leads to the formation of covalent bonds between the photoreactive groups on a substrate and the probe molecule. The photoreactive groups (see left column of 39) can be illuminated to provide localized activation of microarray spots on the support substrate, leading to an addressed grafting of the probes. The photoactivation reactions can be performed under mild conditions, without constraints of pH or temperature, and the reactions can be initiated by irradiation at wavelengths around 350 nm, at which the majority of biomolecules are not affected [29].

Heterobifunctional spacers comprising a photoactivable group and a chemically reactive group may also be used for substrate functionalization, see right column of Table 3 [25]. While the former group is used for immobilization on the support upon irradiation, the latter group is required for coupling to the probe molecules, leading to the probe immobilization.

4.3 Cycloaddition Reactions

4.3.1 Click Chemistry

Click chemistry, which is a specific type of cycloaddition, is based on the reaction between an azide group and an alkyne to form a heterocyclic 1, 2, 3-triazole ring via dipolar cycloaddition. This method has been used for the covalent grafting of enzymes, functionalized with azide groups, on agarose beads bearing acetylene groups [30]. The irreversible formation of the triazole ring is

Table 3
Overview of photoreactive groups and heterobifunctional linkers

Photoreactive groups		Heterobifunctional linkers
Arylazide	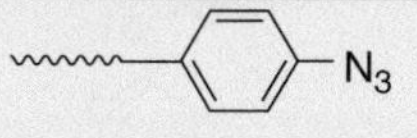	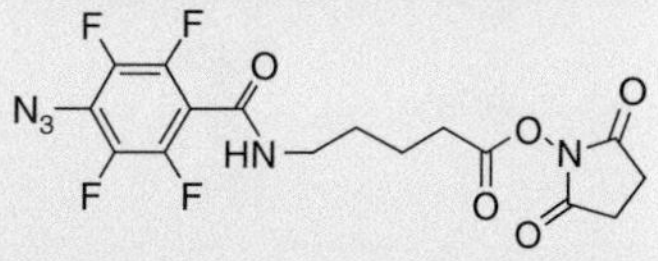
Diazirine	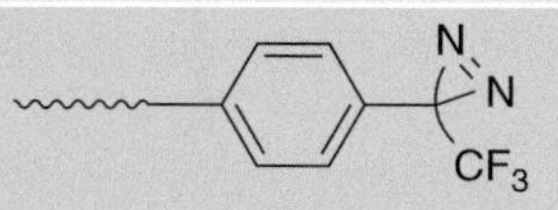	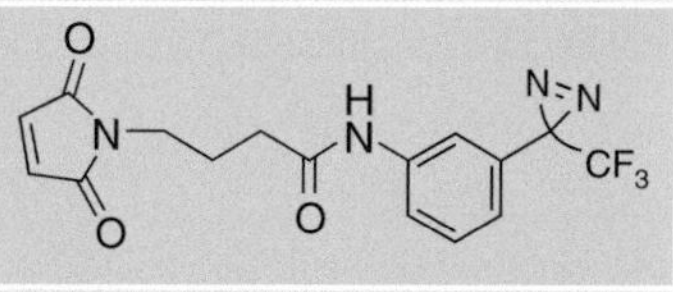
Benzophenone	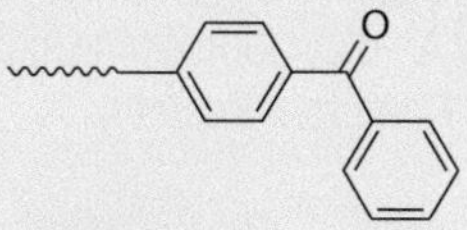	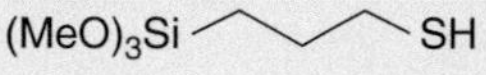
Azidophenylalanine	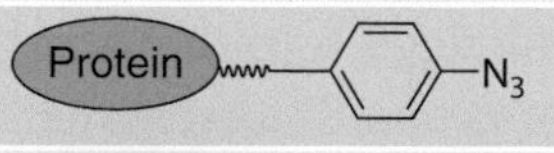	
Disulfide	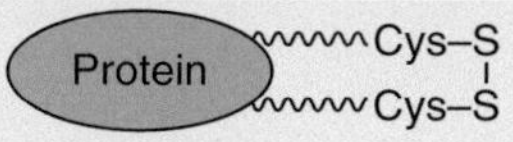	
Nitrobenzyl[a]	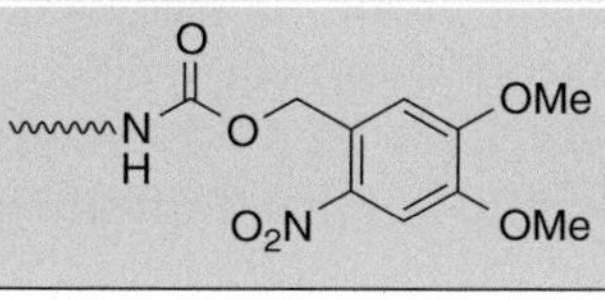	

Reproduced from Ref. 25 with permission from Wiley

[a]Photocleavable protecting group

carried out at room temperature in the presence of copper, which can be a disadvantage for the immobilization of some proteins that are sensitive to copper inhibition.

Waldmann et al. have also used the sulfonamide click reaction between sulfonyl azides and terminal alkynes to immobilize biotin, carbohydrates, peptides, and proteins [31]. In this study, the Ras binding domain (RBD) of the cRafl protein, which was modified with a terminal alkyne functional group, was immobilized on a surface modified with a sulfonyl azide, and the RBD biological activity was conserved.

4.3.2 Diels-Alder Cycloaddition

The Diels-Alder reaction is based on the formation of a cyclohexene from an alkene group and a conjugated diene. The reaction can be performed in an aqueous medium at room temperature, and thus allows the covalent immobilization of protein probes. Houseman et al. used this method for the production of peptide chips dedicated to the evaluation of the activity of kinase proteins [32]. In this study, the peptides of interest that were coupled with

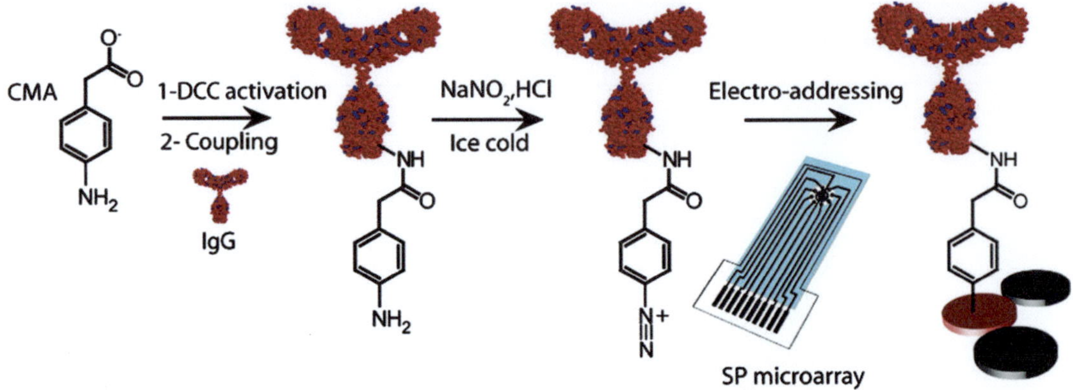

Fig. 6 Direct electro-addressing of modified antibody (IgG) on the graphite electrode surface in an acidic aqueous medium. *CMA* 4-carboxymethylaniline, *DDC N,N′*-dicyclohexylcarbodiimide, *SP* screen-printed (reproduced from Ref. 33) with permission from American Chemical Society)

a cyclopentadiene were immobilized by Diels-Alder cycloaddition with benzoquinone-modified alkanethiols self-assembled on gold surfaces.

4.4 Electrografting

Electrografting of probe molecules is realized on conductive surfaces (i.e., electrodes) on which electrochemical reactions of probe molecules occur. This method, which does not require a deposition robot, permits localized immobilization upon application of the potential on the electrode arrays. This technique can be applied for covalent grafting of a biomolecule (e.g., DNA, protein, polysaccharide) on the surface of an electrode due to the presence of an electro-active group on the biomolecules. By applying a localized potential between the electrode surface and a reference electrode, it is possible to direct the binding of a protein on a specific location of the chip. In 2005, Corgier and colleagues presented an innovative approach to immobilize immunoglobulins using electrochemical immobilization [33]. This approach was achieved by the following steps: (1) conjugation of the proteins to an aniline derivative, 4-carboxymethylaniline (CMA), (2) diazotization of the aniline group, and (3) electro-reduction of the diazonium ion of the conjugated protein individually addressable on the carbon electrode surface, *see* Fig. 6.

4.5 Ligation Reactions

4.5.1 Native Chemical Ligation

This immobilization method is based on the ability of a labile thioester to be acylated. The reaction consists of an exchange between a thioester and the thiol of a cysteine, followed by a rearrangement generating an amide bond. This chemo-selective reaction for surface immobilization of probes has first been described by Falsey et al. for the formation of peptide and small-molecule arrays (Fig. 7). In this study, glass slides were functionalized with glyoxylated derivatives, which allow the formation of thiazolidine

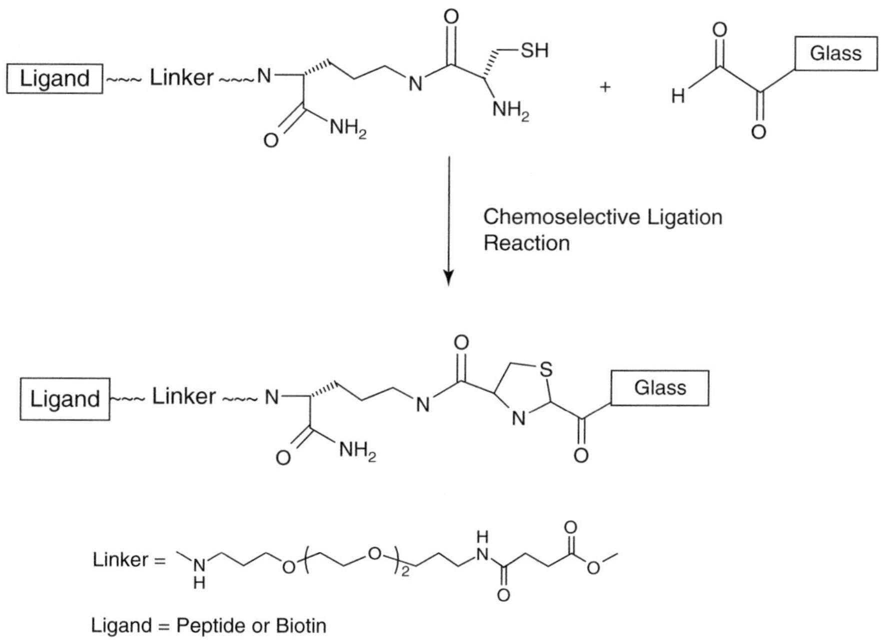

Fig. 7 Covalent attachment of peptides or small molecules to a glass slide via a thiazolidine ring (extract from Ref. 34 with permission from American Chemical Society)

rings with the N-terminal cysteine residue of biotin or peptides. These microarrays have enabled the fluorescent detection of cell adhesion proteins [34].

4.5.2 *Staudinger Ligation*

This method, developed by Bertozzi [35] and Raines [36], allows the formation of an amide bond at the ligation site that involves a phosphinothioester and an azide derivative. A proposed reaction mechanism would involve an iminophosphorane as a chemical intermediate and an acyl transfer to form an aminophosphonium, followed by its hydrolysis. The Staudinger ligation reaction was used in 2003 by the Raines' team [37] for the immobilization of protein on microarrays, and by Waldmann et al. [38] for the formation of small-molecule arrays. In the latter study, small molecules functionalized with an azide group were immobilized on glass supports functionalized with phosphine groups.

References

1. Barbulovic-Nad I, Lucente M, Sun Y et al (2006) Bio-microarray fabrication techniques – a review. Crit Rev Biotechnol 26(4):237–259
2. Xia Y, Whitesides GM (1998) Soft lithography. Angewandte Chemie 37(5):550–575
3. Marquette CA, Corgier BP, Heyries KA et al (2008) Biochips: non-conventional strategies for biosensing elements immobilization. Front Biosci 13(1):382–400
4. Piner RD, Zhu J, Xu F et al (1999) 'Dip-pen' nanolithography. Science 283(5402):661–663
5. Ginger DS, Zhang H, Mirkin CA (2004) The evolution of Dip-Pen nanolithography. Angewandte Chemie 43(1):30–45
6. Avseenko NV, Morozova TY, Ataullakhanov FI et al (2001) Immobilization of proteins in immunochemical microarrays fabricated by electrospray deposition. Anal Chem 73(24):6047–6052

7. Nolan JP, Sklar LA (2002) Suspension array technology: evolution of the flat-array paradigm. Trends Biotechnol 20(1):9–12
8. Sha MY, Walton ID, Norton SM et al (2006) Multiplexed SNP genotyping using nanobarcode particle technology. Anal Bioanal Chem 384(3):658–666
9. Marquette CA, Blum LJ (2004) Direct immobilization in poly(dimethylsiloxane) for DNA, protein and enzyme fluidic biochips. Anal Chim Acta 506(2):127–132
10. Epstein JR, Lee M, Walt DR (2002) High-density fiber-optic genosensor microsphere array capable of zeptomole detection limits. Anal Chem 74(8):1836–1840
11. Epstein JR, Ferguson JA, Lee KH et al (2003) Combinatorial decoding: an approach for universal DNA array fabrication. J Am Chem Soc 125(45):13753–13759
12. Yu X, Xia HS, Sun ZD et al (2013) On-chip dual detection of cancer biomarkers directly in serum based on self-assembled magnetic bead patterns and quantum dots. Biosens Bioelectron 41(1):129–136
13. Piletsky S, Piletska E, Bossi A et al (2003) Surface functionalization of porous polypropylene membranes with polyaniline for protein immobilization. Biotechnol Bioeng 82(1): 86–92
14. Cretich M, Breda D, Damin F et al (2010) Allergen microarrays on high-sensitivity silicon slides. Anal Bioanal Chem 398(4):1723–1733
15. Charles PT, Goldman ER, Rangasammy JG et al (2004) Fabrication and characterization of 3D hydrogel microarrays to measure antigenicity and antibody functionality for biosensor applications. Biosens Bioelectron 20(4):753–764
16. Amine A, Mohammadi H, Bourais I et al (2006) Enzyme inhibition-based biosensors for food safety and environmental monitoring. Biosens Bioelectron 21(8):1405–1423
17. Shi M, Peng Y, Zhou J et al (2006) Immunoassays based on microelectrodes arrayed on a silicon chip for high throughput screening of liver fibrosis markers in human serum. Biosens Bioelectron 21(12):2210–2216
18. Marquette CA, Imbert-Laurenceau E, Mallet F et al (2005) Electroaddressed immobilization of recombinant HIV-1 P24 capsid protein onto screen-printed arrays for serological testing. Anal Biochem 340(1):14–23
19. Rusmini F, Zhong Z, Feijen J (2007) Protein immobilization strategies for protein biochips. Biomacromolecules 8(6):1775–1789
20. Andersson H, Jönsson C, Moberg C et al (2002) Self-assembled and self-sorted array of chemically active beads for analytical and biochemical screening. Talanta 56(2):301–308
21. Andreescu S, Njagi J, Ispas C et al (2009) JEM spotlight: applications of advanced nanomaterials for environmental monitoring. J Environ Monit 11(1):27–40
22. Peijia TY, Xu N, Lu W, Xiao SJ, Liu JN (2011) Covalently derivatized NTA microarrays on porous silicon for multi-mode detection of His-tagged proteins. Sci China Chem 54(3):526–535
23. Wacker R, Niemeyer CM (2004) DDI-μFIA – a readily configurable microarray-fluorescence immunoassay based on DNA-directed immobilization of proteins. ChemBioChem 5(4): 453–459
24. Johnson CP, Jensen IE, Prakasam A et al (2003) Engineered protein A for the orientational control of immobilized proteins. Bioconjug Chem 14(5):974–978
25. Jonkheijm P, Weinrich D, Schröder H et al (2008) Chemical strategies for generating protein biochips. Angew Chem Int Ed 47(50): 9618–9647
26. Patel N, Davies MC, Hartshorne M et al (1997) Immobilization of protein molecules onto homogeneous and mixed carboxylate-terminated self-assembled monolayers. Langmuir 13(24):6485–6490
27. MacBeath G, Schreiber SL (2000) Printing proteins as microarrays for high-throughput function determination. Science 289(5485):1760–1763
28. Fernandez-Lafuente R, Rosell CM, Rodriguez V et al (1993) Preparation of activated supports containing low pK amino groups. A new tool for protein immobilization via the carboxyl coupling method. Enzym Microb Technol 15(7):546–550
29. Blawas AS, Reichert WM (1998) Protein patterning. Biomaterials 19(7-9):595–609
30. Duckworth BP, Xu J, Taton TA et al (2006) Site-specific, covalent attachment of proteins to a solid surface. Bioconjug Chem 17(4): 967–974
31. Govindaraju T, Jonkheijm P, Gogolin L et al (2008) Surface immobilization of biomolecules by click sulfonamide reaction. Chem Commun 32:3723–3725
32. Houseman BT, Huh JH, Kron SJ et al (2002) Peptide chips for the quantitative evaluation of protein kinase activity. Nat Biotechnol 20(3): 270–274

33. Corgier BP, Marquette CA, Blum LJ (2005) Diazonium-protein adducts for graphite electrode microarrays modification: direct and addressed electrochemical immobilization. J Am Chem Soc 127(51):18328–18332
34. Falsey JR, Renil M, Park S et al (2001) Peptide and small molecule microarray for high throughput cell adhesion and functional assays. Bioconjug Chem 12(3):346–353
35. Saxon E, Bertozzi CR (2000) Cell surface engineering by a modified Staudinger reaction. Science 287(5460):2007–2010
36. Nilsson BL, Kiessling LL, Raines RT (2000) Staudinger ligation: a peptide from a thioester and azide. Org Lett 2(13):1939–1941
37. Soellner MB, Dickson KA, Nilsson BL et al (2003) Site-specific protein immobilization by Staudinger ligation. J Am Chem Soc 125(39):11790–11791
38. Köhn M, Wacker R, Peters C et al (2003) Staudinger ligation: a new immobilization strategy for the preparation of small-molecule arrays. Angewandte Chemie 42(47): 5830–5834

Chapter 3

The "Clickable" Method for Oligonucleotide Immobilization Onto Azide-Functionalized Microarrays

Tomasz Ratajczak*, Barbara Uszczyńska*, Emilia Frydrych-Tomczak, and Marcin K. Chmielewski

Abstract

The DNA microarray technique was supposed to help identifying and analyzing the expression level of tens of thousands of genes in the whole genome. But there is a serious problem concerning fabrication of the microarrays by chemical synthesis, such as specific and efficient linking of probes to a solid support. Therefore, we reckon that applying "click" chemistry to covalently anchor oligonucleotides on chemically modified supports may help construct microarrays in applications such as gene identification. Silanization of the glass support with organofunctional silane makes it possible to link azide groups on glass surface and the nucleic acid probe that is equipped with a pentynyl group. This is followed by direct spotting of the nucleic acid on the azide-modified glass support in the presence of copper ions, and this is a frequently applied method of "click" chemistry.

Key words Click chemistry, Oligonucleotides, DNA, Microarrays, Silanization, Hybridization, Fluorescence, Phosphoramidite

1 Introduction

A microarray is defined as a collection of probes (nucleic acids, proteins, lipids, peptides, or carbohydrates) immobilized in an ordered 2-D manner on a solid support, e.g., plates, microchannels, microwells, or particles [1–4]. However, it is the nucleic acid microarrays on glass slides that are first developed and have found major application in biomedical research. The nucleic acid arrays enable assessment of gene expression, analysis of the genome structure, identification of genetic polymorphism, or detection of viral, bacterial, and fungal pathogens. The arrays give an opportunity to analyze thousands of different DNAs or RNAs in parallel with only a small amount of biological material [5]. In practice, there are two

*Author contributed equally with all other contributors.

Paul C.H. Li et al. (eds.), *Microarray Technology: Methods and Applications*, Methods in Molecular Biology, vol. 1368, DOI 10.1007/978-1-4939-3136-1_3, © Springer Science+Business Media New York 2016

technologies for microarray fabrication: by synthesizing oligonucleotides "step by step" in situ or by spotting the previously synthesized oligonucleotides onto the surface of a slide [6, 7]. Currently, in situ synthesis method seems to be the most effective method for preparing high-density microarrays. This method uses a photolabile system of protection/deprotection hydroxyl groups. However, an alternative method for applying thermolabile protecting groups in the "heat-driven" approach has also been reported [8]. Spotting techniques are less effective but simpler, and consequently they are commonly used for the production of customized microarrays [9]. One of the key factors affecting the DNA microarray sensitivity and specificity is the method of probe immobilization on the solid support. Nucleic acid microarrays are being produced via electrostatic interactions or covalent bond formation between probes and solid support. The covalent binding of probes requires DNA to be equipped with a linker, which can form a stable covalent bond with groups located on a solid support [10]. Traditional spotting techniques use spotting pins that may not be small enough and this limitation restricts the size of microarrays. The atomic force microscope tips have been applied in the Dip-Pen Nanolithography (DPN) approach to the miniaturization of DNA nanoarrays [11]. However, the search for effective solutions in biopolymers immobilization on the solid phase is still a challenge in terms of performance and miniaturization of microarrays.

During the 2000s, Sharpless, Kolb, and Finn have introduced to the literature the concept of "click" chemistry [12]. It covers different condensation reactions that characterize rapidity, high efficiency, and specificity [13, 14]. To date, different variants of this "click" reaction are known (Fig. 1) [15], and it has also been

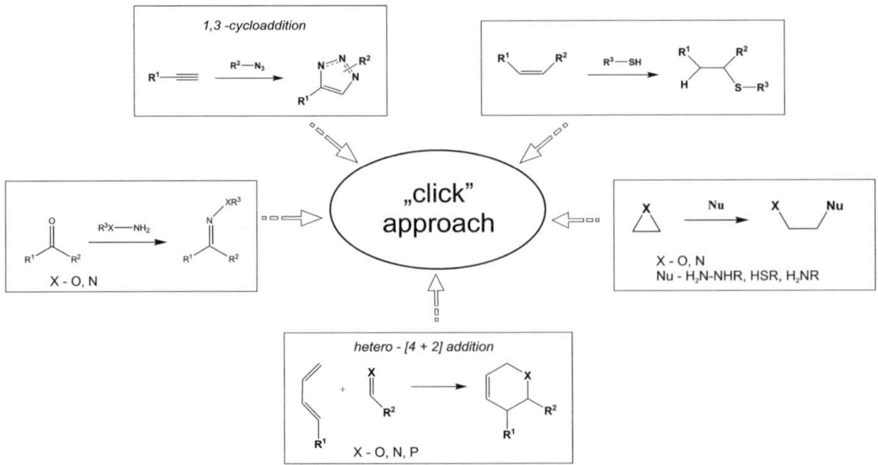

Fig. 1 Examples of reaction types that meet the descriptions of the "click" reaction

demonstrated that the "click" chemistry can be easily combined with microcontact printing [16]. Different types of cycloaddition in the "click" approach can be used for different applications.

The most suitable variant for microarray construction seems to be the copper (Cu(I))-catalyzed variant of Alkyne-Azide Cycloaddition (CuAAC) [17, 18], which was first reported for solid-phase synthesis of peptidotriazoles [19]. This method can be easily applied to attach alkyne or azide-modified oligonucleotides to well-defined surfaces with azide or terminal alkyne group, respectively [20]. The "click" reaction can occur in the presence of various agents, e.g., oxygen, water, and biological molecules [21]. For nucleic acid modifications with azide or alkynyl groups, there are two main strategies. The first one is based on using already modified phosphoramidite building blocks, which can be directly applied in automatic solid-phase DNA synthesis. This strategy provides good results and enables the synthesis of oligomers with an alkynyl group attached to 5′-OH of ribose [22], 2′-OH of ribose [23], the functional groups of nucleobases [24–27], or the internucleotide phosphate backbone [28]. Rozkiewicz et al. have immobilized 5′-alkynyl modified oligonucleotides on azide-bearing slides using microcontact printing. Synthesis in the nanoscale confinement between an elastomeric stamp and a reactive substrate leads to the formation of the desired product under ambient conditions, within a short time, and without a catalyst. The second strategy as applied by Ju et al. [29] is more complicated because the introduction of an azide group to the oligonucleotide during its chemical synthesis is inefficient due to the azide phosphoramidite instability [30]. However, the incorporation of the azide group can take place as a post-synthetic oligonucleotide modification, e.g., a bromine atom can be exchanged by the azide group [31].

We previously reported the application of the "click" reaction for the immobilization of DNA on a solid support [32]. Herein, we present a detailed protocol for this application. To this end, we chose the strategy of binding modified oligodeoxynucleotides (ODNs) (modified by an alkynyl group) to a solid surface covered with azide-functionalized silanes. Figure 2 shows schematically how the ODNs are linked with azide groups on the modified glass surface.

This method is simple and cheap because producing an azide-functionalized silane or a propargyl link is simple and the time required for effective functionalization of the support is short.

The phosphoramidite approach was used to introduce the alkynyl groups to oligodeoxynucleotides (ODN). Pentynyl phosphoramidite was first prepared and it was placed at the 5′-end of the ODN during the last step of chemical synthesis. Owing to the bifunctional nature of the azide-functionalized silane, i.e. 3-azidopropyltrimethoxysilane ($(CH_3O)_3Si(CH_2)_3N_3$) [33–35], it could be used as the "coupling agent" to link the biopolymeric moiety and inorganic silicon-based substrate (e.g., CPG particles or glass microscopic slides) [36].

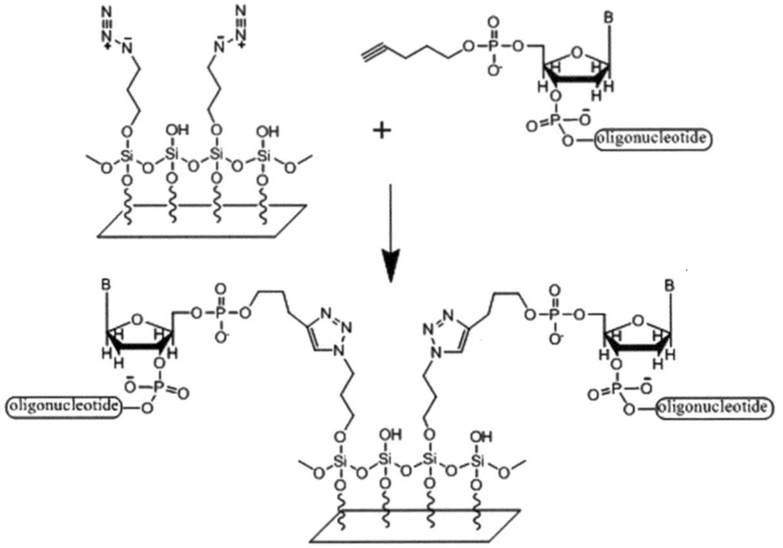

Fig. 2 Immobilization of a DNA oligodeoxynucleotide on the azide-functionalized glass surface (Reproduced from Ref. 32 with permission from Royal Society of Chemistry)

We spotted the modified ODN (with a pentynyl linker) on azide-functionalized slides in the presence of sodium ascorbate, $CuSO_4$, and glycerol. After intensive rinsing of the slide we carried out the hybridization procedure on printed slide with fluorescently labeled ODN targets. As shown in Fig. 3, 0.2 and 0.5 %, but not 0.02 %, of the silane gave hybridization signals that resulted from the formed duplexes.

2 Materials

2.1 Glass Substrate Preparation

1. Glass slides: standard soda-lime microscope slides (26 × 76 × 1 mm).
2. Ultrasonic cleaning units: Elma S60H, ultrasound frequency 37 kHz, ultrasonic power 150 W.
3. Acetone (analytical grade).
4. Microarray High-Throughput Wash Station (Arrayit Co., USA).
5. Magnetic stirrer.
6. Detergent solution: 4 % Trilux detergent in water. Add 250 mL of doubly distilled water to a high-throughput wash station, add 10 mL of Trilux detergent and mix.
7. Nitrogen gas (99.999 %) for drying.

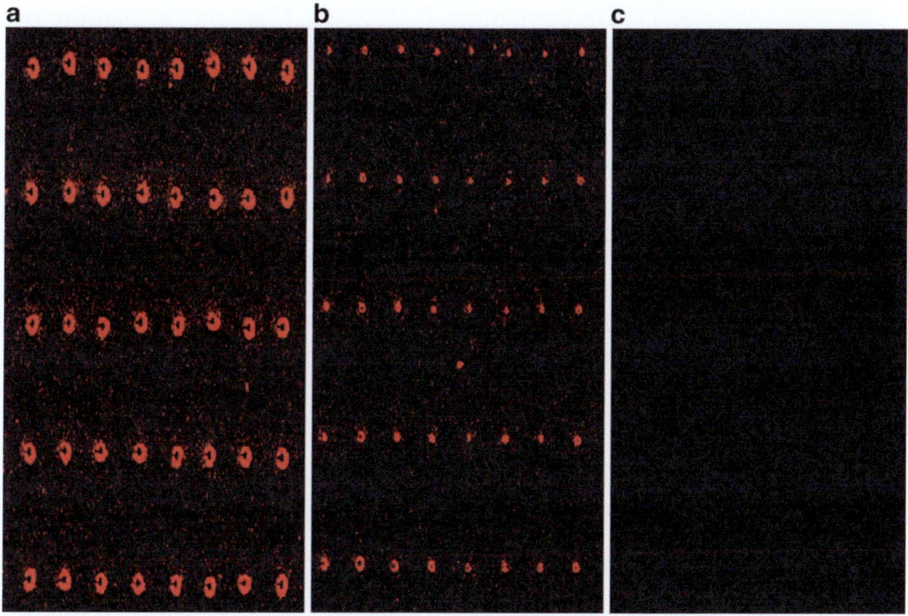

Fig. 3 Hybridization of fluorescently labeled targets with oligodeoxynucleotides that were immobilized via "click" chemistry on glass slides. Different concentrations of azide-functionalized silanes in toluene (v/v) (**a**) 0.5 % v/v, (**b**) 0.2 % v/v, (**c**) 0.02 % v/v were used. The targets are Cy5-labeled xxx. The distance between spots for each slide is 500 μm and the average SNR (signal-to-noise ratio) is 230 (Reproduced from Ref. 32 with permission from Royal Society of Chemistry)

2.2 Glass Substrate Silanization

1. Three Coplin staining jars each holding five slides.
2. Toluene pure.
3. Ethanol, absolut 99,8 % pure.
4. Silane solution for silanization of glass slides (0.02, 0.2, 0.5 % AzPTMS in toluene). Add 40 mL of toluene to three Coplin staining jars and put in 8×10^{-3}, 8×10^{-2}, 0.2 g AzPTMS, respectively, and mix.

2.3 CPG Silanization

1. Controlled pore glass (CPG): pore diameter (dp) = 125–160 μm;
   pore volume (Vp) = 1.06 cm^3 g^{-1}; particle diameter (D) = 94 nm (Cormay, Poland).
2. Round-bottom flask, 25 mL.
3. 3-azidopropyltrimethoxysilane (AzPTMS), synthesized according to the procedure developed by Mader et al. [37].
4. 5 % of AzPTMS in toluene, i.e., 0.25 g of AzPTMS was added to 5 mL of toluene in a round-bottom flask.

2.4 Synthesis of Modified Oligodeoxynucleotides (ODN)

1. 4-pentyl-1-ol, kept for 1 week over activated 3 Å sieves before use.
2. 3-hydroxypropionitrile, freshly distilled under reduced pressure and kept over activated 3 Å sieves before use.

3. Diisopropylamine (DIPEA), dried by distillation over NaH and kept over 3 Å activated sieves for 3 days before use.
4. Phosphorus trichloride (PCl_3) (*see* **Note 1**).
5. Benzene, dried by distillation over NaH and kept over 3 Å activated sieves for 5 days before use.
6. 10 mg of modified CPG solid supports (Biosearch Technologies Inc. USA) (*see* **Note 2**).
7. Oligodeoxynucleotides (ODNs), synthesized according to the procedure developed by *Caruthers* et al. [38] on a DNA synthesizer (K&A Laborgeraete GbR, Germany).
8. Ammonia solution (32 %) for oligonucleotide deprotection.
9. Urea (ultrapure).
10. 100 mL of 10× TBE buffer (tris(hydroxymethyl) aminomethane)/boric acid/EDTA.
11. 20 % polyacrylamide stock solution.
12. 10 % APS prepared by dissolution of 100 mg of ammonium persulfate (APS) in 900 μL milliQ water.
13. For denaturing gel electrophoresis: mix polyacrylamide/*N*,*N′*-methylenebisacrylamide solution (29:1 v:v), 100 mL of 10× TBE buffer, 420 g of urea, and add milliQ water to 1000 mL.
14. For a denaturing gel: mix 60 mL of 20 % stock solution, 750 μL of 10 % APS, and 30 μL of *N*,*N*,*N′*,*N′*-tetramethylethylenediamine (TEMED).
15. Electrophoretic power supply.
16. 0.3 M triethylammonium acetate buffer used for ODN elution, obtained by mixing 300 mL of 1 M triethylammonium acetate buffer and 700 mL of milliQ water and filtrated through a MF™ Membrane filter (0.45 μm).
17. 96 % ethanol.
18. Gel filtration column (NAP™-25, GE Healthcare).
19. 100 mL of autoclaved milliQ water.
20. Dehydrated acetonitrile (max. 30 ppm of a water).
21. Freeze-drier, miVac Concentrator (Genevac, USA) used for lyophilization of probes.

2.5 Spotting of Modified ODNs

1. 40 μM modified ODNs solution.
2. 0.1 M sodium ascorbate solution, 4 mg of sodium ascorbate dissolved in 120 μL autoclaved milliQ water.
3. 0.1 M $CuSO_4$ solution, 6.2 mg of $CuSO_4 \cdot 5H_2O$ dissolved in 250 μL autoclaved milliQ water.
4. 10 mL glycerol.
5. 384-well plate.
6. Microarray spotter (NanoPrint LM60, ArrayIt, USA).

2.6 Hybridization

1. 1 % sodium dodecyl sulfate (SDS) solution: 5 g of SDS dissolved in 80 mL of milliQ water. Next, the solution was heated to 60 °C and stirred until all SDS was dissolved. The volume was adjusted to 550 mL with milliQ water and stirred again at room temperature.
2. milliQ water, T=90–100 °C.
3. 20× SSC buffer: dissolve 173.5 g of NaCl and 88.2 g of sodium citrate in 800 mL of milliQ water. Adjust the pH value to 7.0 with a few drops of 14 M HCl. Adjust the volume to 1 L with milliQ water. Sterilize by autoclaving. The final concentrations are 3.0 M NaCl and 0.3 M sodium citrate [39].
4. 2× SSC+0.1 % SDS: 50 mL of 20× SSC dilute with 300 mL milliQ water and while stirring add 50 mL of 1 % SDS. Adjust the volume to 500 mL with milliQ water.
5. 0.5 % sodium dodecyl sulfate (SDS) solution: 50 mL of 1 % SDS solution was placed in a graduated cylinder and the volume was adjusted to 500 mL with milliQ water.
6. 0.05 % sodium dodecyl sulfate (SDS) solution: 0.5 % SDS solution was diluted ten times with milliQ water in a graduated cylinder and the volume was adjusted to 500 mL.
7. Mixture of modified ODNs (2 nM, 1 μL) in 50 μL SpotQC buffer (Integrated DNA Technologies).
8. Microarray High-Throughput Wash Station (ArrayIt).
9. Microarray High-Speed Centrifuge (ArrayIt).
10. MicroarrayHybridization Chamber (Corning).
11. Microscope coverslips 24×24 mm.
12. MicroarrayHybridization Oven (Thermo Scientific).

2.7 Scanning and Analysis

1. Microarray scanner *GenePix 4200AL* (Molecular Devices).
2. *GenePixPro 6.0* software (Molecular Devices).

3 Methods

3.1 Glass Slide Preparation

1. Place five glass slides in a Coplin staining jar with 40 mL of acetone and sonicate for 10 min (*see* **Note 3**).
2. Dry glass slides under nitrogen gas for 20 min (*see* **Note 4**).
3. Place glass slides in detergent solution in a microarray high-throughput wash station and mix for 20 min (*see* **Note 5**).
4. Rinse twice with doubly distilled water in Microarray high-throughput wash station.
5. Dry in a stream of nitrogen gas for 30 min (*see* **Note 4**).

3.2 CPG Silanization

1. Place 0.1 g CPG and 5 % solution of AzPTMS in toluene in a round-bottom flask. Stir the mixture at room temperature for 24 h (*see* **Note 6**).
2. Filter the silanized CPG using a funnel and wash with 100 ml of toluene.
3. Place the silanized CPG in an oven and dry at 120 °C for 1 h (*see* **Note 7**).

3.3 Glass Slide Silanization

1. Immerse cleaned glass slides in Coplin staining jars containing a toluene solution of the 3-AzPTMS at various concentrations (0.02; 0.2 or 0.5 %) at room temperature for 1 h.
2. Rinse silanized glass slides by sonication in toluene (two times using 40 mL).
3. Dry in a Microarray High-Throughput Wash Station oven at 120 °C for 1 h.
4. After cooling to room temperature, rinse the glass slides with ethanol and dry in a stream of nitrogen.

3.4 Synthesis of Modified Oligodeoxynucleotides (ODN)

1. Use 4-pentyl-1-ol, diisopropylamine, 3-hydroxypropionitrile, PCl_3, and dried benzene (as a solvent) to prepare the linker: 4-pentynyl-(2-cyanoethyl)-N,N-diisopropyl phosphoramidite according to the method of *Pourceau* et al. [40].
2. ODNs were synthesized by following the protocol described by *Caruthers* et al. [3, 7]. The linker was dissolved in a 1 mL of dry acetonitrile (0.2 M) and attached to the 5′ end of each ODN during its automatic ODN synthesis on the DNA synthesizer (*see* **Note 8**).
3. Prepare 20 % polyacrylamide gel for electrophoresis (*see* Subheading 2.3, **item 11**).
4. Run the gel electrophoresis using the settings of 450 V, 10 mA, 25 W.
5. Elute ODNs from a denaturing gel with 0.3 M triethylammonium acetate buffer in two steps. First, crush the gel. Then, swamp it with 1 mL of buffer, shake it overnight at 4 °C, centrifuge, and decant the solution (part I). Put in 1 mL of buffer, shake for 4 h, centrifuge, and decant the solution (part II).
6. Put together solutions from part I and II and add three volumes of 96 % ethanol, keep it for 1 h at 4 °C, centrifuge at 18,514 × *g* force for 30 min. Decant the supernatant and wash the precipitate with 500 μL of 70 % ethanol solution. Centrifuge it at 18,514 × *g* force for 15 min at 4 °C, decant the supernatant, and dry the modified ODN precipitate.
7. The modified ODNs are desalted. Dissolve the ODNs obtained from gel in 500 μL autoclaved milliQ water. Introduce them on

a NAP™ -25 column, and elute the ODN with autoclaved milliQ water. Lyophilize the ODNs using a freeze-drier afterwards.

3.5 Spotting of Modified ODNs

1. Mix in a tube: 250 μL $CuSO_4$ solution (0.04 M), 1 μL of modified ODN (40 μM), 50 μL of glycerol, and 120 μL sodium ascorbate (0.1 M) (*see* **Note 8**).
2. Place 20 μL of the mixture into a single well of 384-well plate. In total, 100 wells per modified ODN were filled with the mixture (5 columns × 20 rows). Apply the microarray spotter pin configuration: four pins in a row (*see* **Note 9**).
3. Incubate the slides for 1 h in the spotter humidity chamber (70–80 %) at room temperature (*see* **Note 10**).
4. Using the modified ODNs in the 384-well plate, perform spotting on the slides.

3.6 Hybridization

1. Immerse the spotted slides in the High-Throughput Wash station filled with 1 % SDS solution (500 mL) at room temperature and stir for 1 min.
2. Next, fill the station with milliQ H_2O (500 mL, T = 90–100 °C) and immerse the slides in it and stir for 30 s.
3. Take out the slides and dry them by centrifugation (10 s) (*see* **Note 11**).
4. Place the slides in the MicroarrayHybridization Chamber. Place 50 μL mixture of fluorescently labeled ODN in SpotQC buffer on the slide. Cover the slide surface with coverslip and place 10 μL of milliQ H_2O in the well in the Microarray Hybridization Chamber (*see* **Note 12**).
5. Lock the MicroarrayHybridization Chamber and incubate it for 30 min at 40 °C.
6. Next, remove slides from Hybridization Chambers and immerse it in the High-Throughput Wash station filled with 2× SSC + 0.1 % SDS solution (500 mL) at room temperature and stir for 1 min.
7. Fill the station with 0.5 % SDS solution (500 mL) and immerse slides in it and stir for 1 min at room temperature.
8. Lastly, wash the slides by immersing them in 0.5 % SDS solution (500 mL) with stirring for 1 min at room temperature. Then dry them by centrifugation (10 s).

3.7 Scanning and Analysis

1. To detect fluorescence of hybridized probes, scan the microarrays using a microarray scanner (*see* **Note 13**).
2. Perform quantitative analysis using the *GenePixPro 6.0* software. As a result, convert the signal intensities into numbers and save them as .gpr files (*GenepixPro results*). This file format is used for many programs enabling advanced microarray data analysis.

4 Notes

1. The reagents: 4-pentyl-1-ol, benzene, and DIPEA must be dried prior to use, otherwise we have to use more PCl_3, which consumes water in reaction mixture, i.e., in a small excess than in the method described by *Pourceau* et al.
2. For modified CPG, we used the 1000 Å of 5'-DMT-X-Suc-CPG where X refers to nucleotides: T, dG(iBu), dC(Bz), and dA(Bz) (where iBu and Bz are isobutyryl and benzoyl exocyclic amine protecting groups, respectively).
3. Acetone removes organic contaminants from the glass surface.
4. The glass slides are dried in a desiccator through which a stream of nitrogen gas flows.
5. The use of a detergent solution is sufficient to clean the surface of the glass slides. This removes any stains that have background fluorescence to interfere with the microarray image on the glass slide.
6. The CPG silanization process is carried out in a round-bottom flask. The flask must be shaken to keep the CPG particles in constant motion. To this end, one may use an automatic shaker or mechanical stirrer. In the latter case, it is vital to keep the stir bar off the CPG particles in order to avoid destruction of their structure.
7. Thermal curing of the film which, in cross-linking the free silanol groups, reduces the effect of hydrolysis of one or more of the siloxane linkages on the glass surface. Post-silanization curing of the substrate has been shown to improve the stability of silane films by cross-linking of free silanols.
8. The time of the coupling stage was extended to 200 s to maximize the reaction yield. The presence of linker at the 5′ end of the ODNs was confirmed chromatographically after their purification.
9. Copper (I) sulfate (formed by the reduction of copper (II) sulfate by sodium ascorbate) is present to catalyze the "click" reaction. Glycerol reduces water evaporation in small liquid volumes of the spotted droplets and enables more effective covalent attachment of probes to the solid surface.
10. While using the contact microarray spotter, it is important to optimize the contact distance between the pins and the glass slide. In this way, the solution is precisely located and spread on the slide surface. The liquid spread on the slide also depends on the hydrophilic character of the surface.
11. To enable a stable attachment of the probes on the solid surface via the "click" reaction, it is crucial to keep humidity at ~70–80 % in order to decrease evaporation of the solvent.

12. The use of centrifugation to dry the slides helps to reduce the fluorescent background of the microarray image.
13. The application of nitrogen gas, especially during hybridization, results in a reduction in the fluorescent background.

Acknowledgment

This research was supported by the European Regional Development Fund within the Innovative Economy Programme [Grant No. POIG.01.03.01-30-045].

References

1. Zhang W (2004) 2nd ed. Wiley-Liss, New York, p 106
2. Simon R (2003) 2nd ed. New York, p 37–40
3. Blalock EMA (2003) Kluwer Academic, New York, p 15
4. Baldi P (2002) 2nd ed. Cambridge University Press, p 117
5. Hardiman G (2003) The Nuts and Bolts Series. Independent Publishers Group
6. Angenendt P, Glokler J, Sobek J, Lehrach H, Cahill DJ (2003) Next generation of protein microarray support materials: evaluation for protein and antibody microarray applications. J Chromatogr A 1009:97–104
7. Beaucage SL (2001) Strategies in the preparation of DNA oligonucleotide arrays for diagnostic applications. Curr Med Chem 8:1213–1244
8. Grajkowski A, Cieslak J, Chmielewski MK, Marchan V, Phillips LR, Wilk A, Beaucage SL (2003) Conceptual "heat-driven" approach to the synthesis of DNA oligonucleotides on microarrays therapeutic oligonucleotides. Ann N Y Acad Sci 1002:1–11
9. Schaeferling M, Schiller S, Paul H, Kruschina M (2002) Application of self-assembly techniques in the design of biocompatible protein microarray surfaces. Electrophoresis 23:3097–3105
10. Venkatasubbarao S (2004) Microarrays: status and prospects. Trends Biotechnol 22:630–637
11. Piner RD, Zhu J, Xu F, Hong S, Mirkin CA (1999) "Dip-pen" nanolithography. Science 283:661–663
12. Kocalka P, El-Sagheer AH, Brown T (2008) Rapid and efficient DNA strand cross-linking by click chemistry. Chem Bio Chem 9:1280–1285
13. Kolb HC, Finn MG, Sharpless KB (2001) Click chemistry: diverse chemical function from a few good reactions. Angew Chem Int Ed 40:2004–2021
14. Lutz JF (2007) 1,3-dipolar cycloadditions of azides and alkynes: a universal ligation tool in polymer and materials science. Angew Chem Int Ed 46:1018–1025
15. Kolb HC, Sharpless KB (2003) The growing impact of click chemistry on drug discovery. Drug Discov Today 8:1128–1137
16. Rozkiewicz DI, Janczewski D, Verboom W, Ravoo BJ, Reinhoudt DN (2006) "Click" chemistry by microcontact printing. Angew Chem Int Ed 45:5292–5296
17. El-Sagheer AH, Brown T (2010) Click chemistry with DNA. Chem Soc Rev 39: 1388–1405
18. Chen L, Rengifo HR, Grigoras C, Li X, Li Z, Ju J, Koberstein JT (2008) Spin-on end-functional diblock copolymers for quantitative DNA immobilization. Biomacromolecules 9:2345–2352
19. Tornes CW, Christensen C, Meldal M (2002) Peptidotriazoles on solid phase: [1,2,3]-triazoles by regiospecific copper(I)-catalyzed 1,3-dipolar cycloadditions of terminal alkynes to azides. J Org Chem 67:3057–3064
20. Brase S, Gil C, Knepper K, Zimmermann V (2005) Organic azides: an exploding diversity of a unique class of compounds. Angew Chem Int Ed 44:5518–5240
21. Jewetta JC, Bertozzi CR (2010) Cu-free click cycloaddition reactions in chemical biology. Chem Soc Rev 39:1272–1279
22. Rozkiewicz DI, Gierlich J, Burley G, Gutsmiedl K, Carell TR, Bart J (2007) Transfer printing of DNA by "click" chemistry. ChemBioChem 16:1997–2002
23. Jawalekar AM, Meeuwenoord N, Cremers JGO, Overkleeft HS, van der Marel GA, Rutjes FPJT, van Delft FL (2008) Conjugation of nucleosides and oligonucleotides by [3+2] cycloaddition. J Org Chem 73:287–290

24. Seela F, Sirivolu VR (2007) Nucleosides and oligonucleotides with diynyl alkyne-azide 'click' cycloaddition. Helv Chim Acta 90:535–552
25. Gramlich PME, Warncke S, Gierlich J, Carell T (2008) Click-click-click: single to triple modification of DNA. Angew Chem Int Ed 47:3442–3444
26. Nakane M, Ichikawa S, Matsuda A (2008) Triazole-linked dumbbell oligodeoxynucleotides with NF-κB binding ability as potential decoy molecules. J Org Chem 73:1842–1851
27. Qing G, Xiong H, Seela F, Sun T (2010) Spatially controlled DNA nanopatterns by "click" chemistry using oligonucleotides with different anchoring sites. J Am Chem Soc 43:15228–32
28. Lietard J, Meyer A, Vasseur JJ, Morvan F (2008) New strategies for cyclization and bicyclization of oligonucleotides by click chemistry assisted by microwaves. J Org Chem 73:191–200
29. Seo TS, Bai X, Ruparel H, Li Z, Turro NJ, Ju J (2004) Photocleavable fluorescent nucleotides for DNA sequencing on a chip constructed by site-specific coupling chemistry. PNAS 15:5488–93
30. Fujino T, Yamazaki N, Guillot-Nieckowski M (2009) Convergent synthesis of oligomers of triazole-linked DNA analogue (TLDNA) in solution phase. Tetrahedron Lett 50:4101–4103
31. Geci I, Filichev VV, Pedersen EB (2007) Stabilization of parallel triplexes by twisted intercalating nucleic acids (TINAs) incorporating 1,2,3-triazole units and prepared by microwave-accelerated click chemistry. Chem Eur J 13:6379–6386
32. Uszczyńska B, Ratajczak T, Frydrych E, Maciejewski H, Figlarowicz M, Markiewicz WT, Chmielewski MK (2012) Application of click chemistry to the production of DNA microarrays. Lab Chip 12:1151–1156
33. Buder W, Karl A (1983) US Patent No. 3,705,911
34. Keogh MJ (1985) US Patent No. 3,697,551
35. Dow Corning (1971) GB Patent No. 1,377,214
36. Haensch C, Hoeppener S, Schubert SU (2008) Self-assembled nanoscale architecture of TiO2 and application for dye-sensitized solar cells. Nanotechnology 19:1–7
37. Mader HS, Link M, Achatz DE, Uhlmann K, Li X, Wolfbeis OS (2010) Surface-modified upconverting microparticles and nanoparticles for use in click chemistries. Chem Eur J 16:5416–5424
38. Caruthers MH, Barone AD, Beaucage SL, Dodds DR, Fisher EF, McBride LJ, Matteucci M, Stabinsky Z, Tang JY (1987) Chemical synthesis of deoxyoligonucleotides by the phosphoramidite method. Methods Enzymol 154:287–313
39. Sambrook J, Russell DW (2001) Molecular cloning: a laboratory manual, 3rd edn. Cold Spring Harbor Laboratory Press, New York
40. Pourceau G, Meyer A, Vasseur JJ, Morvan F (2009) Azide solid support for 3′-conjugation of oligonucleotides and their circularization by click chemistry. J Org Chem 74:6837–6842

Chapter 4

Microarray Developed on Plastic Substrates

María-José Bañuls, Sergi B. Morais, Luis A. Tortajada-Genaro, and Ángel Maquieira

Abstract

There is a huge potential interest to use synthetic polymers as versatile solid supports for analytical microarraying. Chemical modification of polycarbonate (PC) for covalent immobilization of probes, microprinting of protein or nucleic acid probes, development of indirect immunoassay, and development of hybridization protocols are described and discussed.

Key words Synthetic polymers, Plastic substrates, Polycarbonate, Surface functionalization, Probe covalent immobilization, Immunoassay, Hybridization assay, Optical detection

1 Introduction

Microarray techniques have been developed to increase the work capacity, to miniaturize assay format, and to improve the performances of different analytical approaches. Three steps are key for the microarraying method, and they are probe immobilization, probe-target biorecognition, and detection. There are many microarraying methods, but all of them use a support material where the bioanalytical assay is carried out. The performance of the different assays depends highly on the nature of the support, what determines probe immobilization strategy, bioreceptor properties, and detection mode.

There are three main types of solid supports: inorganic materials such as silicon and its derivatives, oxides of elements such as tantalum, indium, and aluminum, and metals such as gold or silver. Other support families are carbon and its composites, or more sophisticated substances such as nanocrystals.

Synthetic polymers or plastics are another large group of solid supports for microarraying. Besides the own nature of each support material, plastics have key properties such as diversity, versatility, and fabrication in different presentations at a very competitive cost. Especially for disposable medical devices, this type of plastic supports is the gold standard if marketing biosensing systems is the goal.

Paul C.H. Li et al. (eds.), *Microarray Technology: Methods and Applications*, Methods in Molecular Biology, vol. 1368, DOI 10.1007/978-1-4939-3136-1_4, © Springer Science+Business Media New York 2016

Polymers have exceptional properties as microarraying supports, mainly related to their immobilization strategies and their interaction between targets and capture receptors (proteins in general, antibodies, carbohydrates, lipids, nucleic acids, semi-synthetic and synthetic substances). The polymers allow different ways for probe anchoring, high throughput and compatibility with practically all types of detection principles, particularly the optical ones. Other advantages of polymeric supports are the ease for integrating microfluidic structures to develop multiple steps of the analytical protocol.

The proper choice of the surface material properties is also key to obtain good results. Normally, polymer surfaces are modified in the fabrication procedure, and it is important to note that even the change due to the extrusion process to manufacture bulk polymer alters significantly the behavior of the microarray related to probe attachment. In addition, surface properties of the polymeric materials affect the immobilization density of the probes and their bioavailability. At the same time, nonspecific interaction should be reduced for maintaining the assay reproducibility and reliability.

Among the different polymers, polycarbonate (PC), polymethylmethacrylate (PMMA), nylon, and polystyrene have been extensively used. Some general approaches in using these substances for microarray solid support are found in the literature [1, 2].

Polycarbonate (PC) possesses interesting physical properties, molding facility, and low cost, and it is a very good candidate to be applied as the microarray platform in optical detection devices. Moreover, chemical surface functionalization has been developed for plastic substrates and their applications in microarraying and biosensing have been demonstrated.

Several microarraying strategies developed on PC substrates are presented and they are employed in competitive immunoassay for low molecular weight analytes and in DNA hybridization assay for pathogenic bacteria (Fig. 1).

2 Materials

Prepare all aqueous solutions using ultrapure water and analytical grade reagents. Work in a well-ventilated fume hood when handling hazardous reagents. Discard the waste by following the safety recommendations for hazardous reagent disposal.

Before use, all the solutions must be filtered through a 0.22-μm pore size nitrocellulose membrane filter.

2.1 Reagents

1. Nitric acid 69 %, sulfuric acid 98 %.
2. Glutaraldehyde (25 %, aqueous solution), sodium borohydride, $NaBH_4$.
3. Dicyclohexyl carbodiimide (DCC), dimethylformamide (DMF).

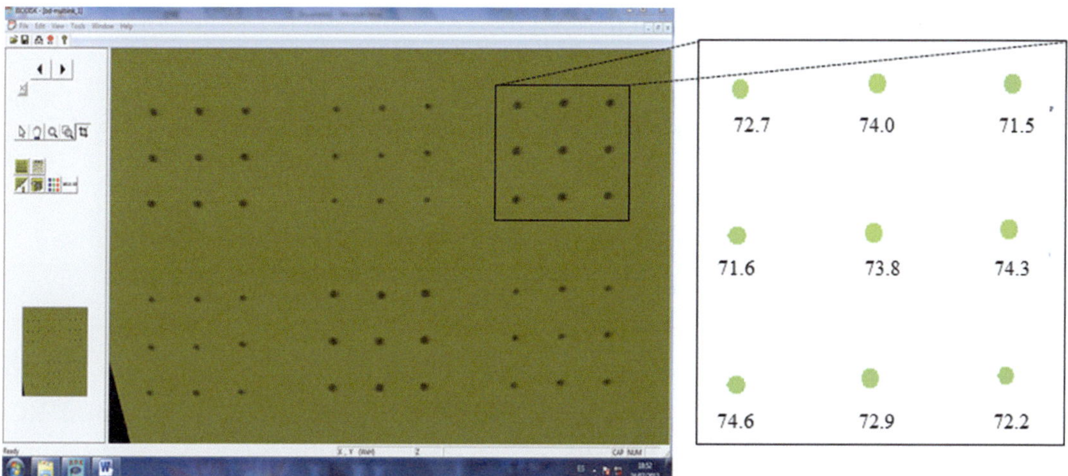

Fig. 1 Representative image for groups of 3 × 3 microarrays for the determination of low molecular weight compounds. The data analysis (signal-to-noise ratios) is represented in the *inset*

4. Ethylene carbodiimide (EDC), *N*-hydroxysuccinimide (NHS).
5. KCN, ninhydrin, pyridine, *t*-butanol.
6. Acetic acid, *p*-anisaldehyde.
7. Crystal violet dye, HCl (0.1 M), NaOH (0.1 M).

2.2 Buffers and Working Solutions

1. Coating buffer: 50 mM sodium carbonate/bicarbonate buffer (CBB), pH 9.6.
2. Phosphate buffered saline (PBS) 1× buffer: 0.15 M NaCl, 10 mM sodium phosphate, pH 7.4. Use as the immobilization buffer.
3. PBST 10× buffer: PBS 10× buffer containing Tween 20, i.e., 1.5 M NaCl, 0.1 M sodium phosphate, pH 6.8, 0.5 % (v/v) Tween 20.
4. PBST 1× buffer: PBS 1× buffer containing 0.05 % (v/v) Tween 20, pH 7.4.
5. MES buffer: 0.5 M 2-(*N*-morpholino)ethanesulfonic acid, pH 5.0. Use as an activation buffer.
6. Silver enhancement solutions A and B. Prepare just before use. Use for signal amplification.
7. Hybridization buffer (HB): 150 mM sodium chloride and 15 mM sodium citrate, containing 25 % formamide (v/v), pH 7.0. Sterilize the solution by autoclaving. Use: DNA hybridization.

8. Mineral oil (8 %): used to reduce the risk of evaporation of HB during hybridization.
9. Enzymatic substrate: 3,3′,5,5′-Tetramethylbenzidine (TMB). Use for signal amplification. Store at 4 °C.

2.3 DNA Amplification Reagents

1. Primers (10 μM).
2. Digoxigenin-11-dUTP (1 mM).
3. dNTPs (1 mM).
4. Template DNA.
5. Taq DNA Polymerase (1 U/μL).
6. PCR buffer 10×: 750 mM Tris–HCl pH 9.0, 500 mM KCl, 2 mM $MgCl_2$.

2.4 Immunoreagents

1. Ovalbumin from chicken egg white (OVA).
2. An analog of the target analyte (such as hapten) conjugated to a protein (*see* **Note 1**).
3. Analyte standard at 1 mg/L in methanol.
4. Anti-analyte rabbit polyclonal antibody solution (*see* **Note 2**).
5. Gold-coated anti-rabbit IgG (whole molecule) antibody (*see* **Note 3**).
6. Affinity-purified streptavidin from *Streptomyces avidinii*.
7. Oligonucleotides: 5′ end-modified (biotin, amine, and digoxigenin).
8. Peroxidase-labeled anti-digoxigenin monoclonal antibody.
9. Sephadex G-25 medium.

2.5 Apparatus

1. PC slides are provided by Grupo Condor, Calatayud, Spain.
2. Crystallizing dish.
3. Hot plate.
4. Microwave plasma reactor (PVA Tepla 200 plasma system).
5. Spotter.

3 Methods

3.1 Conjugation of Haptens to Carrier Proteins

Bring all the solutions to room temperature and carry out all procedures at room temperature, unless otherwise specified.

1. Prepare 100 μL fresh solutions of *N*-hydroxysuccinimide (NHS; 100 μmol), dicyclohexylcarbodiimide (DCC; 100 μmol), and the analog of the target analyte (hapten; 25 μmol) in anhydrous dimethylformamide (DMF).

2. Pipette 25 μL of each NHS and DCC solutions and dispense them to the hapten solution (100 μL) prepared in a 1-mL glass vial to activate it.
3. Stir the solution by magnetic stirrer for 3 h at room temperature.
4. Centrifuge the mixture for 2 min at 8161 ×*g* to remove the acyl urea precipitate.
5. Take 35 μL of the supernatant and make it up to 100 μL with DMF.
6. Weigh in a 2-ml polypropylene tube 10 mg ovalbumin (OVA) and dissolve it with 900 μL sodium carbonate/bicarbonate buffer (CBB).
7. Add the activated hapten (100 μL) dropwise to the OVA solution and stir for 4 h at room temperature for conjugation.
8. Purify the hapten–OVA conjugate by gel-exclusion chromatography on a Sephadex G-25 medium (*see* **Note 4**), using PBS 1× as the elution buffer.
9. Store the conjugate at −20 °C until use (*see* **Note 5**).

3.2 Amine-Modified Polycarbonate (PC) Surface

1. Add 20 mL of nitric acid to a crystallizing glass dish.
2. Add 5 mL of sulfuric acid to the dish (*see* **Note 6**).
3. Heat the sulfuric-nitric acid mixture at 65 °C on a hot plate (*see* **Note 7**).
4. Immerse a PC slide into the hot acid mixture.
5. Heat for 10 min (*see* **Note 8**).
6. Pour the acid mixture carefully into a beaker containing 200 mL of water (*see* **Note 9**).
7. Wash the PC slides three times with deionized water and blow dry.
8. Weight 2.5 g of sodium borohydride ($NaBH_4$) and dissolve it in 25 mL of PBS 1× (*see* **Note 10**).
9. Immerse the PC slides in the $NaBH_4$ solution and stir for 6 h at room temperature.
10. Take out the slides which are aminated (Fig. 3a). Wash with PBS 1×. Rinse with deionized water, wash again with ethanol, and blow dry. The surface can be tested for the presence of amine groups (*see* **Note 11**).

3.3 Aldehyde-Modified PC Surface

1. Mix 5 mL of glutaraldehyde with 20 mL of PBS 1× in a dish (*see* **Note 12**).
2. Immerse the aminated PC slide in the glutaraldehyde solution and incubate at room temperature for 2 h (*see* **Note 13**).

3. Take out the slides which are aldehyde-modified (Fig. 3b), wash with PBS 1×, water and blow dry. The surface can be tested for the presence of aldehyde groups (*see* **Note 14**).

3.4 Carboxylate-Modified PC Surface

1. Wash the aldehyde-modified PC slides with ethanol, then with deionized water and blow dry.
2. Put the slides inside the microwave plasma reactor.
3. Set the oxygen pressure inside the reactor to 120 Pa and operate the microwave radiation at 2.45 GHz and with continuous power, 100 W, for 30 s (*see* **Note 15**). Take out the slides which are carboxylate-modified (Fig. 3c). The surface can be tested for the presence of carboxylate groups (*see* **Note 16**).
4. Store the slides under vacuum until its use (*see* **Note 17**) [3].

3.5 General Method for Spotting Probes on PC Surfaces

The procedure for printing probe solutions in microarray format on polycarbonate surfaces [4, 5] is described as follows.

1. Transfer 50 μL of probe solutions (e.g., oligonucleotides, haptens, protein–hapten conjugates) to the wells of a 384-microwell plate (*see* **Note 18**).
2. Place the 384-microwell plate into the spotting robot's plate holder (*see* **Note 19**).
3. Place the slide into the spotting robot's slide holder. Be sure that activated surface is faced up.
4. Set the printing layout programming for probe spotting (i.e., 24 arrays of 5 × 5 spots each). An example is shown in Fig. 2 (*see* **Note 20**).

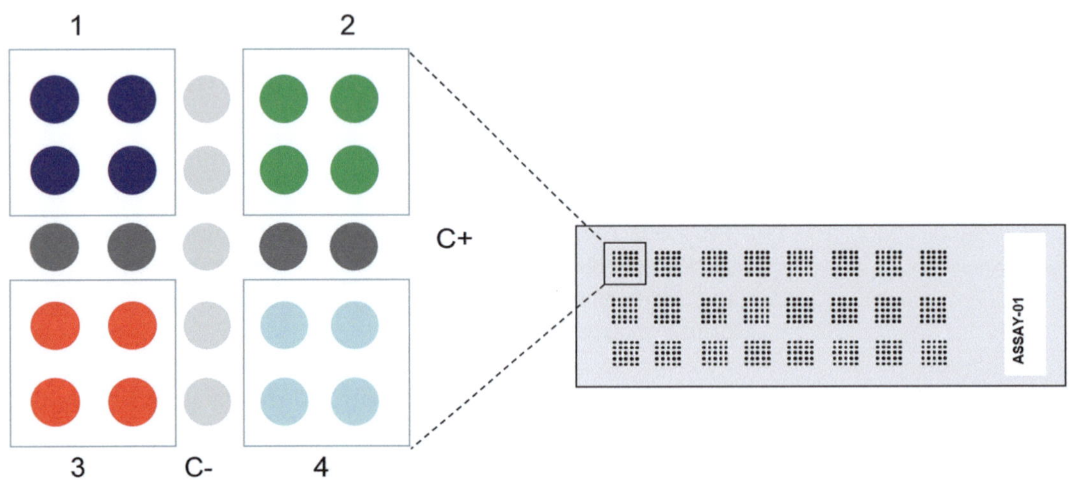

Fig. 2 Microarray layout: 24 arrays of 25 dots each (5 × 5). Spot-distance is 1 mm. Controls: C+ for positive (four replicates) and C– for negative (five replicates). *Boxes* 1–4: targeted probes (four replicates)

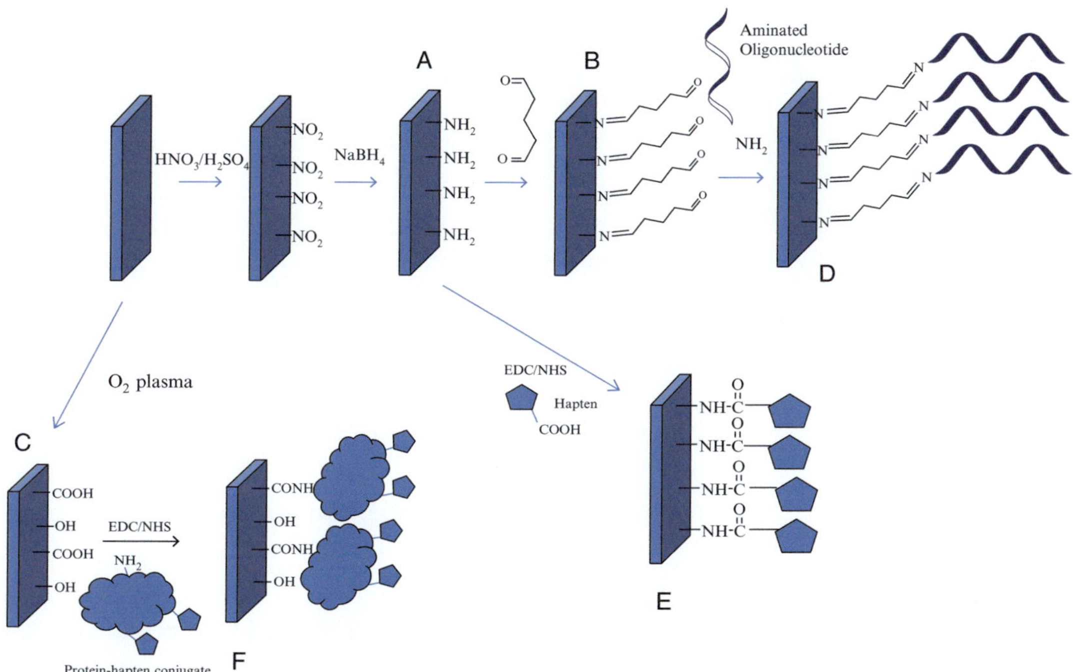

Fig. 3 Schemes of the different probe immobilization approaches conducted on the PC surface: *A* for amine-modified surface, *B* for aldehyde-modified surface, *C* for carboxylate-modified surface, *D* for oligonucleotides coated on the aldehyde-modified surface, *E* for haptens coated on the amine-modified surface, *F* for hapten–protein conjugate coated on the carboxylate-modified surface

5. Set the printing conditions: humidity 90 % and temperature 25 °C.
6. Adjust the height of slides (*see* **Note 21**).
7. Test that the spotting pins are all printing well and the arrayer positioning is properly calibrated by performing test prints (*see* **Note 22**).
8. Print probe solutions on PC slides. Repeat the process for replicate slides (*see* **Note 23**).

3.6 Covalent Attachment of Oligonucleotides on Aldehyde-Modified PC Surface

1. For printing oligonucleotides, aldehyde-modified PC slides are used (Fig. 3b).
2. Prepare a solution of aminated oligonucleotide 5 μM in CBB.
3. Transfer 50 μL of the aminated oligonucleotide solution to a 384-well plate.
4. Follow **steps 2–8** of Subheading 3.5 using oligonucleotide probes.
5. Incubate the arrayed surface at 42 °C for 1 h (*see* **Note 24**). Perform the incubation in a dark humid chamber (*see* **Note 25**).
6. Wash the slide with CBB, then with water, and blow dry (Fig. 3d) (*see* **Note 26**).

7. Prepare a solution of OVA of 1 % (w/v) in CBB.
8. Dispense 35 μL of OVA solution on the microarray slide to block the surface.
9. Incubate the slide at room temperature in the humid chamber for 30 min (*see* **Note 27**).
10. Wash the slide with PBST, then with water, and blow dry [6].
11. Store the slides at 4 °C until use.

3.7 Covalent Attachment of Haptens on Amine-Modified PC Surface

1. For printing haptens that consist of carboxylate group, amine-modified PC slides are used (Fig. 3a).
2. Dissolve 50 μmol of hapten, 50 μmol (5.8 mg) of NHS, and 50 μmol of DCC (10.3 mg) in 0.4 mL anhydrous DMF to prepare the hapten solution.
3. Incubate the slide in the hapten solution in the dark, at room temperature, for 2 h with gently stirring.
4. Centrifuge at 5223 × *g* for 15 min.
5. Take the supernatant and discard the acyl urea precipitate.
6. Prepare a 1/100 dilution of the supernatant in PBS 1× (*see* **Note 28**).
7. Transfer 50 μL of the solution to the wells of a 384-well plate.
8. Follow **steps 2–8** of Subheading 3.5 using the hapten solution.
9. Incubate the arrayed surface at room temperature for 16 h under controlled humidity conditions (*see* **Note 25**).
10. Wash the slide with PBS 1×, then with water, and blow dry (Fig. 3e) [7].
11. Store the slides at 4 °C until use.

3.8 Covalent Attachment of Protein–Hapten Conjugate on Carboxylate-Modified Polycarbonate Surface

1. For printing protein–hapten conjugates, carboxylate-modified PC slides are used (Fig. 3c).
2. Prepare a solution of 0.2 M NHS and 0.2 M EDC in MES buffer (pH 5.0) (*see* **Note 29**).
3. Immerse the carboxylate-modified PC slide in the solution and incubate for 30 min at room temperature.
4. Wash the slide with MES buffer.
5. Prepare a 1 mg/mL of a solution of protein–hapten conjugate (e.g., streptavidin-biotin conjugate) in MES buffer (*see* **Note 30**).
6. Transfer 50 μL of the protein–hapten conjugate solution to the wells of a 384-well plate.
7. Follow **steps 2–8** of Subheading 3.5. Incubate at 4 °C for 16 h in humid, dark chamber (*see* **Note 31**).
8. Wash the slide with MES buffer, then with water and blow dry (Fig. 3f).
9. Store the slides at 4 °C until use.

3.9 Competitive Immunoassays on Plastic Surfaces for Determination of Low Molecular Weight Analytes

1. Prepare standards of analytes (e.g., chlorpyriphos) of 85 μL in 0.2-mL polypropylene tubes by making fourfold serial dilutions, ranging from 0.016 to 16.0 μg/L of analyte in ultrapure water.
2. Add 10 μL PBST 10× to each standard to condition it.
3. For competitive immunoassay, the analyte is first mixed with the antibody which will be incubated with the protein–hapten conjugate attached on the microarray slide surface. To do this, add 5 μL rabbit anti-analyte antibody solution to the conditioned analyte standards and vortex the mixture for 5 s (standard solutions).
4. For sample analysis, repeat **steps 1–3** by using the sample analytes. Pipette 85 μL of samples into polypropylene tubes and proceed as before for sample conditioning and antibody addition.
5. Dispense 25 μL of standard solutions or samples onto the sensing zones, covering the microarray slide under a coverslip, and incubate the solutions for 15 min. The reaction of the analyte to the protein–hapten conjugate and the subsequent detection are depicted schematically in Fig. 4a.

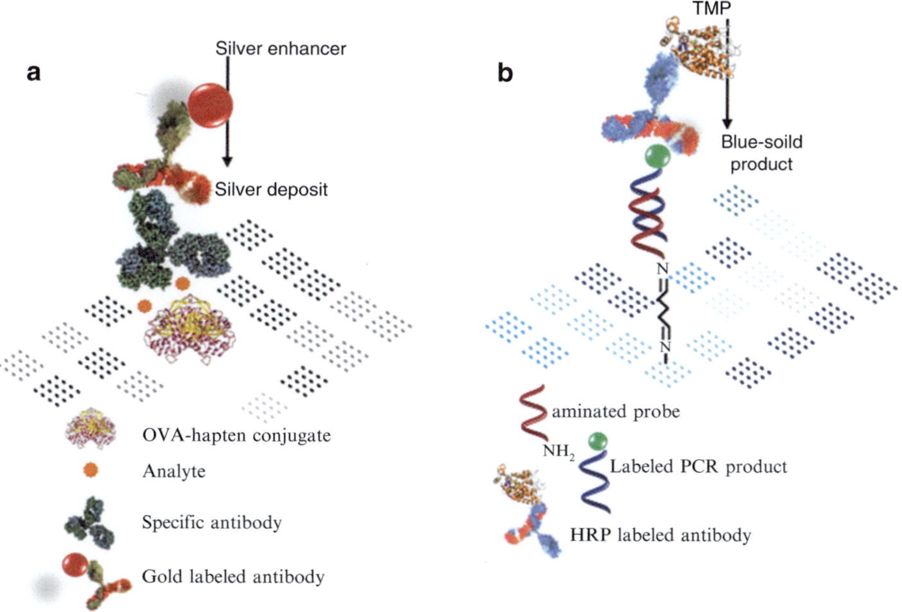

Fig. 4 Schemes of the assays conducted on the PC surface. (**a**) Whole process of a competitive immunoassay. The assay consists on a competitive step between the analyte (sample) and the hapten conjugate for the binding sites of the specific antibody. The immuno-interaction is indirectly detected using gold-labeled secondary antibody and silver enhancement signal amplification, rendering a metallic silver deposit. (**b**) DNA hybridization assay on an aldehyde-modified PC surface

6. Wash the slide with deionized water to remove unbound analytes.
7. Dispense 25 μL of 5-nm gold-labeled goat anti-rabbit antibody solution (1:100 dilution in detection buffer) onto each sensing zone and incubate it for 15 min.
8. Wash the slide with deionized water to remove unbound labeled antibody.
9. Prepare the silver enhancement solution by pipetting solutions A and B (500 μL each) in a polypropylene tube (1.5 mL) and vortex for 2 s.
10. Dispense the silver enhancement solution (25 μL) onto each sensing zone and incubate it for 8 min. The silver enhancement amplification produces a black precipitate whose optical density in inversely proportional to the analyte concentration.
11. Stop the amplification reaction by washing the slide with water.
12. Detect the black precipitate generated on the plastic surface with the naked eye, document scanner, smartphone camera, or an optical disc drive when standard optical discs are used as supports [8].

3.10 DNA Hybridization Assays on Plastic Surfaces for Detection of Pathogenic Bacteria

1. For PCR amplification, prepare the PCR mix (25 μL) in a 0.2-mL tube by adding 2.5 μL of PCR buffer 10×, 0.5 μL dNTP-stock solution, 1.0 μL of each forward and reverse primers, and 1.0 μL of digoxigenin-11-dUTP (1 μM). Keep the tubes on ice (*see* **Note 32**).
2. Add 5.5 μL template DNA, 2.5 μL Taq DNA Polymerase, and 11 μL nuclease-free water (*see* **Note 33**).
3. Gently mix the PCR mixture. Collect all liquid to the bottom of the tube by a quick spin, if necessary (*see* **Note 33**).
4. Remove the PCR tubes from ice and transfer them to a PCR machine with the heat block pre-heated to 95 °C and begin thermocycling (*see* **Note 34**).
5. Prepare 100 μL of tenfold serial dilutions of the amplified PCR product in the hybridization buffer.
6. Dispense 25 μL of each dilution onto each sensing zone and incubate the solutions at 37 °C, for 30 min (s).
7. Wash the sensing surface with deionized water.
8. Dispense 25 μL of horseradish peroxidase-labeled anti-digoxigenin monoclonal antibody solution (1/1000 dilution in PBST 1×) onto each sensing zone to detect the hybridization event. Incubate the solution for 15 min.
9. Wash the slide with deionized water to remove unbound labeled antibody.

10. Dispense 25 μL of TMB liquid substrate system onto each sensing zone and incubate for 10 min for the HRP-catalyzed TMB conversion to the oxidized form, which is a dark-blue precipitate.
11. Stop the enzymatic reaction by washing the slide with water.
12. Detect the colored insoluble product generated onto the plastic surface with the naked eye, document scanner, smartphone camera, or a disc drive when a standard optical disc was used [9].

4 Notes

1. Ovalbumin (OVA) is one of the recommended protein for the conjugation of low molecular weight target analytes to be immobilized on the microarray slide surface.
2. Other than polyclonal rabbit anti-analyte antibody, specific mouse monoclonal antibody can be used.
3. For detecting monoclonal, rather than polyclonal, antibodies, gold-labeled anti-mouse instead of anti-rabbit, IgG antibody (whole molecule) produced in goat should be used.
4. Follow the supplier recommendations. The Sephadex G-25 column specifications are exclusion limit: MW 5000; bed volume: 8–10 mL; sample volume: 1 mL.
5. Prepare 50 μL aliquots of the hapten–OVA conjugate for storage. This is to avoid repeated cycle freeze-thaw of the conjugate.
6. Add sulfuric acid very carefully to nitric acid and in small portions, since the solutions will heat up locally and pose the risk of splashing. The mixture will release acidic vapors, so the procedure must be conducted in a fume hood.
7. It is important not to exceed 65 °C by too much in order to avoid polymer degradation.
8. Shake the dish slightly for several times during heating. This step can be extended to a maximum of 30 min.
9. The dilution of the sulfonitric mixture in water results in a very exothermic reaction and local over-heating can produce undesired liquid splashing.
10. When dissolving $NaBH_4$, gas bubbles will be generated continuously in the solution. The gas generated is hydrogen, and so perform this step in a fume hood and stay away from ignition sources. If there is some $NaBH_4$ residue, it must be deactivated (hydrolyzed) before discarding it to the waste bin. Don't put the solid residue in contact with dry paper, as there is a risk of fire.

11. To confirm the presence of functional groups on the surface, a qualitative ninhydrin test for amine-modified surfaces is detailed as follows [10].

    Dissolve 33 mg KCN in 50 mL water. Dilute 2 mL of the KCN solution up to 100 mL with pyridine. This is reagent A.

    Dissolve 2.5 g ninhydrin in 50 mL ethanol (reagent B).

    Dissolve 80 g phenol in 20 mL *t*-butanol (reagent C).

    Cut a small piece of the polycarbonate (PC) slide and place it inside a dish. Add a few drops (2–3) of each reagent (A, B, and C) to the dish. Heat at 100 °C for 5 min. The test is positive for the presence of amine if the PC slide turns purple-violet. A control test should be performed using untreated PC slide.

12. It is important to maintain the pH of the glutaraldehyde mixture to be around 7.0. A higher pH induces the dialdehyde self-polymerization and the polycarbonate surface will lose its transparency.
13. Swirl the dish occasionally several times during the reaction.
14. The *p*-anisaldehyde test for aldehyde-modified surfaces:

    Prepare a solution with sulfuric acid (9 mL), ethanol (88 mL) and acetic acid (1 mL), and p-anisaldehyde (2.55 mL). This solution can be stored at 0 °C for several days. Cut a small piece of the PC slide and put it in a dish. Add a few drops of the solution to cover fully the surface and heat at 110 °C for 5 min. The test for the presence of aldehyde is positive if the surface turns dark purple.

15. Time can be increased to about 1 min.
16. The crystal violet test for carboxylate-modified surfaces:

    Prepare a 1 mM aqueous solution of crystal violet dye in 0.01 M NaOH.

    Cut a small portion of the PC slide and place it in a dish. Add a few drops of the dye solution until it fully covers the slide surface. Incubate at room temperature for 5 min. Remove the slide and wash it thoroughly with water. Introduce the slide in 80 % ethanol (i.e., 80 % ethanol and 20 % water), and then in 0.1 M HCl. If the slide surface turns violet, the test for the presence of carboxylate is positive. Alternatively, assuming 1:1 stoichiometry binding between the carboxylate group and the dye, the absorbance can be measured for quantification. The amount of carboxylate groups on the PC surface can be determined through the corresponding calibration curve and upon normalization for the slide surface area [11].

17. It is highly recommended to perform the bioreceptor conjugation to the surface by spotting immediately after finishing the carboxylate functionalization steps.
18. The solution volume depends on the number of slides to be spotted. Here 50 μL of solution is used for 16 slides.
19. Different spotters can be used for microarray spotter [12]. A non-contact printer is recommended (e.g., AD 1500 BioDot).
20. For a standard 25 mm × 75 mm e slide, a recommended layout consists of 24 arrays of 25 dots each (5 × 5): four spots for each analyte (four in total), four positive controls, and five negative controls. Other layouts can be designed, depending on the surface area and geometry [13]. Spots are distributed with a center-to-center distance of 1 mm. This layout is designed for a gasket applied to introduce 3 × 8 wells.
21. The height of polycarbonate slides can differ from that of slides made of other materials. Adjust the *Z*-position of the spotting nozzle using the software that controls the robot.
22. All pins should be able to print several hundred consecutive spots. In case of improper printing, rinse the pins with dilute acid solutions and water. Put the pins in a bath containing the dilute acid and sonicate them extensively, if necessary [14].
23. The dispensing of 50 nL of printing solutions by non-contact system onto PC slides produces spots of approximately 500 μm in diameter. In the recommended configuration, the array density is 1 spot/mm^2.
24. Alternatively, incubation can be conducted for 4 h at room temperature.
25. Humid conditions can be reached by placing a moisturized filter paper on the bottom of the container.
26. To determine the immobilization yield, fluorescent labeled oligonucleotides can be used. To this end, 1 μl of known concentrations of oligonucleotides is spotted on the surface, and fluorescence detected with a CCD camera or a fluorescence scanner.
27. The blocking step minimizes nonspecific adsorption, but it can also diminish the hybridization efficiency. Generally, for hybridization assays a blocking step is not required under the described conditions.
28. The dilution described here is only an example. However, with other carboxylate-modified haptens, it is recommended to test different dilution factors in order to reach the optimal value.
29. It is important to carefully maintain the pH at 5.0 in order to stabilize the active ester formed.
30. Alternatively, PBS1× pH 7.0 can be employed instead.

31. If the protein can tolerate the temperature, room temperature, instead of 4 °C, can be used.
32. The optimal concentration of the master mix components, including magnesium ion, must be determined empirically for each specific application. For *Salmonella* spp. detection, the PCR mix is prepared by adding 2.5 μL 10 × Taq standard polymerase reaction buffer, 0.5 μL dNTP-stock solution (with 10 mM of each dNTP), 1.0 μL of each forward and reverse primers (10 μM), and 1.0 μL of digoxigenin-11-dUTP (1 μM).
33. For detection of *Salmonella* spp., the primers pair consisting of LHNS 531 (5′-TACCAAAGCTAAACGCGCAGCT-3′) and RHNS 682 (5′-TGATCAGGAAATCTTCCAGTTGC-3′) were used to amplify a 152-bp region of the *hns* gene. The probe sequences were Probe 1: 5′-T10-GCTCGTCCGGCTA AATATAGCTATGTTG-3′, Probe 2: 5′- T10-CGAAAACGG TGAAACTAAAACCTGGAC-3′, Probe 3: 5′- T10-AGGGTC GTACACCGGCTGTAATCAAA-3′, Probe 4: 5′- T10-GCAA TGGAAGAACAAGGTAAGCAACTG-3′, C+: 5′- T10-GTCA TGGGCCTCGTGTCGGAAAACC-digoxigenin-3′,C-:5′-T10-TAGAGACTTAAAGAGGGAGCCCGGG-3′
34. Use high-quality and purified DNA templates. The recommended amounts of DNA template range from 1 ng to 1 μg. Mix all reagents in a PCR tube kept on ice. When needed, quickly transfer the tube to a thermocycler.
35. For *Salmonella* spp. detection, the PCR program starts with a denaturation step of 7 min at 95 °C, followed by 35 cycles at 95 °C (30 s), 62 °C (30 s), and 72 °C (30 s) and a final cycle at 72 °C for 4 min.

Acknowledgements

This research has been funded through FEDER Projects: GVA-PROMETEO II/2014/040 (Generalitat Valenciana) and MINECO CTQ/2013-45875-R.

References

1. Hook DG, AAnderson DG, Langer R, Williams P, Davies MC, Alexander MR (2010) High throughput methods applied in biomaterial development and discovery. Biomaterials 31:187–198
2. Bhattacharyya D, Fakirov S (2012) Synthetic polymer–polymer composites. Carl Hanser Verlag, Munich, Germany
3. Tamarit-López J, Morais S, Puchades R, Maquieira A (2011) Oxygen plasma treated interactive polycarbonate DNA microarraying platform. Bioconjug Chem 22:2573–2580
4. Campàs M, Katakis I (2004) DNA biochip arraying, detection and amplification strategies. Trends Anal Chem 23:49–62
5. Miller MB, Tang Y-W (2009) Basic concepts of microarrays and potential applications in clinical microbiology. Clin Microbiol Rev 22:611–633
6. Bañuls MJ, García-Piñón F, Puchades R, Maquieira A (2008) Chemical derivatization of

compact disc polycarbonate surfaces for SNPs detection. Bioconjug Chem 19:665–672

7. Tamarit-López J, Morais S, Bañuls MJ, Puchades R, Maquieira A (2010) Development of hapten-linked microimmunoassays on polycarbonate discs. Anal Chem 82:1954–1963
8. Morais S, Tamarit Lopez J, Carrascosa J, Puchades R, Maquieira A (2008) Analytical prospect of compact disk technology in immunosensing. Anal Bioanal Chem 391:2837–2844
9. Arnandis-Chover T, Morais S, Tortajada LA, Puchades R, Maquieira A, Olabarria G, Berganza J (2012) Detection of food-borne pathogens with DNA arrays on disk. Talanta 101:405–412
10. Vazquez J, Qushair G, Albericio F (2003) Qualitative colorimetric tests for solid phase synthesis. Methods Enzymol 369:21–27
11. Xu F, Datta P, Wang H, Gurung S, Hashimoto M, Wei S, Goettert J, McCarley RL, Soper SA (2007) Polymer microfluidic chips with integrated waveguides for reading microarrays. Anal Chem 79:9007–9013
12. Zubritsky E (2000) Spotting a microarray system. Anal Chem 72:761–767
13. Morais S, Tortajada-Genaro LA, Arnandis-Chover T, Puchades R, Maquieira A (2009) Multiplexed microimmunoassays on a Digital Versatile Disk. Anal Chem 81:5646–5654
14. McWilliam I, Kwan MC, Hall D (2011) Inkjet printing for the production of protein microarrays. In Protein Microarrays. Methods and Protocols. Korf U (Ed.). 345–361. Humana Press

Chapter 5

Detection and Quantification of MicroRNAs by Ligase-Assisted Sandwich Hybridization on a Microarray

Ryo Iizuka, Taro Ueno, and Takashi Funatsu

Abstract

Extracellular microRNAs (miRNAs) in body fluids have been identified as promising biomarkers for different human diseases. The high-throughput, multiplexed detection and quantification of these miRNAs are highly beneficial for the rapid and accurate diagnosis of diseases. Here, we developed a simple and convenient microarray-based technique, named ligase-assisted sandwich hybridization (LASH), for the detection and quantification of miRNAs. The LASH assay involves the hybridization of capture and detection probe pairs with the target miRNA to form a double-stranded structure which is then nick-sealed by T4 DNA ligase. Using this assay, we successfully demonstrated the multiplexed detection and quantification of different miRNAs in total RNA samples derived from blood obtained within 3 h. Here, we provide a detailed protocol for the LASH assay to detect a specific miRNA, as a model for the detection and quantification of extracellular miRNAs.

Key words microRNA (miRNA), Biomarker, Diagnosis, Sandwich hybridization, T4 DNA ligase

1 Introduction

MicroRNAs (miRNAs) are single-stranded non-coding RNAs of 18–25 nucleotides in length, and they are found in various organisms, including plants, and animals and viruses. The miRNAs bind to the complementary regions on one or more messenger RNAs (mRNAs), thereby promoting their degradation and inhibiting their translation. As a result, miRNAs play fundamental roles in different cellular processes such as cell proliferation, differentiation, and apoptosis, as well as in disease processes [1–3]. Previous studies have focused on extracellular miRNAs in body fluids such as serum, plasma, saliva, breast milk, and urine [4–6]. Despite high RNase activity in the body fluid [4, 5], these miRNAs are remarkably stable in it because they are complexed with proteins (e.g., Argonaute 2, high-density lipoprotein) [7, 8] and encapsulated in

Paul C.H. Li et al. (eds.), *Microarray Technology: Methods and Applications*, Methods in Molecular Biology, vol. 1368,
DOI 10.1007/978-1-4939-3136-1_5, © Springer Science+Business Media New York 2016

extracellular lipid-bilayer vesicles (e.g., exosomes and microvesicles) [9–11]. Recent studies have indicated that extracellular miRNAs are associated with various pathological conditions, including cancer [4–6, 12], and so these biomolecules are considered to be promising biomarkers for cancer and other diseases. Therefore, a method to facilitate the high-throughput detection and quantification of numerous miRNAs in clinical practice is required.

In general, microarrays are very useful for the high-throughput, multiplexed detection and quantification of miRNAs. However, in the conventional microarrays used for the detection and quantification of miRNA, there are various challenges, such as stoichiometric labeling with fluorophores (i.e., fluorophore-labeling bias), discrimination among highly homologous sequence variants, and discrimination between mature miRNAs and precursors (e.g., pri-miRNAs and pre-miRNAs). To address these challenges, we developed a microarray-based assay, namely ligase-assisted sandwich hybridization (LASH). Using this assay, we successfully demonstrated within 3 h the multiplexed detection and quantification of different miRNAs in total RNA samples derived from blood [13].

Figure 1 shows a schematic representation of the LASH assay used for miRNA detection. The LASH assay employs two different types of DNA oligonucleotide probes. The first one is C-probe, a

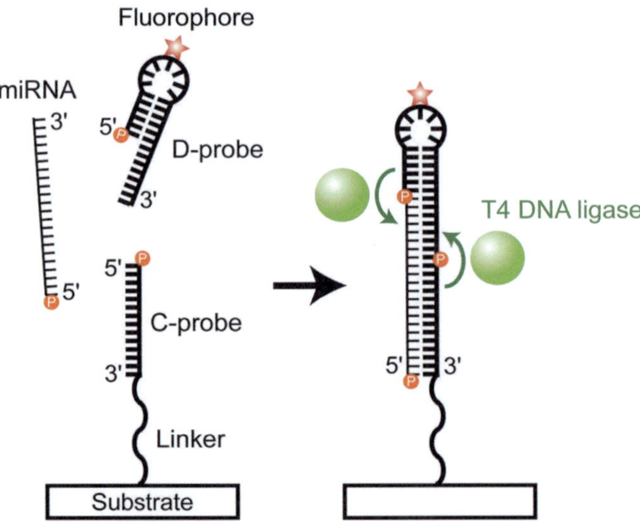

Fig. 1 Schematic representation of the ligase-assisted sandwich hybridization assay (LASH assay) for detecting miRNAs. A target miRNA hybridizes to complementary C-probe and D-probe, resulting in a tertiary complex. Both C-probe and D-probe are each capped with a phosphate group at the 5′ end and a hydroxyl group at the 3′ end. This permits the T4 DNA ligase to ligate the 5′ end of C-probe to the 3′ end of D-probe, and to ligate the 5′ end of D-probe to the 3′ end of the target miRNA

DNA oligonucleotide that is complementary to the 5′ half of the target miRNA, and the probe is immobilized on a glass surface via a flexible linker. The linker enhances the hybridization of the target miRNA to an immobilized probe on the surface [14, 15]. The other is D-probe, a DNA oligonucleotide with a stem-loop structure labeled with a fluorophore and the probe sequence is complementary to that of the 3′ half of the miRNA. Both probes carry a phosphate group at the 5′ end. The combination of the target miRNA, C-probe, and D-probe results in the formation of a double-stranded structure with nicks. The nicks can be sealed by T4 DNA ligase which works on dsDNA, but not on ssDNA (*see* **Note 1**). Consequently, the target miRNA can be detected fluorescently without labeling the miRNA that involves complex sample preparation procedures and variability in the labeling efficiency. In addition, the LASH procedure can be completed within 3 h, whereas conventional microarray experiments used for miRNA detection usually require 24 h.

The LASH assay is highly specific because T4 DNA ligase will only join nicks that are perfectly aligned on the oligonucleotide (i.e., no gaps or mismatches at the junction). Furthermore, the hybridization of the precursors of miRNAs (i.e., pre-miRNAs and pri-miRNAs) to the D-probe is prevented by its stem-loop structure, which acts as a size discriminator [16]. Note that miRNA is not released from C-probe and D-probe after ligation. Thus, miRNA cannot hybridize to another C-probe and D-probe pair, preventing multiple counts of hybridization to the same miRNA.

In the LASH assay, homologous miRNAs can be detected separately using D-probes labeled with different fluorescent dyes (Fig. 2). Here, we demonstrate an example of the detection of two human miRNAs, hsa-miR-141-3p and hsa-miR-200a-3p, which can serve as biomarkers for cancer detection [13, 17, 18] (*see* **Note 2**). There are differences of two nucleotides between hsa-miR-141-3p and hsa-miR-200a-3p (indicated as boxed nucleotides in Fig. 2a). The C-probe was designated to hybridize to the identical sequences of both miRNAs (indicated as underlined sequences), whereas D-probes were targeted to the rest of the sequences. D-probes targeted to hsa-miR-141-3p and hsa-miR-200a-3p were labeled with Alexa647 and Alexa532, respectively (Fig. 2a). The fluorescent signals were detected separately in the presence of target miRNAs (Fig. 2b). On the other hand, the fluorescent signals were only slightly detectable in the presence of non-target miRNAs (Fig. 2b). These results indicate that each D-probe individually hybridized to the target miRNA and that two-color detection enabled us to simultaneously quantify hsa-miR-141-3p and hsa-miR-200a-3p without significant cross-hybridization.

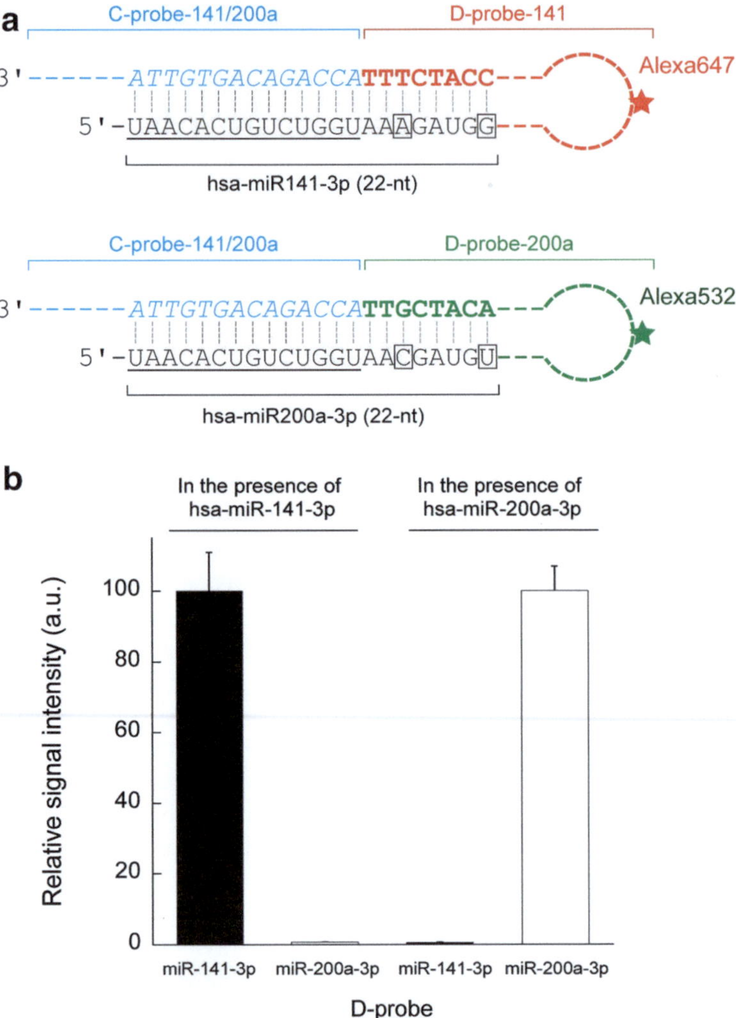

Fig. 2 Two-color detection of homologous miRNAs. (**a**) Design of probe sets used for the detection of hsa-miR-141-3p and hsa-miR-200a-3p. There are differences of two nucleotides between hsa-miR-141-3p and hsa-miR-200a-3p (indicated as *boxed* nucleotides), which are located in the sequence that binds to D-probes. In addition, the *underlined* sequences of both miRNAs are identical, and they can bind to the common C-probe (indicated in *italics*). (**b**) Detection of hsa-miR-141-3p and hsa-miR-200a-3p in the presence of the target and non-target D-probes. Data are presented as mean ± SD ($n = 3$)

2 Materials

miRNAs are readily degraded by RNase [5], and so RNase contamination should be avoided. Use RNase-free reagents, buffers, and water [e.g., diethylpyrocarbonate (DEPC)-treated water or Milli-Q water filtered through a BioPak ultrafiltration membrane] when preparing solutions.

2.1 Construction of Oligonucleotide Microarrays

1. C-probe for hsa-miR-143-3p (synthetic DNA oligonucleotide): 5′-(phosphate)GTGCTTCATCTCAACAACAACAACAACAACAACA(6-FAM)(SH)-3′. The oligonucleotide is phosphorylated at the 5′ end and modified with 6-FAM and a thiol group at the 3′ end (*see* **Note 3**). The underlined sequence is complementary to hsa-miR-143-3p (*see* **Note 4**). The synthesized oligonucleotide should be purified by HPLC. Dissolve the oligonucleotide in DEPC-treated water to obtain a concentration of 20 μM. Store the aliquots at −20 °C, and avoid multiple cycles of thawing and freezing.
2. DEPC-treated water or nuclease-free water.
3. 1 M Tris–HCl, pH 8.0.
4. 1 M dithiothreitol (DTT). Store at −20 °C.
5. 2× reduction buffer: 167 mM Tris–HCl (pH 8.0), 80 mM DTT.
6. Silane-PEG-Maleimide (silane-PEG-Mal) (MW 5000; Nanocs Inc., cat. no. PG2-MLSL-5k).
7. NAP-5 column (GE Healthcare, cat. no. 17-0853-01).
8. Non-stick RNase-free 1.5-mL microcentrifuge tube.
9. UV–visible spectrophotometer.
10. 20× SSC buffer: 3 M NaCl, 300 mM sodium citrate, pH 7.0.
11. 3× SSC buffer: 450 mM NaCl, 45 mM sodium citrate, pH 7.0.
12. Betaine solution.
13. Inkjet spotter.
14. Coverslips (e.g., 18 × 18 mM, 0.12–0.17 mm thick).
15. Humidity chamber (plastic petri dish containing wet Bemcot paper soaked in DEPC-treated water or Milli-Q water).
16. Parafilms.
17. Coverslip staining rack.
18. Glass petri dish (Φ60 × 90H mM).
19. W1 buffer: 0.1 % SDS, 300 mM NaCl, 30 mM sodium citrate, pH 7.0. This solution should be freshly prepared each time.
20. Benchtop reciprocal shaker (e.g., desktop shaking water bath).
21. Milli-Q water.
22. Compressed-gas blower.

2.2 Construction of the Hybridization Chamber

1. Metal disks (custom-made, 11 mm in diameter, 5 mm in height).
2. Square petri dish (e.g., Sterile 100 mm square petri dishes in a non-compartmentalized format).
3. Adhesive agent (e.g., Araldite AR-S30).

4. Polydimethylsiloxane (PDMS).
5. Disposable weighing boat.
6. Vacuum desiccator.
7. Water aspirator pump.
8. Oven.

2.3 Detection of miRNA

1. D-probe for hsa-miR-143-3p (synthetic DNA oligonucleotide): 5′-(phosphate)*CTCAACTGGTGTCG*TGGA(Alexa647)GT*CGGCA*AT*TCAGTTGAG*GAGCTACA-3′.

   The oligonucleotide is phosphorylated at the 5′ end and internally modified with Alexa647. The underlined sequence is complementary to hsa-miR-143-3p (*see* **Note 4**). The sequence in italics is a self-complementary sequence that can form a stem-loop structure. The synthesized oligonucleotide should be purified by HPLC. Dissolve the oligonucleotide in DEPC-treated water or nuclease-free water to obtain a concentration of 20 μM. Store the aliquots at −20 °C and avoid multiple cycles of thawing and freezing.
2. hsa-miR-143-3p (synthetic RNA oligonucleotide): 5′-UGAGAUGAAGCACUGUAGCUC-3′.

   The synthesized oligonucleotide should be purified by HPLC. Dissolve the oligonucleotide in DEPC-treated water or nuclease-free water to obtain a concentration of 100 μM. Store the aliquots at −20 °C and avoid multiple cycles of thawing and freezing.
3. 1 M Tris–HCl, pH 7.5.
4. 1 M $MgCl_2$.
5. 100 mM ATP. Store at −20 °C.
6. 50 mg/mL BSA. Use RNase-inactivated or RNase-free BSA. Store at −20 °C.
7. 50 % PEG6000 (e.g., Hampton Research, cat. no. HR2-533).
8. 1.5× hybridization buffer: 100 mM Tris–HCl (pH 7.5), 15 mM $MgCl_2$, 225 mM NaCl, 150 μM ATP, 15 mM DTT, 0.75 mg/mL BSA, 15 % PEG6000, 150 nM D-probe.
9. T4 DNA ligase.
10. 0.2× SSC buffer: 30 mM NaCl, 3 mM sodium citrate, pH 7.0.
11. Microplate shaker.
12. Paper towels.
13. Binder clips.
14. Metal plate.
15. Hybridization incubator.

2.4 Observation of Microarray Spots

1. Fluorescence microscope with a xenon lamp as the light source.
2. Objective lens [e.g., UPlanApo 20×, NA 0.70].
3. Excitation filter for 6-FAM detection [e.g., BP460-490 (Olympus)].
4. Dichroic mirror for 6-FAM detection [e.g., FF495-Di03-25 × 36 (Semrock)].
5. Emission filter for 6-FAM detection [e.g., HQ535/50m (Chroma Technology)].
6. Excitation filter for Alexa647 detection [e.g., FF01-628/40 (Semrock)].
7. Dichroic mirror for Alexa647 detection [e.g., Q660lp (Chroma Technology)].
8. Emission filter for Alexa647 detection [e.g., FF01-692/40 (Semrock)].
9. Electron-multiplying charge-coupled device camera.

2.5 Washing the Hybridization Chamber

1. W2 buffer: 1 % SDS, 300 mM NaCl, 30 mM sodium citrate, pH 7.0. This solution should be freshly prepared each time.

3 Methods

RNase contamination should be avoided by adopting the following precautions.

- Always wear gloves when handling reagents, materials, and equipment. Change the gloves after coming in contact with RNase-contaminated surfaces.
- Avoid using equipment and work areas that have been exposed to RNases. Clean the equipment and work surfaces with ethanol or commercially available RNase decontamination solutions (e.g., RNaseZap).

3.1 Preparation of the Oligonucleotide Microarray

In this protocol, C-probes are immobilized on coverslips via a PEG linker (Fig. 1). A custom-made DNA microarray can also be used [e.g., eArray (Agilent Technologies)]. A long spacer sequence [e.g., 15 repeats of trinucleotide (ACA)] should be incorporated on the 3′ end of C-probe. In addition, the 5′ ends of C-probes must be phosphorylated by T4 polynucleotide kinase prior to ligation.

1. Prepare 100 μL of the 2× reduction buffer. This solution is used to reduce the disulfide bonds and free the thiol groups for preparing the C-probe (*see* **Note 3**).
2. Mix 100 μL of C-probe solution (20 μM) with 100 μL of the 2× reduction buffer and incubate the mixture at room temperature for 1 h.

3. Remove DTT in the mixture using a NAP-5 column equilibrated with 3× SSC buffer. Determine the concentration of C-probe by measuring the absorbance.
4. Mix the C-probe with silane-PEG-Mal (the linker) at a molar ratio of 1:3 in the 3× SSC buffer for reaction and incubate at room temperature for 1 h.
5. Quench the reaction by adding DTT (10 mM) to the solution and incubate at room temperature for 1 h.
6. Prepare the probe printing solution by mixing 3 μM C-probe-PEG-silane with 750 mM betaine in the 3× SSC buffer (*see* **Note 5**).
7. Spot 1 nL of the printing solution on the coverslips using an inkjet spotter. Mark the coverslip to indicate its side that is printed with C-probes.
8. To ensure the quality of spot morphology and reproducibility of spot size, place the coverslip in a humidity chamber immediately after spotting, seal the chamber with parafilm, and then incubate in the dark at room temperature for 2 h.
9. Place the coverslips on a coverslip staining rack. Handle them carefully (*see* **Note 6**). Immerse the rack in a glass petri dish containing approximately 100 mL of W1 buffer for 10 min with gentle shaking using a benchtop reciprocal shaker. This step is to remove C-probes nonspecifically bound to the glass surface.
10. Rinse the coverslips with Milli-Q water (approximately 100 mL) while in a glass petri dish. Repeat this process three times (*see* **Note 6**).
11. Dry the coverslips using a compressed-gas duster.
12. Observe the microarray spots using a fluorescence microscope.
13. Measure the fluorescence intensities of five randomly selected spots in the absence and presence of C-probe. Calculate the average fluorescence intensities after background subtraction and determine the amount of immobilized C-probe using a standard curve (*see* **Note 7**).

3.2 Construction of the Hybridization Chamber

1. The hybridization chamber will be constructed by PDMS casting. First, prepare the mold of the chamber. Use an adhesive agent to stick metal disks at uniform intervals of approximately 20 mm on a square petri dish.
2. Measure 100 g of the PDMS base and 10 g of the curing agent (at a ratio of 10:1 w/w) in a disposable weighing boat and mix them thoroughly to prepare the PDMS prepolymer mixture.

3. Degas the mixture in a vacuum desiccator which is connected to a water aspirator pump until no bubbles remain on the surface of the PDMS prepolymer.
4. Slowly pour the degassed prepolymer mixture on the mold and heat it in an oven at 65 °C for 2 h.
5. Carefully peel off the casted PDMS from the mold.

3.3 Detection of miRNA

1. For experiment on each coverslip, prepare 200 μL of 1.5× hybridization buffer in non-stick RNase-free 1.5-mL microcentrifuge tubes. For more experiments, the total volume of the buffer can be made to be $200 \times (X + 0.5)$ where X is the number of coverslips.
2. Prepare 100 μL of hsa-miR-143-3p each of a fourfold serial dilution series (i.e., 20 fM–20 nM) in non-stick RNase-free 1.5-mL microcentrifuge tubes.
3. Add 200 μL of 1.5× hybridization buffer to the hsa-miR-143-3p solutions and then add T4 DNA ligase (final concentration of 5 U/μL).
4. Transfer the mixtures into the wells of the hybridization chamber, and cover them with the coverslips that are immobilized with the C-probes.
5. Fix the hybridization chamber on a microplate shaker using binder clips, as shown in Fig. 3.
6. Agitate the hybridization chamber at 1,200 rpm in an oven at 30 °C for 2 h.
7. Carefully place the coverslips on a staining rack (*see* **Note 6**). Immerse it in a glass petri dish containing 0.2× SSC buffer (approximately 100 mL) for 10 min with gentle shaking using a benchtop reciprocal shaker. This step will remove any D-probes nonspecifically bound to the glass surface. Repeat this process twice.
8. Dry the coverslips using a compressed-gas duster.
9. Measure the fluorescence intensities of five randomly selected spots in the absence and presence of C-probe. Calculate the average fluorescence intensities after background subtraction and determine the amount of D-probe ligated to C-probe and hsa-miR143-3p using a standard curve (*see* **Note 7**). An example of the results obtained is shown in Fig. 4.

3.4 Washing the Hybridization Chamber

The hybridization chamber can be used repeatedly after washing the wells.

1. Introduce 300 μL of W2 buffer into the wells of the hybridization chamber, and cover them with coverslips.
2. Fix the hybridization chamber on a microplate shaker, as shown in Fig. 3c.

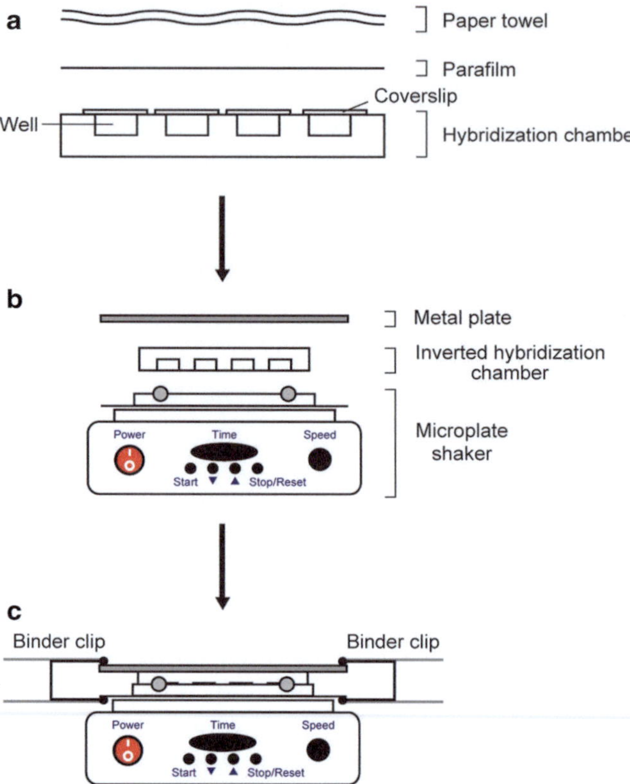

Fig. 3 Fixing the hybridization chamber on a microplate shaker. (**a**) Place the parafilm and paper towels on the hybridization chamber sealed with coverslips. (**b**) Sandwich the inverted hybridization chamber between a metal plate and a microplate shaker. (**c**) Clamp the edge of the metal plate to the microplate shaker using binder clips

3. Agitate the chamber on the shaker at 1,200 rpm in an oven at 37 °C overnight.
4. Rinse the wells with Milli-Q water, and dry them.

4 Notes

1. T4 DNA ligase catalyzes the formation of a phosphodiester bond between the adjacent 3′-hydroxyl and 5′-phosphate ends of double-stranded DNA, but it does not join single-stranded nucleic acids. This enzyme can also seal nicks in DNA–RNA hybrids with RNA strands via the 3′-hydroxyl end [19, 20].
2. In the present study, we assigned the names of the miRNAs according to the nomenclature defined by the miRBase database for miRNAs (http://www.mirbase.org/) [21]. In the case of hsa-miR141-3p, the first three letters (hsa) denote

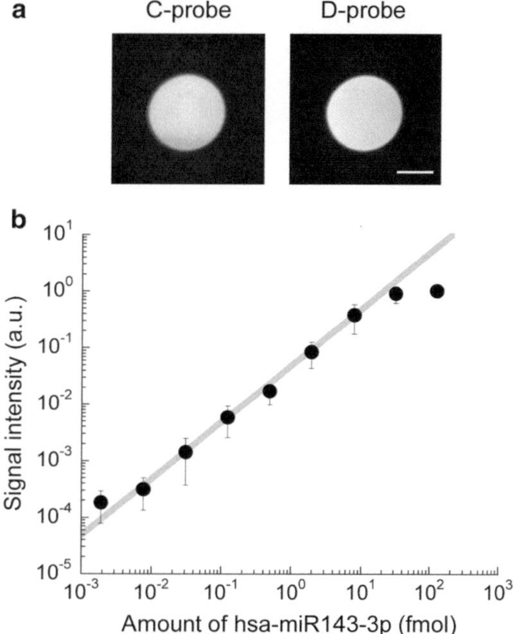

Fig. 4 Detection of hsa-miR-143-3p by the LASH assay. (**a**) Fluorescence images of 6-FAM-labeled C-probe (*left*) immobilized on a substrate and Alexa647-labeled D-probe (*right*) bound to C-probe in the presence of 1.25 nM hsa-miR-143-3p. Scale bar represents 100 μm. (**b**) The linearity of the plot is maintained between 10 amol and 10 fmol of the input hsa-miR-143-3p. The upper limit of the linearity depends on the amount of C-probe immobilized on the substrate. Data are presented as mean ± S.E. ($n = 5$)

Homo sapiens, the next three letters (miR) refer to mature miRNA, the number represents the gene designation, and the last two letters (3p) indicate that the miRNA is generated from the 3′ arm of the precursor miRNA.

3. In general, thiol-modified oligonucleotides are supplied in their oxidized form as a disulfide bond to protect the thiol group from undesired side reactions. Thus, it is necessary to reduce the disulfide bond in order to free the thiol group before use.
4. D-probe has eight nucleotides complementary to the 3′ end of the miRNA, and C-probe has 13 nucleotides complementary to the 5′ end of the target. In the detection of hsa-miR143-3p, this combination has yielded the highest sensitivity [13].
5. This probe printing solution can be stored for up to 1 month at 4 °C. However, 6-FAM is light-sensitive and thus the solution should be protected from exposure to light.
6. Handle the coverslips carefully. It is important to ensure that they do not stick to each other.

7. Spot a dilution series of C-probe and D-probe (1 nL per spot) on coverslips, and dry them. Measure the fluorescence intensities of the entire spots using a fluorescence microscope and create standard curves. Use the respective standard curves to determine the amounts of immobilized and ligated probes.

Acknowledgments

This research was partly supported by the Funding Program for World-Leading Innovative R&D on Science and Technology (FIRST Program) from the Japan Society for the Promotion of Science (JSPS) and by the Center of Innovation Program from Japan Science and Technology Agency (JST).

References

1. Bushati N, Cohen SM (2007) microRNA functions. Annu Rev Cell Dev Biol 23:175–205
2. Sayed D, Abdellatif M (2011) MicroRNAs in development and disease. Physiol Rev 91:827–887
3. Iorio MV, Croce CM (2012) MicroRNA dysregulation in cancer: diagnostics, monitoring and therapeutics. A comprehensive review. EMBO Mol Med 4:143–159
4. Chen X, Ba Y, Ma L, Cai X, Yin Y, Wang K, Guo J, Zhang Y, Chen J, Guo X, Li Q, Li X, Wang W, Zhang Y, Wang J, Jiang X, Xiang Y, Xu C, Zheng P, Zhang J, Li R, Zhang H, Shang X, Gong T, Ning G, Wang J, Zen K, Zhang J, Zhang CY (2008) Characterization of microRNAs in serum: a novel class of biomarkers for diagnosis of cancer and other diseases. Cell Res 18:997–1006
5. Mitchell PS, Parkin RK, Kroh EM, Fritz BR, Wyman SK, Pogosova-Agadjanyan EL, Peterson A, Noteboom J, O'Briant KC, Allen A, Lin DW, Urban N, Drescher CW, Knudsen BS, Stirewalt DL, Gentleman R, Vessella RL, Nelson PS, Martin DB, Tewari M (2008) Circulating microRNAs as stable blood-based markers for cancer detection. Proc Natl Acad Sci U S A 105:10513–10518
6. Weber JA, Baxter DH, Zhang S, Huang DY, Huang KH, Lee MJ, Galas DJ, Wang K (2010) The microRNA spectrum in 12 body fluids. Clin Chem 56:1733–1741
7. Arroyo JD, Chevillet JR, Kroh EM, Ruf IK, Pritchard CC, Gibson DF, Mitchell PS, Bennett CF, Pogosova-Agadjanyan EL, Stirewalt DL, Tait JF, Tewari M (2011) Argonaute2 complexes carry a population of circulating microRNAs independent of vesicles in human plasma. Proc Natl Acad Sci U S A 108:5003–5008
8. Vickers KC, Palmisano BT, Shoucri BM, Shamburek RD, Remaley AT (2011) MicroRNAs are transported in plasma and delivered to recipient cells by high-density lipoproteins. Nat Cell Biol 13:423–433
9. Valadi H, Ekström K, Bossios A, Sjöstrand M, Lee JJ, Lötvall JO (2007) Exosome-mediated transfer of mRNAs and microRNAs is a novel mechanism of genetic exchange between cells. Nat Cell Biol 9:654–659
10. Hunter MP, Ismail N, Zhang X, Aguda BD, Lee EJ, Yu L, Xiao T, Schafer J, Lee ML, Schmittgen TD, Nana-Sinkam SP, Jarjoura D, Marsh CB (2008) Detection of microRNA expression in human peripheral blood microvesicles. PLoS One 3:e3694
11. Kosaka N, Iguchi H, Yoshioka Y, Takeshita F, Matsuki Y, Ochiya T (2010) Secretory mechanisms and intercellular transfer of microRNAs in living cells. J Biol Chem 285:17442–17452
12. Cortez MA, Bueso-Ramos C, Ferdin J, Lopez-Berestein G, Sood AK, Calin GA (2011) MicroRNAs in body fluids—the mix of hormones and biomarkers. Nat Rev Clin Oncol 8:467–477
13. Ueno T, Funatsu T (2014) Label-free quantification of microRNAs using ligase-assisted sandwich hybridization on a DNA microarray. PLoS One 9:e90920
14. Southern EM, Mir KU, Shchepinov MS (1999) Molecular interactions on microarrays. Nat Genet 21:5–9
15. Chou CC, Chen CH, Lee TT, Peck K (2004) Optimization of probe length and the number

of probes per gene for optimal microarray analysis of gene expression. Nucleic Acids Res 32, e99

16. Chen C, Ridzon DA, Broomer AJ, Zhou Z, Lee DH, Nguyen JT, Barbisin M, Xu NL, Mahuvakar VR, Andersen MR, Lao KQ, Livak KJ, Guegler KJ (2005) Real-time quantification of microRNAs by stem-loop RT-PCR. Nucleic Acids Res 33:e179
17. Taylor DD, Gercel-Taylor C (2008) MicroRNA signatures of tumor-derived exosomes as diagnostic biomarkers of ovarian cancer. Gynecol Oncol 110:13–21
18. Li A, Omura N, Hong SM, Vincent A, Walter K, Griffith M, Borges M, Goggins M (2010) Pancreatic cancers epigenetically silence SIP1 and hypomethylate and overexpress miR-200a/200b in association with elevated circulating miR-200a and miR-200b levels. Cancer Res 70:5226–5237
19. Bullard DR, Bowater RP (2006) Direct comparison of nick-joining activity of the nucleic acid ligases from bacteriophage T4. Biochem J 398:135–144
20. Lohman GJ, Tabor S, Nichols NM (2011) DNA ligases. Curr Protoc Mol Biol 94:3.14.1–3.14.7
21. Griffiths-Jones S, Grocock RJ, van Dongen S, Bateman A, Enright AJ (2006) miRBase: microRNA sequences, targets and gene nomenclature. Nucleic Acids Res 34:D140–D144

Chapter 6

Probe Design Strategies for Oligonucleotide Microarrays

Nicolas Parisot, Eric Peyretaillade, Eric Dugat-Bony, Jérémie Denonfoux, Antoine Mahul, and Pierre Peyret

Abstract

Oligonucleotide microarrays have been widely used for gene detection and/or quantification of gene expression in various samples ranging from a single organism to a complex microbial assemblage. The success of a microarray experiment, however, strongly relies on the quality of designed probes. Consequently, probe design is of critical importance and therefore multiple parameters should be considered for each probe in order to ensure high specificity, sensitivity, and uniformity as well as potentially quantitative power. Moreover, to assess the complete gene repertoire of complex biological samples such as those studied in the field of microbial ecology, exploratory probe design strategies must be also implemented to target not-yet-described sequences. To design such probes, two algorithms, KASpOD and HiSpOD, have been developed and they are available via two user-friendly web services. Here, we describe the use of this software necessary for the design of highly effective probes especially in the context of microbial oligonucleotide microarrays by taking into account all the crucial parameters.

Key words KASpOD, HiSpOD, DNA microarrays, Probe design, Explorative probe

1 Introduction

The advancement of microarray technology (e.g., in situ synthesis technologies coupled to the evolution of microarrays slide formats) led oligonucleotide microarrays to become high-throughput molecular tools. Holding millions of probes on a single glass slide, the high-density oligonucleotide microarrays can monitor the presence and/or the expression, of thousands of genes, combining qualitative and quantitative aspects in only one experiment [1–3].

The success of a microarray experiment strongly depends on the determination of the best probe set while taking the biological question into account. For instance, transcriptome arrays or whole-genome arrays (WGAs) target a single organism whose genome is sequenced and known, whereas environmental DNA microarrays aim to study complex microbial mixtures of known and unknown microorganisms. Probe design is thus one of the most critical steps

Paul C.H. Li et al. (eds.), *Microarray Technology: Methods and Applications*, Methods in Molecular Biology, vol. 1368,
DOI 10.1007/978-1-4939-3136-1_6, © Springer Science+Business Media New York 2016

because the selected oligonucleotide probe set will have to provide (a) sensitivity (i.e., probes should detect low abundance targets in complex mixtures), (b) specificity (i.e., probes should not cross-hybridize with non-target sequences), and (c) uniformity (i.e., probes should display similar hybridization behaviors) [4, 5].

Improvements in the qualitative and quantitative aspects in the next-generation sequencing methods produce a substantial volume of sequence information and these methods will continue to accumulate large amounts of sequence data sets in public databases. It is now possible to take advantage of such sequencing data to develop comprehensive microarrays by using explorative probe design strategies. Such strategies offer the opportunity to survey both known and unknown sequences [6]. Explorative probe design strategy uses the sequence variability within the targeted sequences to predict new combinations that are potentially present in natural environments but have not yet been described and deposited in public databases.

The software is available (*see* Table 1) to help the user solve the current bottlenecks in the choice of high-quality probe sets. Each software program has its own advantages and drawbacks, and the choice of programs must be made in total accordance with the nature of projects and the content of the scientific question.

More recently, microarrays were adapted in a flexible and easy-to-use manner for profiling environmental communities in the area of microbial ecology [5, 7]. Indeed, designing oligonucleotide microarrays that can be used to survey the extreme diversity of microorganisms residing in various ecosystems represents a great challenge in the field of microbial ecology. Ever since several whole-genome arrays (WGAs) have been developed in the last few years, phylogenetic oligonucleotide arrays (POAs) and functional gene arrays (FGAs) are the two major approaches to assess diversity of microbial communities in the environment [5]. The POAs target the SSU rRNA genes, while the FGAs target key genes encoding enzymes involved in metabolic processes. These arrays are described in more detail in subsequent sections.

1.1 Phylogenetic Oligonucleotide Arrays (POAs)

To rapidly characterize the members of microbial communities present in complex environments, numerous phylogenetic oligonucleotide arrays (POAs) have been developed, and they target the SSU rRNA biomarker [8–15]. Manual approaches and fully automated software have both been developed to design POAs (Table 1).

Such probe design strategies are generally based on aligned input sequence data or on performing a multiple sequence alignment as the first step of the algorithm. To design probes with an optimal coverage of the target group, multiple sequence alignments are usually converted into consensus sequences that account for the sequence variability at each position. Then, the design programs search in conserved regions to select the best oligonucleotide probes.

Table 1
Comparison of oligonucleotide probe design software

Software	Reference	Application	Availability	URL
ARB (v 5.5)	[26]	POA	Downloadable, stand-alone GUI (L, M)	http://www.arb-home.de/
CaSSiS (v 0.5.0)	[27]	POA	Downloadable, command-line (L)	http://cassis.in.tum.de
PhylArray	[16]	POA	Web interface	http://g2im.u-clermont1.fr/serimour/phylarray
SSPD	[28]	POA	Web interface	http://ieg.ou.edu/SSPD/
ORMA	[29]	POA, FGA	Matlab Script	Upon request
KASpOD	[17]	POA, FGA, WGA-ORF	Web interface or command-line (L)	http://g2im.u-clermont1.fr/kaspod/
DEODAS (v 0.1.0)	[30]	FGA	Downloadable, GUI (L)	http://deodas.sourceforge.net/
ProDesign	[31]	FGA	Web interface	http://www.uhnresearch.ca/labs/tillier/ProDesign/ProDesign.html
ArrayOligoSelector (v 3.8.4)	[32]	FGA, WGA-ORF	Downloadable, command-line (L)	http://arrayoligosel.sourceforge.net
BOND	[33]	FGA, WGA-ORF	Downloadable, command-line (L, W, M)	www.csd.uwo.ca/~ilie/BOND/
CommOligo (v 2.0)	[34]	FGA, WGA-ORF	Downloadable, stand-alone GUI (W)	http://ieg.ou.edu/software.htm
HiSpOD	[3]	FGA, WGA-ORF	Web interface	http://g2im.u-clermont1.fr/hispod/
MProbe (v 2.0)	[35]	FGA, WGA-ORF	Downloadable, GUI (W)	http://www.biosun.org.cn/mprobe/
OligoArray (v 2.1)	[36]	FGA, WGA-ORF	Downloadable, command-line (L)	http://berry.engin.umich.edu/oligoarray2_1/
OligoPicker (v 2.3.2)	[37]	FGA, WGA-ORF	Downloadable, command-line (L)	http://pga.mgh.harvard.edu/oligopicker/
OligoWiz (v 2.3.0)	[38]	FGA, WGA-ORF	Downloadable client program, GUI (L, W, M)	http://www.cbs.dtu.dk/services/OligoWiz2

(continued)

Table 1 (continued)

Software	Reference	Application	Availability	URL
PRIMEGENS (v 2.0)	[39]	FGA, WGA-ORF	Web interface or command-line stand-alone (L, W)	http://primegens.org/
UPS 2.0	[40]	FGA, WGA-ORF	Web interface	http://array.iis.sinica.edu.tw/ups/
Mprime	[41]	WGA-ORF	Web interface	http://kbrin.a-bldg.louisville.edu/Tools/OligoDesign/MPrime.html
OliD	[42]	WGA-ORF	Downloadable, command-line (L)	Upon request
PICKY (v 2.2)	[43]	WGA-ORF	Downloadable, stand-alone GUI (L, W, M)	http://www.complex.iastate.edu/download/Picky/index.html
ProbeSelect	[44]	WGA-ORF	Available upon request, command-line (L)	http://stormo.wustl.edu/src/probeselect-src.tar
ChipD	[45]	WGA-tiling	Web interface	http://chipd.uwbacter.org/
MAMMOT (v 1.21)	[46]	WGA-tiling	Downloadable, local server	http://www.mammot.org.uk/
MOPeD	[47]	WGA-tiling	Web interface	http://moped.genetics.emory.edu/newdesign.html
OligoTiler	[48]	WGA-tiling	Web interface	http://tiling.gersteinlab.org/OligoTiler/oligotiler.cgi
PanArray (v 1.0)	[49]	WGA-tiling	Available upon request, command-line (L)	Upon request
Teolenn (v 2.0.1)	[50]	WGA-tiling	Downloadable, command-line (L)	http://transcriptome.ens.fr/teolenn/

POA phylogenetic oligonucleotide array, *FGA* functional gene array, *WGA-ORF* open-reading-frame oriented whole-genome array, *WGA-tiling* tiling of whole-genome array, *GUI* graphical user interface, *L* Linux, *M* MacOS, *W* Windows

The first software program dedicated to POAs that offered the possibility of designing explorative probes was the PhylArray program [16]. PhylArray was developed to survey whole microbial communities, including both known and unknown microorganisms, in complex environments. A degenerate consensus sequence is deduced from a multiple alignment of targeted SSU rRNA sequences. Degenerate candidate probes are then selected along the consensus sequence, and all the nondegenerate combinations deduced from the consensus sequence are checked for cross-hybridizations against a SSU rRNA database. Among the combinations derived from each degenerate probe, some correspond to sequences that have not yet been deposited in public databases, namely explorative probes. Such probes should, therefore, allow the detection of undescribed microorganisms belonging to the targeted taxon.

Even if PhylArray was used to account for all of the sequence variability within the targeted sequences, this software is limited in its ability to manage large input datasets due to the fact that it relies on initial multiple sequence alignments. Consequently, new probe design strategies are needed to define explorative probes based on large databases. KASpOD was developed to overcome this limitation, see the following section.

1.1.1 Introducing KASpOD

KASpOD stands for "K-mer based Algorithm for highly Specific and explorative Oligonucleotide Design" [17]. The user first provides two sequence datasets that correspond to the target group and the non-target group. KASpOD will subsequently search probes that cover the target group minimizing the cross-hybridization with the non-target sequences. This software consists of three computational stages (Fig. 1). The first stage of this algorithm is the extraction of every *k*-mer from the target and the non-target groups. Every *k*-mer found in both the target and the non-target groups is removed from the list of known probe candidates. In the second stage, a clustering step is then performed to gather probe sequences from the same genomic location. For each cluster, a degenerate consensus probe is deduced that accounts for the sequence variability within the cluster. Among the sequence combinations derived from each degenerate oligonucleotide, some correspond to sequences not previously included in the target group and therefore represent explorative probes. Finally, the last stage of the KASpOD algorithm consists of assessing the coverage (i.e., percentage of sequences within the target group matching with the oligonucleotide probe) and specificity of each degenerate consensus *k*-mer against the target and non-target groups.

1.1.2 Probe Design Parameters Using KASpOD

KASpOD defines group-specific signatures based on two input FASTA files: one for the target group and the second one for the non-target sequences. The user can choose the oligonucleotide

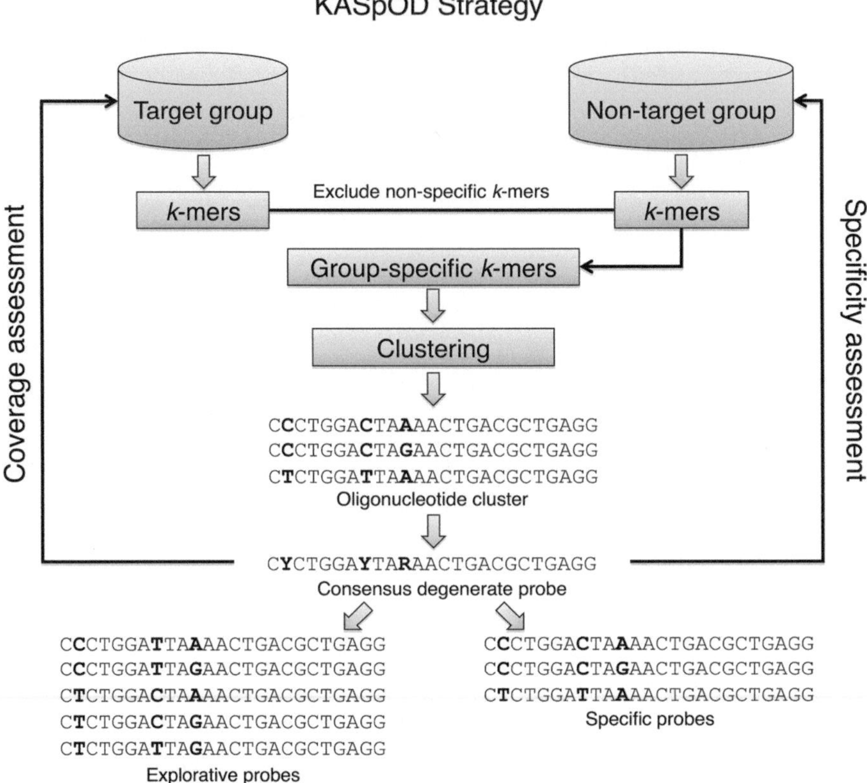

Fig. 1 Overview of the KASpOD algorithm

length (between 18 and 31 nucleotides) as well as the edit distance to perform both coverage and specificity assessments. The edit distance is defined as the upper limit of tolerated differences (gaps and/or mismatches) between the probe and its target (or non-target). For a classical POA design we suggest 25 mers probes and an edit distance threshold set to 2. KASpOD is also not restricted to construct POAs and can be used to design probes for other microarrays (WGAs or FGAs).

1.1.3 Probe Design Results Using KASpOD

Once the probe design is completed, KASpOD provides a downloadable CSV file containing the results. Each row represents a candidate probe and the columns correspond to: probe sequence, number of sequences in the target group, start position, end position, coverage (%), number of sequences in the non-target group, and the number of cross-hybridizations. Start and end positions are given for guidance since they are only defined according to the probe's best match in the target group.

1.2 Functional Gene Arrays (FGAs)

High-density oligonucleotide functional gene arrays (FGAs) provide the best high-throughput molecular tools to access the tremendous functional diversity of ecosystems [18]. Although most tools are limited to the determination of probes that target specific gene sequences within a single genome dataset, few strategies offer the opportunity to design probes that permit broad coverage of multiple sequence variants for a given gene family [6, 19] (Table 1).

1.2.1 Introducing HiSpOD

The HiSpOD (High Specific Oligo Design) program was developed [3] in this context of microbial ecology. HiSpOD (Fig. 2) allows the design of both gene-specific and sequence-specific probes. Gene-specific probes are computed using consensus sequences obtained after multiple alignments of several nucleic sequences belonging to the same gene family. All combinations deduced from the degenerate probes are then divided into regular probes for sequences available in databases, and explorative probes that represent putative new genetic signatures that do not correspond to any previously described sequence. On the other hand, sequence-specific probes can also be designed through HiSpOD by using nondegenerate nucleic acid sequences corresponding to a unique gene.

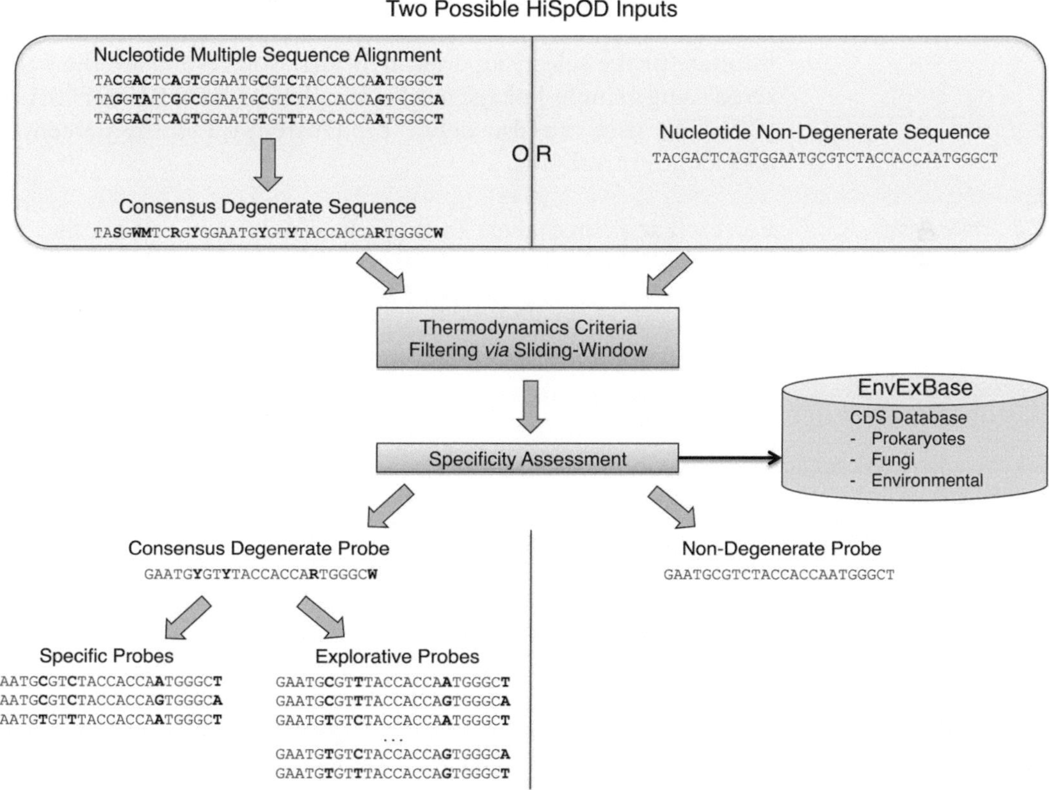

Fig. 2 Overview of the HiSpOD program workflow

To limit cross-hybridizations, the specificity of all selected probes is checked against a large formatted database dedicated to microbial communities, i.e., the EnvExBase (Environmental Expressed sequences dataBase), which is composed of 13,697,580 coding DNA sequences (CDSs) from the prokaryotic (PRO), fungal (FUN), and environmental (ENV) taxonomic divisions of the EMBL databank.

1.2.2 Probe Design Parameters Using HiSpOD

Unlike KASpOD, the HiSpOD program only needs a single input FASTA file containing at least one sequence. This file may be either degenerate or nondegenerate sequences. Thus, the user can submit (a) consensus sequences obtained after multiple alignments of nucleic sequences, or (b) separate nondegenerate nucleic sequences.

HiSpOD offers the classical parameters for the design of effective probes, including probe length, melting temperature, and complexity, and adds supplemental properties that were not considered by previous programs. Indeed the HiSpOD program performs specificity tests using the software called BLAST [20] with an expectation value defined by the user. The identity percentage threshold and the maximal continuous stretch of nucleotides between the probe and a non-target sequence to detect putative cross-hybridizations are also user-defined parameters. In order to facilitate probe selection, cross-hybridization results are then clustered using a single-linkage method implemented in BLASTCLUST [20]. The user can also define the clustering identity percentage and length thresholds.

1.2.3 Probe Design Results Using HiSpOD

For each sequence given in the submitted input FASTA file, two result files will be generated through HiSpOD: a .probes file providing the designed probes in FASTA format, and a .result file that summarizes the design results. The second file contains the probe sequence, its position on the sequence, and the clustered cross-hybridization results.

2 Materials

Both KASpOD and HiSpOD algorithms are provided through a web service, and a computer that can access the internet is therefore needed. There are requirements for the input files listed below (*see* **Note 1**).

2.1 KASpOD Requirements

1. A FASTA file containing the targeted nucleic sequences (e.g., the 16S rDNA sequence of the *Borrelia* genus).
2. A FASTA file containing the non-targeted nucleic sequences (e.g., the 16S rDNA sequence of all the *Spirochaetes* phylum except *Borrelia*) (*see* **Note 2**).

2.2 HiSpOD Requirements

A FASTA file containing (a) a consensus nucleic sequence for a gene-specific design or (b) several nondegenerate nucleic sequences for a sequence-specific design (*see* **Note 3**).

1. According to the IUPAC nomenclature, the allowed characters in the DNA sequence are: A, C, G, T, R, Y, M, K, W, S, B, D, H, V, and N (e.g., Y = {C,T}, H = {A,T,C}, and N = {A,C,G,T}).
2. HiSpOD specificity tests are performed against a comprehensive CDS database; users are therefore recommended to use CDSs as input for the probe design.
3. Consensus sequences can be obtained using the following strategy:
   - Multiple sequence alignment using tools such as ClustalW2 [21] or Muscle [22].
   - Consensus creation using stand-alone software such as Seaview [23], or web services (e.g., Consensus Maker http://www.hiv.lanl.gov/content/sequence/CONSENSUS/consensus.html) or custom scripts.

3 Methods

3.1 How to Design Group-Specific Probes Using KASpOD

This section summarizes the steps the user has to go through to perform a group-specific probe design using KASpOD.

1. Connect to the KASpOD web service by going to the KASpOD website: http://g2im.u-clermont1.fr/kaspod. Create an account ("Create account" tab on the left menu) or log in using your login and password information.
2. Start a new probe design job by clicking on the "New Job" tab on the left menu.
3. Select input files (*see* **Note 4**) by (a) selecting the target sequences file in FASTA format using the "Browse" button, and (b) browsing the non-target sequences file in FASTA format using the "Browse" button.
4. Customize probe design parameters (Fig. 3).
   (a) Set the oligonucleotide length from 18 to 31 nucleotides (*see* **Note 5**) [Default: 25].
   (b) Input the Edit distance by setting the maximal number of differences (mismatches/gaps) allowed between the probe and its target or non-target [Default: 2].
5. Launch the probe design query by pressing the "Launch" button.

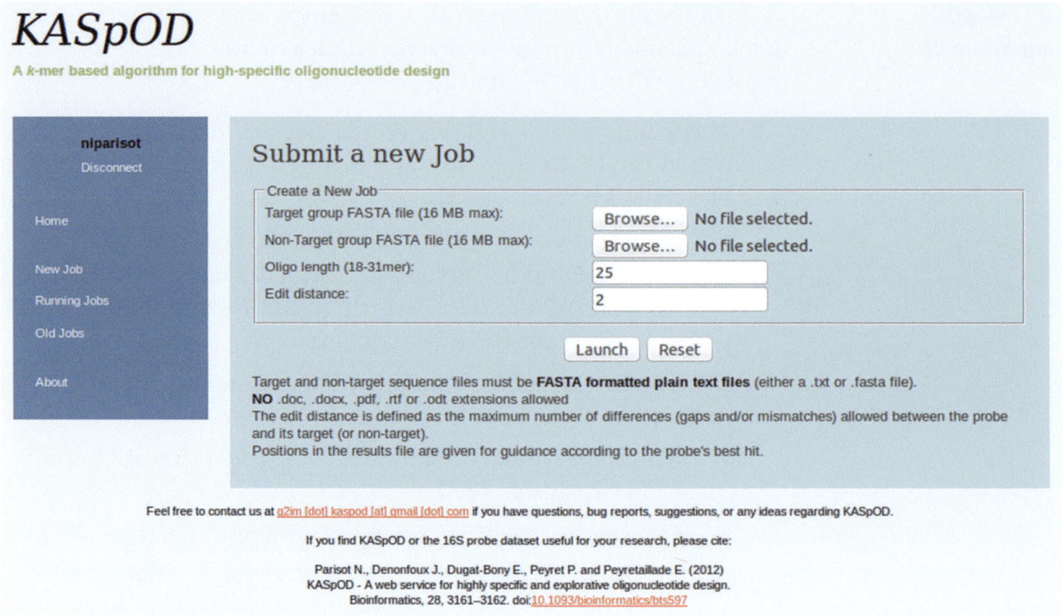

Fig. 3 Screenshot of the KASpOD "new design" tab

6. Wait for the server to finish processing the query. The status of the processing can be seen by clicking on the "Running jobs" tab on the left menu. Job status could be: "running" if the job is currently running on a CPU node, "queue" that means that the job will be launched as soon as a CPU node is free, or "waiting" if the job has not been submitted yet to the queue management system. The computation times range from hours to days depending on the size of both the target and the non-target sequences files (*see* **Note 6**).

7. Once the processing has completed, the job will appear in the "Old jobs" tab with a "done" status (Fig. 4). Otherwise, if the probe design encountered an error, the status will be "failed".

   To download the results, click on the "Old jobs" tab on the left menu. Then, click on the green arrow to download the .csv result file.

8. Load the data file. Double-click on the downloaded file to open it in a spreadsheet program such as Excel. The table can be sorted by decreasing coverage to help the selection of the best probes. We strongly recommend the users to select multiple probes per group in order to avoid misleading interpretation of hybridization data (e.g., cross-hybridizations, secondary structures, etc.)

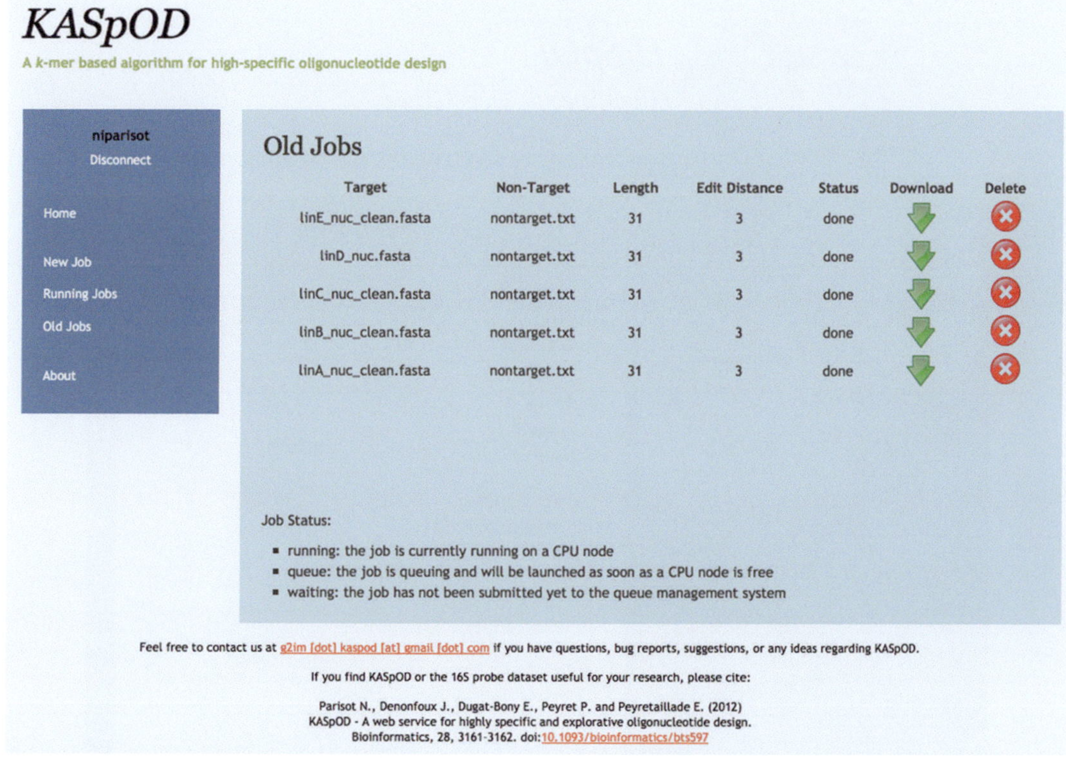

Fig. 4 Screenshot of the KASpOD "old jobs" tab

3.2 How to Define Protein-Coding Gene-Specific Probes Using HiSpOD

This section summarizes the different steps to perform a probe design for a gene that codes a protein using HiSpOD.

1. Connect to the HiSpOD web service. Go to the HiSpOD website: http://g2im.u-clermont1.fr/hispod/ and create an account ("Create Account" tab on the left menu) or log in using your login and password information.
2. Start a new probe design job by clicking on the "New design" tab on the left menu.
3. Customize "oligonucleotide generation" parameters (Fig. 5).
   (a) Oligonucleotide length: Set the probe length from 18 to 120 nucleotides in the first box [Default: 50 nucleotides].
   (b) Melting temperature (T_m) range: Set the T_m range (in °C) in the next two boxes. Melting temperature is computed using the following formula in which [*Na*+] is fixed by default at 0.5 M, where *w*, *x*, *y*, and *z* are the number of bases A, T, G, and C, respectively:

$$T_m = 79.8 + 18.5 \times \log_{10}([Na^+]) + 58.4 \times (yG + zC) / (wA + xT + yG + zC) + 11.8 \times (yG + zC)^2 / (wA + xT + yG + zC) - 920 / (wA + xT + yG + zC)$$

[Default: 64–79 °C].

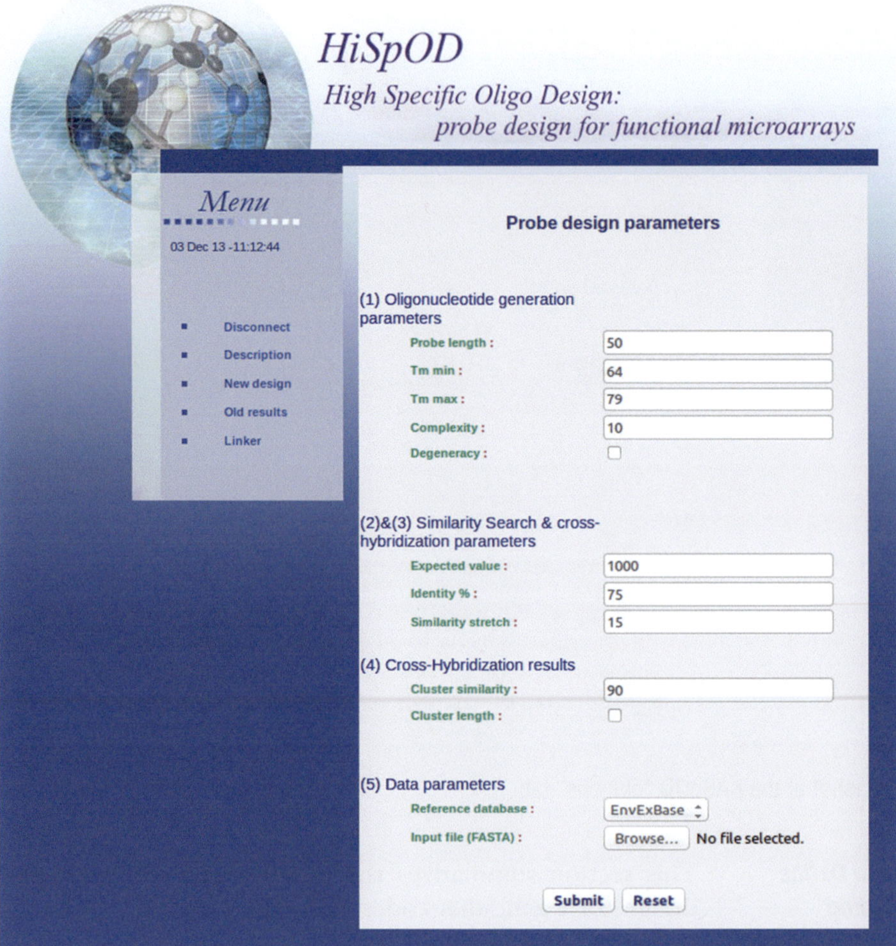

Fig. 5 Screenshot of the HiSpOD "new design" tab

(c) Complexity: click on the checkbox to set the complexity as the maximal number of successive identical nucleotides allowed in the candidate probes (from zero to the probe length). For instance, for a low-complexity oligonucleotide: "AGATGC**AAAAAAAAAA**GCTGACGTA", the complexity is 10. [Default: 10].

(d) Degeneracy: Click on the checkbox to set the maximal degeneracy for the candidate probe to a value from 1 to 32. For example, the following probe "GATGAT**YC**GTA **HG**TAGCTA**NC**TGAC" contains three degenerate oligonucleotides according to the IUPAC code. The degeneracy of this probe is therefore equal to $2 \times 3 \times 4 = 24$ [Default: 1].

4. Customize "Similarity search and cross-hybridization" parameters.

(a) Expectation value: Set the expectation value threshold for the specificity test using BLAST (from 0 to 40,000) [Default: 1000].

(b) Identity threshold: Set the minimal percentage identity threshold between the probe and a target to consider a cross-hybridization (from 0 to 100) [Default: 75 %].

(c) Identical nucleotide stretch: Set the minimal number of successive identical nucleotides between the probe and a target to consider a cross-hybridization (from 0 to probe length) [Default: 15 nucleotides]. Kane and colleagues [24] showed that for a given oligonucleotide any non-target sequence harboring at least 75 % identity over 50 nucleotides or sharing at least 15 contiguous identical bases will contribute to the overall signal intensity and may therefore lead to misleading interpretation of hybridization data.

5. Customize "Cross-hybridization" results.

   (a) Cluster similarity threshold: Set the percentage identity threshold to cluster cross-hybridizing sequences (from 0 to 100). [Default: 90 %].

   (b) Cluster length threshold: Click on the checkbox to set the sequence length threshold to cluster cross-hybridizing sequences (from 0 to 100). A value of 0 corresponds to a clustering without taking into account the sequence length. A value of 90 will allow clustering sequences that met the percentage identity threshold over an area covering 90 % of the length of each sequence [Default: 0].

6. Select input file (*see* **Note 7**) by browsing the target sequences file in FASTA format using the "Browse" button.

7. Launch the design query by pressing the "Submit" button. Verify your design parameters, and then click on the "Launch HiSpOD" button.

8. Wait for the server to finish processing the query. The status of the processing job can be seen by clicking on the "Old results" tab on the left menu. The run time for a nondegenerate single sequence (e.g., about 1000 nucleotides) is a few hours. An increasing number of sequences and degeneracy may lead to a computation time of over a few days.

9. Download the results. Once the processing job has completed, it will appear in the "Old results" tab with a "done" status.

   (a) Click on the "Old results" tab on the left menu.

   (b) Click on the "Download" link next to your job.

   (c) Click on the "Download all" link at the bottom of the page to retrieve all the files in a compressed archive (.tar.gz extension). The user may also right-click and select "Save the target as…" on each file you want to download.

10. Load the data file. Double-click on the downloaded .result file to open it in a text editor. Results are sorted by decreasing number of cross-hybridizations. We strongly recommend the users to select multiple probes per group to test in order to avoid misleading interpretation of hybridization data (e.g., cross-hybridizations, secondary structures, etc.). The position field of the result file allows the users to select probes from different regions of the gene.

4 Notes

1. Problems related to input data: The most common source of problems with running KASpOD or HiSpOD is problems with input data. Make sure that the data is in FASTA formatted plain text files. Notice that the file must be a text-only file, either a .txt or .fasta file for instance (an otherwise correctly formatted FASTA file within a MS-Word document will NOT work). NO .doc, .docx, .pdf, .rtf or .odt extensions allowed.
2. For the KASpOD program, make sure that the first file contains only the targeted sequences. The second file should contain only the non-targeted sequences. The presence of the same sequence in both files will lead to misleading probe design results.
3. For the HiSpOD program, please make sure that the input file contains the sequences of the genes which should be targeted. Submitting a file with a single large DNA sequence representing an entire microbial genome will not work (for comments on how to design a chromosomal tiling array *see* Ref. 25).
4. Large datasets problems. The web service is not dedicated to very large datasets.

   Input files size is limited to 16 MB for the KASpOD program. Since the KASpOD web service is not suitable for large datasets, a stand-alone version of KASpOD is available for download (64-bits GNU/Linux version). Registered users can download it through the "About" tab on the left menu.
5. Make sure that the input sequences are of a sufficient length. Entries that are shorter than the minimum probe length will be discarded.
6. Computations are distributed on a cluster (135 CPUs) and each job cannot be processed more than 30 days.
7. The web service is not dedicated to very large datasets. Input files size is limited to 2 MB for HiSpOD.

Acknowledgements

This work was supported by the French "Direction Générale de l'Armement" (DGA) and the programme Investissements d'avenir AMI 2011 VALTEX.

References

1. Tiquia SM, Wu L, Chong SC et al. (2004) Evaluation of 50-mer oligonucleotide arrays for detecting microbial populations in environmental samples, BioTechniques. 36: 664–70– 672–674–5
2. Marcelino LA, Backman V, Donaldson A et al (2006) Accurately quantifying low-abundant targets amid similar sequences by revealing hidden correlations in oligonucleotide microarray data. Proc Natl Acad Sci U S A 103: 13629–13634
3. Dugat-Bony E, Missaoui M, Peyretaillade E et al (2011) HiSpOD: probe design for functional DNA microarrays. Bioinformatics 27:641–648
4. Loy A, Bodrossy L (2006) Highly parallel microbial diagnostics using oligonucleotide microarrays. Clin Chim Acta 363:106–119
5. Wagner M, Smidt H, Loy A et al (2007) Unravelling microbial communities with DNA-microarrays: challenges and future directions. Microb Ecol 53:498–506
6. Dugat-Bony E, Peyretaillade E, Parisot N et al (2012) Detecting unknown sequences with DNA microarrays: explorative probe design strategies. Environ Microbiol 14:356–371
7. Zhou J (2003) Microarrays for bacterial detection and microbial community analysis. Curr Opin Microbiol 6:288–294
8. Loy A, Lehner A, Lee N et al (2002) Oligonucleotide microarray for 16S rRNA gene-based detection of all recognized lineages of sulfate-reducing prokaryotes in the environment. Appl Environ Microbiol 68:5064–5081
9. Wilson KH, Wilson WJ, Radosevich JL et al (2002) High-density microarray of small-subunit ribosomal DNA probes. Appl Environ Microbiol 68:2535–2541
10. Brodie EL, DeSantis TZ, Joyner DC et al (2006) Application of a high-density oligonucleotide microarray approach to study bacterial population dynamics during uranium reduction and reoxidation. Appl Environ Microbiol 72:6288–6298
11. Palmer C, Bik EM, Eisen MB et al (2006) Rapid quantitative profiling of complex microbial populations. Nucleic Acids Res 34:e5
12. Brodie EL, DeSantis TZ, Parker JPM et al (2007) Urban aerosols harbor diverse and dynamic bacterial populations. Proc Natl Acad Sci U S A 104:299–304
13. DeSantis TZ, Brodie EL, Moberg JP et al (2007) High-density universal 16S rRNA microarray analysis reveals broader diversity than typical clone library when sampling the environment. Microb Ecol 53:371–383
14. Hazen TC, Dubinsky EA, DeSantis TZ et al (2010) Deep-sea oil plume enriches indigenous oil-degrading bacteria. Science 330:204–208
15. Hazen TC, Rocha AM, Techtmann SM (2013) Advances in monitoring environmental microbes. Curr Opin Biotechnol 24:526–533
16. Militon C, Rimour S, Missaoui M et al (2007) PhylArray: phylogenetic probe design algorithm for microarray. Bioinformatics 23:2550–2557
17. Parisot N, Denonfoux J, Dugat-Bony E et al (2012) KASpOD--a web service for highly specific and explorative oligonucleotide design. Bioinformatics 28:3161–3162
18. He Z, Van Nostrand JD, Wu L et al (2008) Development and application of functional gene arrays for microbial community analysis. Trans Nonferrous Metals Soc China 18:1319–1327
19. Lemoine S, Combes F, Le Crom S (2009) An evaluation of custom microarray applications: the oligonucleotide design challenge. Nucleic Acids Res 37:1726–1739
20. Altschul SF, Gish W, Miller W et al (1990) Basic local alignment search tool. J Mol Biol 215:403–410
21. Thompson JD, Higgins DG, Gibson TJ (1994) CLUSTAL W: improving the sensitivity of progressive multiple sequence alignment through sequence weighting, position-specific gap penalties and weight matrix choice. Nucleic Acids Res 22:4673–4680
22. Edgar RC (2004) MUSCLE: multiple sequence alignment with high accuracy and high throughput. Nucleic Acids Res 32(5):1792–1797
23. Gouy M, Guindon S, Gascuel O (2010) SeaView version 4: a multiplatform graphical user interface for sequence alignment and phylogenetic tree building. Mol Biol Evol 27(2): 221–224

24. Kane MD, Jatkoe TA, Stumpf CR et al (2000) Assessment of the sensitivity and specificity of oligonucleotide (50mer) microarrays. Nucleic Acids Res 28:4552–4557
25. Parisot N, Denonfoux J, Dugat-Bony E et al. (2014) Software tools for the selection of oligonucleotide probes for microarrays. In Microarrays: current technology, innovations and applications, Z. He, ed. (Norfolk, UK: Caister Academic Press)
26. Ludwig W, Strunk O, Westram R et al (2004) ARB: a software environment for sequence data. Nucleic Acids Res 32:1363–1371
27. Bader KC, Grothoff C, Meier H (2011) Comprehensive and relaxed search for oligonucleotide signatures in hierarchically clustered sequence datasets. Bioinformatics 27: 1546–1554
28. Tu Q, He Z, Deng Y et al (2013) Strain/species-specific probe design for microbial identification microarrays. Appl Environ Microbiol 79(16):5085–5088
29. Severgnini M, Cremonesi P, Consolandi C et al (2009) ORMA: a tool for identification of species-specific variations in 16S rRNA gene and oligonucleotides design. Nucleic Acids Res 37:e109
30. Fredrickson HL, Perkins EJ, Bridges TS et al (2001) Towards environmental toxicogenomics — development of a flow-through, high-density DNA hybridization array and its application to ecotoxicity assessment. Sci Total Environ 274:137–149
31. Feng S, Tillier ER (2007) A fast and flexible approach to oligonucleotide probe design for genomes and gene families. Bioinformatics 23:1195–1202
32. Bozdech Z, Zhu J, Joachimiak MP et al (2003) Expression profiling of the schizont and trophozoite stages of Plasmodium falciparum with a long-oligonucleotide microarray. Genome Biol 4:R9
33. Ilie L, Mohamadi H, Golding GB et al (2013) BOND: basic oligonucleotide design. BMC Bioinformatics 14:69–69
34. Li X, He Z, Zhou J (2005) Selection of optimal oligonucleotide probes for microarrays using multiple criteria, global alignment and parameter estimation. Nucleic Acids Res 33: 6114–6123
35. Li W, Ying X (2006) Mprobe 2.0: computer-aided probe design for oligonucleotide microarray. Appl Bioinforma 5:181–186
36. Rouillard J-M, Zuker M, Gulari E (2003) OligoArray 2.0: design of oligonucleotide probes for DNA microarrays using a thermodynamic approach. Nucleic Acids Res 31:3057–3062
37. Wang X, Seed B (2003) Selection of oligonucleotide probes for protein coding sequences. Bioinformatics 19:796–802
38. Wernersson R, Nielsen HB (2005) OligoWiz 2.0 – integrating sequence feature annotation into the design of microarray probes. Nucleic Acids Res 33:W611–W615
39. Xu D, Li G, Wu L et al (2002) PRIMEGENS: robust and efficient design of gene-specific probes for microarray analysis. Bioinformatics 18:1432–1437
40. Chen SH, Lo CZ, Su SY et al (2010) UPS 2.0: unique probe selector for probe design and oligonucleotide microarrays at the pangenomic/genomic level. BMC Genomics 11(Suppl 4):S6
41. Rouchka EC, Khalyfa A, Cooper NGF (2005) MPrime: efficient large scale multiple primer and oligonucleotide design for customized gene microarrays. BMC Bioinformatics 6:175
42. Talla E, Tekaia F, Brino L et al (2003) A novel design of whole-genome microarray probes for Saccharomyces cerevisiae which minimizes cross-hybridization. BMC Genomics 4:38
43. Chou HH, Hsia AP, Mooney DL et al (2004) Picky: oligo microarray design for large genomes. Bioinformatics 20:2893–2902
44. Li F, Stormo GD (2001) Selection of optimal DNA oligos for gene expression arrays. Bioinformatics 17:1067–1076
45. Dufour YS, Wesenberg GE, Tritt AJ et al (2010) chipD: a web tool to design oligonucleotide probes for high-density tiling arrays. Nucleic Acids Res 38:W321–W325
46. Ryder E, Jackson R, Ferguson-Smith A et al (2006) MAMMOT--a set of tools for the design, management and visualization of genomic tiling arrays. Bioinformatics 22:883–884
47. Patel VC, Mondal K, Shetty AC et al (2010) Microarray oligonucleotide probe designer (MOPeD): a web service. Open Access Bioinformatics 2:145–155
48. Bertone P, Trifonov V, Rozowsky JS et al (2006) Design optimization methods for genomic DNA tiling arrays. Genome Res 16:271–281
49. Phillippy AM, Deng X, Zhang W et al (2009) Efficient oligonucleotide probe selection for pan-genomic tiling arrays. BMC Bioinformatics 10:293
50. Jourdren L, Duclos A, Brion C et al (2010) Teolenn: an efficient and customizable workflow to design high-quality probes for microarray experiments. Nucleic Acids Res 38:e117

Part III

Applications of Microarray Technology to Nucleic Acid Assays

Chapter 7

Analyzing Illumina Gene Expression Microarray Data Obtained From Human Whole Blood Cell and Blood Monocyte Samples

Alexander Teumer, Claudia Schurmann, Arne Schillert, Katharina Schramm, Andreas Ziegler, and Holger Prokisch

Abstract

Microarray profiling of gene expression is widely applied to studies in molecular biology and functional genomics. Experimental and technical variations make not only the statistical analysis of single studies but also meta-analyses of different studies very challenging. Here, we describe the analytical steps required to substantially reduce the variations of gene expression data without affecting true effect sizes. A software pipeline has been established using gene expression data from a total of 3358 whole blood cell and blood monocyte samples, all from three German population-based cohorts, measured on the Illumina HumanHT-12 v3 BeadChip array. In summary, adjustment for a few selected technical factors greatly improved reliability of gene expression analyses. Such adjustments are particularly required for meta-analyses of different studies.

Key words Gene expression analysis, Transcriptomics, Human whole blood cells, Human monocytes, Microarray, ILLUMINA HumanHT-12, BeadChip

1 Introduction

Global gene expression studies are widely conducted for research in molecular biology and functional genomics [1]. These studies have successfully provided new insights into the etiology of common diseases [2]. Although gene expression studies have been successfully applied to a wide range of clinical issues, these studies are often criticized for low robustness and lack of reproducibility [3, 4]. Other concerns include improper statistical analysis or validation, insufficient control of false positives, and inadequate reporting of methods [1].

When the results of multiple independent gene expression studies are available, the combined analysis or even a meta-analysis of the data can increase their reliability and generalizability [1].

Paul C.H. Li et al. (eds.), *Microarray Technology: Methods and Applications*, Methods in Molecular Biology, vol. 1368, DOI 10.1007/978-1-4939-3136-1_7,

However, meta-analyses of global gene expression studies remain challenging, which include problems that are common to traditional meta-analyses [5] such as differences in study design as well as concerns that are specific for analyzing gene expression data [1, 6]. One of the latter concerns is related to the technology used. Specifically, different types of microarrays vary in several important aspects, such as length of probes, range of probes' measurement values or coefficients of variation [1, 6], and therefore cross-platform comparisons are difficult to perform.

Even if several independent studies use the same microarray technology, there may be study-specific variations, originating from differences in experimental procedures, laboratory protocols, sample preparation [1, 6], or the type of tissue samples used [7]. In addition, different preprocessing of the data as well as batch effects (e.g., due to grouping of samples being processed) may lead to differences in measured gene expression values in large sample size studies.

Here, we describe the analytical steps to process gene expression data in terms of mRNA abundances measured on the Illumina HumanHT-12 v3 BeadChip array to circumvent the problems mentioned. First, we compared the $\log_2$ transformation of intensity values [8] with the recently proposed variance-stabilizing transformation (VST) [9]. It turned out that both methods are valid, but the commonly applied $\log_2$ transformation was simpler to implement. Next, we searched for main factors correlating with the overall expression values. Since intra-study variation is often corrected for by adjusting for the eigen-genes of the study obtained by a principal component analysis [10, 11], we analyzed the correlation between the eigen-genes and technical as well as biological factors. Our data demonstrate that the variation of gene expression signal intensities can be reduced by appropriate technical covariate adjustment. Previously, doubts have been raised about the suitability of using probes containing single nucleotide polymorphisms (SNPs) in gene expression studies [11]. We therefore investigated to what extent the signal intensities were affected by mismatch alleles of SNPs within probes. In summary, we propose the adjustment for selected technical covariates minimizing the technical variation of gene expression data without affecting true effect sizes, thus reducing the number of false negatives in the association results.

Analytical steps implemented into a software pipeline have been established using gene expression data from a total of 3358 whole blood cell and blood monocyte samples, from three German population-based cohorts, i.e., the Study of Health in Pomerania (SHIP-TREND) [12], the Cooperative Health Research in the Region of Augsburg (KORA) [13], and the Gutenberg Health Study (GHS) [14]. These three cohorts are part of the MetaXpress (Meta-Analysis of Gene Expression) Consortium within the German Center for Cardiovascular Disease (DZHK). For details regarding the analyses that led to this software pipeline, see Ref. 15.

2 Materials

Perform sample preparation and mRNA array hybridization according to manufacturer's guidelines. An example is given below. This project was conducted using fasting samples only. The workflow for processing samples from whole blood cells and monocytes is shown in Fig. 1.

2.1 RNA Isolation and Quality Control

1. PAXgene tubes (BD, Heidelberg, Germany).
2. PAXgene Blood miRNA Kit (Qiagen, Hilden, Germany). The kit is designed for the purification of total RNA longer than ~18 nucleotides.
3. QIAcube (Qiagen).
4. NanoDrop ND-1000 UV-Vis Spectrophotometer (Thermo Scientific, Hennigsdorf, Germany).
5. 2100 Bioanalyzer and RNA 6000 Nano Lab Chips (both from Agilent Technologies, Santa Clara, CA, USA).

2.2 Amplification, Labeling, and Hybridization

1. 96-well plates.
2. Illumina TotalPrep-96 RNA Amplification Kit (Ambion, Darmstadt, Germany).
3. biotin-UTP.
4. HumanHT-12 v3 Expression BeadChip array (Illumina, San Diego, USA).
5. Illumina Bead Array Reader.

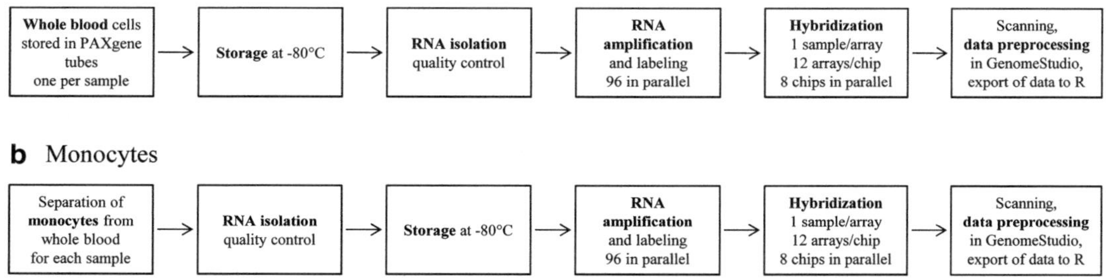

Fig. 1 Workflow—from blood sampling to measured mRNA intensities for whole blood cells and monocytes. (**a**) From *left* to *right*: Whole blood was taken and stored in PAXgene tubes until all blood samples had been collected. The processing continued with isolation of RNA from whole blood. (**b**) Monocytes were separated from whole blood and RNA was isolated from monocytes within 24 h after blood sampling; subsequently the isolated RNA samples were stored until all RNA samples had been collected. The samples were processed in 96-well plates both after isolation and amplification of the RNA. The corresponding plate layouts were called (**a**) *RNA isolation batch* and (**b**) *RNA amplification batch*. Finally, the RNA was hybridized and the arrays were scanned, and the expression data was processed and analyzed (reproduced from Ref. 15 with permission from Creative Commons Attribution (CC BY) license)

2.3 Preparation of Data Analysis

1. Illumina's GenomeStudio software and the Gene Expression Module. To install the software, a PC with Windows XP/Vista/7 64-bit, at least 8 GB RAM and a 100 GB hard drive is required.
2. R statistics software [16] including the lumi package [8] is open-source software that can be obtained from Bioconductor (http://www.bioconductor.org).

3 Methods

3.1 RNA Isolation and Quality Control

1. Collect and store whole blood samples in PAXgene tubes at −80 °C until processing. Use the PAXgene Blood miRNA Kit to isolate RNA. This isolation can be achieved manually or by using a QIAcube for automated isolation according to protocols provided by the manufacturer. Alternatively, isolate RNA from monocytes within 24 h after blood sampling to ensure rapid sample processing. Separate monocytes from whole blood and prepare RNA as previously described [14].
2. Determine purity and concentration of RNA using the NanoDrop Spectrophotometer.
3. Ensure a consistently high RNA quality by analyzing all preparations using a 2100 Bioanalyzer and RNA 6000 Nano Lab Chips according to the manufacturer's instructions. As quality control, exclude samples exhibiting a RNA integrity number (RIN) less than seven from further analyses.

3.2 Amplification, Labeling, and Hybridization

1. Process RNA samples in 96-well plates and reverse transcribe 500 ng of RNA into cRNA (200 ng in case of monocyte samples) using the Illumina TotalPrep-96 RNA Amplification Kit, thereby labeling the cDNA with biotin-UTP (*see* **Note 1**).
2. Hybridize 3000 ng of labeled cRNA (700 ng in case of monocyte samples) to the Illumina HumanHT-12 v3 Expression BeadChip array, and subsequently perform washing steps as described in the Illumina protocol.
3. Scan the BeadChips using the Illumina Bead Array Reader.

3.3 Preparation of Data Analysis

1. Install the Illumina's GenomeStudio software and the Gene Expression Module on a PC. For using alternative open-source software (*see* **Note 2**).
2. Install R statistics software including the lumi package from Bioconductor (http://www.bioconductor.org/) for subsequent data normalization, data analysis, and association testing.

3.4 Quality Control and Data Exporting

1. In GenomeStudio, create a new project. Import the scanned microarray data.
2. Perform imputation which replaces missing values using the default setting of the Gene Expression Module. There should be less than ten of the 48,000 probes per array that need to be imputed.
3. For quality control, exclude samples with less than 6000 significantly detected probes (detection p-value <0.01).
4. Export the sample-probe profile of the gene expression data as final report (including the average signal intensity, bead standard error, number of beads, and detection p-value). For further analysis, such as normalization and association, load this data into the R statistics software (*see* **Note 2**).

3.5 Data Normalization and Transformation

If not stated otherwise, all steps in this subsection are performed using the R statistical software.

1. Perform quantile normalization on the gene expression data for each microarray, e.g., by using the lumi package.
2. Perform a log transformation to base two (log2 transformation) on the normalized data. This step can also be accomplished using the lumi package. Variance-stabilizing transformation (VST) can be used as an alternative transformation (*see* **Note 3** and Fig. 2).
3. Estimate the sex of the samples by performing a principal component analysis on the gene expression intensities of the probes that have DNA sequences located on the X and Y chromosomes (Fig. 3).
4. Exclude samples that have a mismatch between the estimated gender and the reported (true) gender.
5. Save the processed data as RData or a text file (*see* **Note 4**) with probe names as rownames and sample IDs as columnnames. Set the Illumina Array_Address_Id as the probe identifier.
6. Other than Array_Address_Id, the Nucleotide Universal Identifier (nuID) could be used (*see* **Note 5**).
7. Use the saved $\log_2$ transformed gene expression data for subsequent association analyses. This data has been stored in a text table having each probe (gene) in rows and the samples in columns (*see* **Note 6**).

3.6 Covariate and Phenotype Selection

1. Select the following covariates that should be included in every analysis to reduce variations due to sample processing: (a) *RNA amplification batch* (each 96-well plate on which the samples were processed together for RNA amplification and labeling) (*see* **Note 7**), (b) RNA Integrity number (*RIN)*

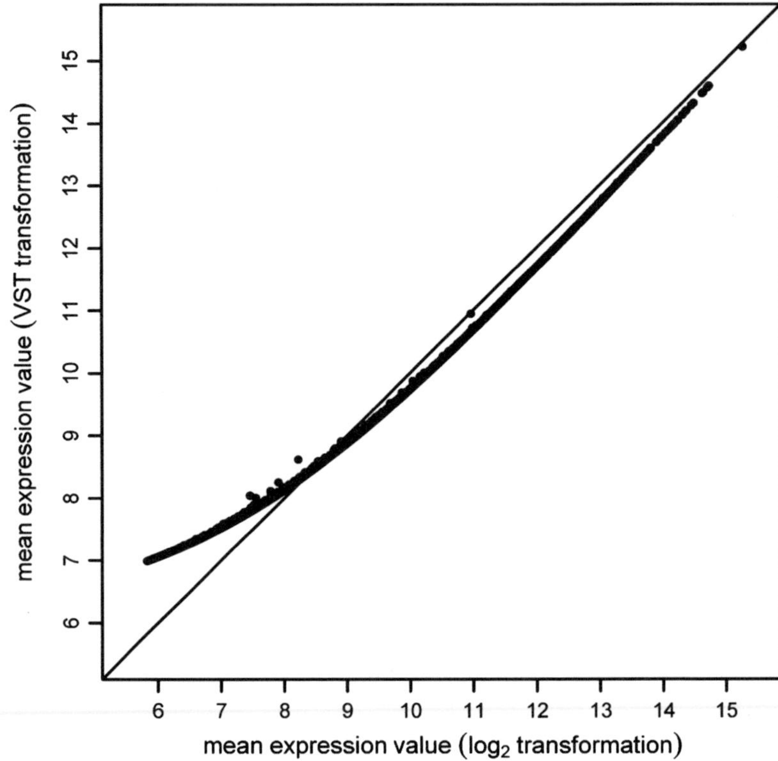

Fig. 2 Log_2 transformation (L2T) versus variance-stabilizing transformation (VST): comparison of mean expression values. The mean VST gene expression values (*y*-axis) are plotted against the mean L2T expression values (*x*-axis) for each probe. If all probes would have the same mean expression value in both transformation methods, all *dots* (*dark line*) would be located on the *light diagonal line*. The mean expression values of L2T probe intensities were highly correlated with the mean expression values of VST probe intensities for probes with mean expression values greater than 2^9. This correlation was recognizably smaller for probes with lower mean expression values (reproduced from Ref. 15 with permission from Creative Commons Attribution (CC BY) license)

(*see* **Note 8**), and (c) sample *storage time* (for whole blood samples, this is the time between blood donation and RNA isolation; for monocyte samples, this is the time between RNA isolation and RNA amplification; they usually range from several months to a few years). However, the proportion of technical variation explained by each of these covariates (*see* **Note 9**) may vary in case a different sample processing workflow was applied in a study.

2. Select the phenotype.
3. Select additional covariates for the analysis depending on the specific analytical question (*see* **Note 10**).

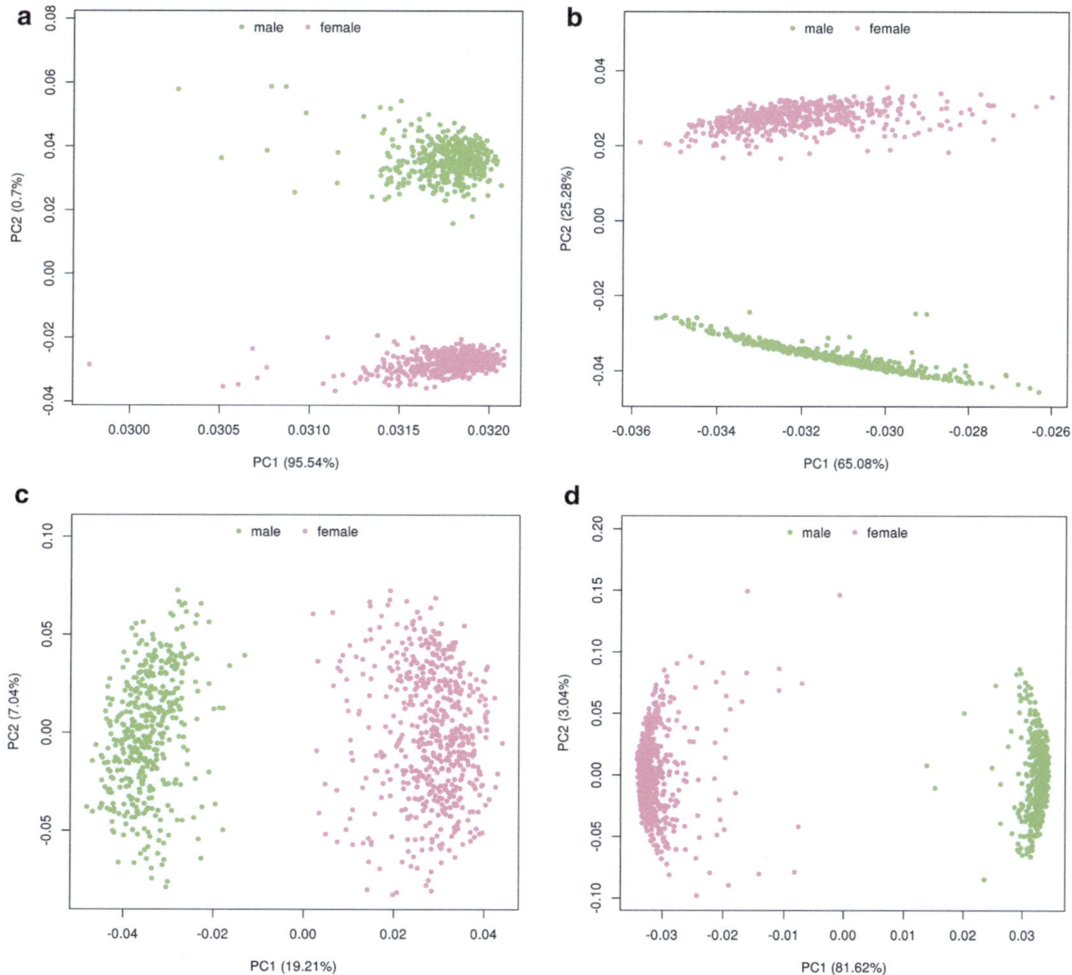

Fig. 3 Principal Component Analysis (PCA) based on allosomal probes for sex determination. (**a**) and (**b**) represent the PCA using the unadjusted intensities and (**c**) and (**d**) represent the PCA using the intensity values adjusted for technical variables. The first two eigenvectors (PCs) of the PCA shown are based on the normalized linear intensity values of the X- and Y-chromosomal probes (**a** and **c**) and Y-chromosomal probes only (**b** and **d**). The *rose* and *light green* colored data points represent the true gender of female and male individuals, respectively. The percentage of variance in the probe's gene expression values explained by the corresponding PC is displayed in *brackets*. PC1 separates the male from female samples after adjusting for technical covariates (PC2 using unadjusted data). In this example, no mismatches between estimated and reported gender occurred. (Reproduced from Schurmann, C (2013) Analysis and Integration of Complex Omics Data of the SHIP Study, Ph.D. thesis with permission from the author)

3.7 Association Analysis

1. Select the appropriate association model for analysis. The correct model depends on several factors, like the type of data need to be analyzed (i.e., the number of repeated measurement per sample) and the sample structure (i.e., possible relatedness) (*see* **Note 11**).

2. For linear regression, include the *RIN* and *storage time* as fixed effects. Include the *RNA amplification batch* as a factor: dichotomize the n different values of the *RNA amplification batch* to n-1 dummy variables using one selected value as the reference category and add these dummy categories as fixed effects to the model (i.e., using the "as.factor()" function in R) (*see* **Note 12**).
3. For each probe, regress the normalized $\log_2$ transformed gene expression values on the covariates and the phenotype (*see* **Note 13**). The association results obtained using the VST are comparable (*see* **Note 3** and Fig. 4).

4 Notes

1. Carefully select the samples placed on a 96-well plate for amplification and labeling to be able to correct for the *RNA amplification batch* without concurrently correcting for the analyzed phenotype. Each treatment category and sample group (i.e., males and females, diabetics and nondiabetics, normal weight and obese individuals, or age groups) should be equally distributed over several 96-well plates. This is particularly important when preparing gene expression arrays of population-based cohorts whereas the specific parameter for analysis is not defined at time of array preparation and will be selected afterwards.

Fig. 4 Association results of $\log_2$ transformation (L2T) versus variance-stabilizing transformation (VST). The panels show the association results for a simulated phenotype without true associations (**a–c**) and for body mass index (BMI) (**d–f**) on each mRNA probe adjusted for sex, age, *RNA amplification batch*, *RNA integrity number* (*RIN*), and the sample *storage time* based on L2T expression values (*x*-axis) and on VST values (*y*-axis). The association results in the *upper panels* (**a, d**) show the per unit phenotype increase of the transformed expression values (betas), the *middle panels* (**b, e**) show the standard errors of the betas (SEs), and the *lower panels* (**c, f**) show the negative $\log_{10}$ association *p*-values. The values on the axis represent the values of the betas, SEs, and *p*-values, respectively. The corresponding squared Pearson product-moment correlation coefficient (R^2) between the plotted values is given in the *upper right corner* of each plot. Each *dot* represents a probe and is *colored* according to its mean L2T expression value from all samples. The *color code* is given in the legend located in the *lower right corner* of each plot, whereas the values from 6 to 15 represent the probes' $\log_2$ expression values associated with the color. Although betas and SEs differ between both transformations, the association *p*-values are highly correlated (reproduced from Ref. 15 with permission from Creative Commons Attribution (CC BY) license)

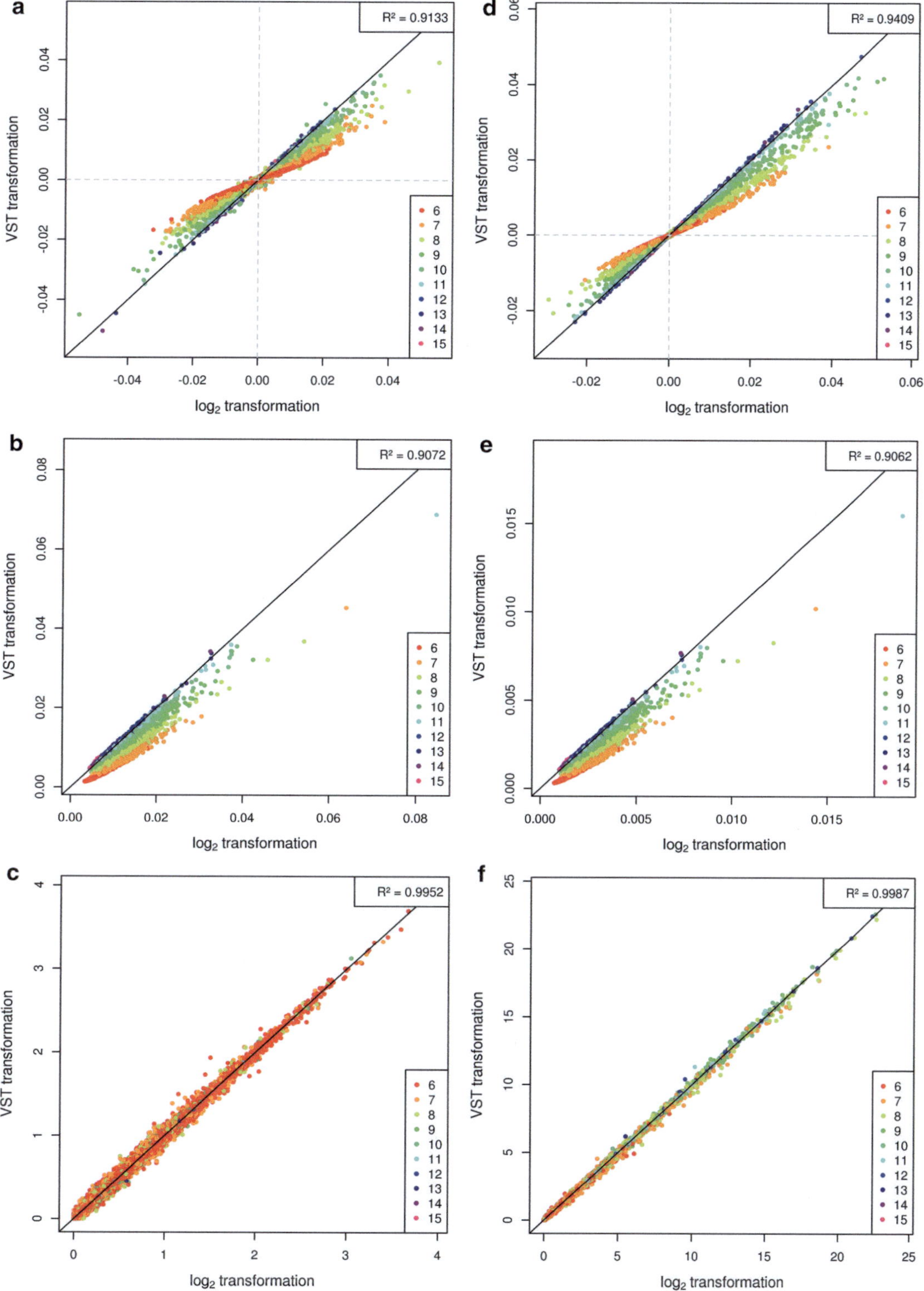
a
R² = 0.9133
VST transformation
log2 transformation
6
7
8
9
10
11
12
13
14
15
b
R² = 0.9072
c
R² = 0.9952
d
R² = 0.9409
e
R² = 0.9062
f
R² = 0.9987

2. An alternative to exporting the complex final reports created from GenomeStudio is to solely export the measured gene expression intensities to other software. Additionally, to circumvent the use of the GenomeStudio software, the Illumina raw data files (*.idat) can be imported into R using the illuminaio package [17]. Array data at bead level can be imported using the beadarray package [18].
3. Instead of using the $\log_2$ transformation of gene expression data, alternative methods like variance-stabilizing transformation (VST) were proposed [9] (Fig. 2). With respect to association results, both methods differ slightly regarding the effect estimates and their standard errors of low expressed probes. However, p-values of the association are virtually the same (Fig. 4).
4. Storing microarray data in RData format, instead of a text table, saves a substantial amount of disk space and processing time.
5. Alternatively to the Array_Address_Id as the probe identifier, the Nucleotide Universal Identifier (nuID) could be used [19] in theory. The nuID is unique for each probe's sequence and can be used to identify the same probe sequences across different gene expression array types. However, a duplicate nuID due to duplicate probe sequences was found on the Illumina HumanHT-12 v3 array. These probes could still be distinguished by their corresponding Array_Address_Id.
6. Some remarks on the data import if analyses are performed in R. Since the Array_Address_Id is coded by a number on the Illumina HumanHT-12 array, leading zeros may be omitted depending on the software used. This can lead to problems when joining different data sources (i.e., regression results) with gene annotation tables. When reading the Illumina gene annotation file, the parameters *quote=""*, *comment.char=""* should be added to the "read.table()" function to avoid problems with respect to the gene description columns. The parameter *check.names=F* added to the "read.table()" function can be useful to avoid automatically re-naming of the column names. This is particularly the case when importing the gene expression table in text format whereas the sample names start with a number or contain underscores, minus signs, etc. If using this parameter, the user has to ensure uniqueness of the column names and avoid problematic characters like whitespaces in the sample identifiers.
7. The *Illumina chip*, which comprises 12 arrays, had the largest influence on the variation of gene expression values. However, adjustment for the *Illumina chip* often led to analytical problems regarding the regression model, whereas most of the chip's variation was accounted by the *RNA amplification batch*. Furthermore, the parameter *RNA isolation batch*

(performed on 96-well plates) had a large effect on the gene expression variation. However, since the same group of samples were also processed together in the *RNA amplification batch*, adjustment for the latter factor corrected also for the *RNA isolation batch* in our workflow.

8. The influence of the *RIN* on the variation of gene expression values is comparably small [15]. This might seem surprising, but is most probably due to the fact that samples with degraded mRNA (RIN < 7) were excluded from the study and not further processed.
9. Although additional adjustment for the technical parameters *signal-to-noise ratio* or the *number of detected genes* reduced the variation of the expression data, both parameters are correlated with clinical endpoints, e.g., body mass index (BMI) which might therefore falsify the association results.
10. Although the specific covariates included in the association analysis are project-dependent, adjustment for sex, age, BMI, and blood cell composition parameters (where applicable) will reduce phenotypic variance and confounding effects, thus decreasing the number of false positive and false negative association results. Blood cell composition parameters include the white and red blood cell counts, platelets, and hematocrit, as well as the percentage of the following white blood cell types: lymphocytes, neutrophils, monocytes, basophils, lymphocytes, and eosinophils.
11. Most of the technical variation can be removed by adjusting for the eigenvectors (EVs), the so-called eigen-genes [20], obtained from a principal component analysis on the sample's expression profiles. However, since the EVs also correlate with many phenotypes and thus diminish true association, this type of adjustment is only suitable for a few traits of interest, like eQTL analyses. Adjustment for the first 40 to 50 EVs performs well in this case [15, 21].
12. Adding the *RNA amplification batch* as a random factor to a mixed effect association model did not change the results markedly compared to the proposed factor adjustment.
13. For SNPs located within probes, the number of transcripts associated with decreased expression signal intensity per mismatch allele of a SNP was significantly higher than the number of transcripts associated with increased signal intensity [15]. Surprisingly, in almost 45 % of the associations a positive effect per mismatch allele on expression signal intensity was observed (Fig. 5). Additionally, taking the relatively small effects of these SNPs into account, it is most unlikely that these SNPs will lead to false positive associations with respect to the analyzed phenotype. However, special care should be taken when performing eQTL analyses.

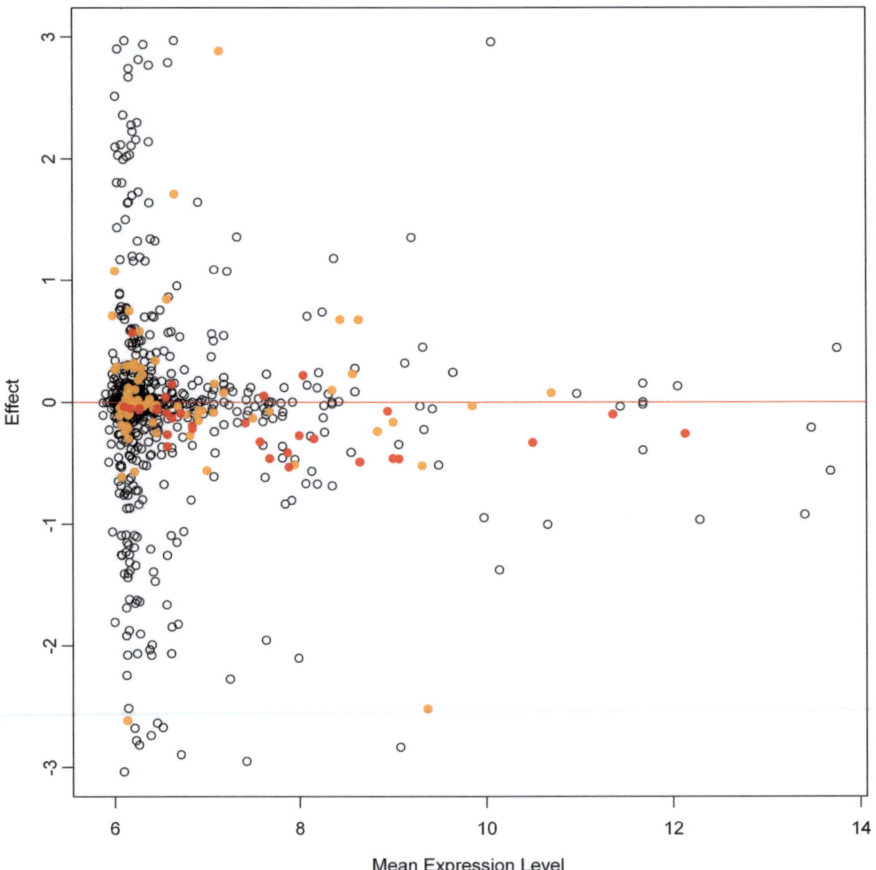

Fig. 5 Effect sizes of SNPs within probes on signal intensities. The change on measured $\log_2$ transformed (L2T) gene expression values per number of mismatch alleles of SNPs located within probes (*y*-axis) is plotted against the mean L2T expression value for each probe (*x*-axis). Each dot represents a SNP-probe combination (*black open circle*); significant associations of mismatch alleles and expression values after Bonferroni correction ($p < 2.3 \times 10^{-5}$) are filled with *red* and association *p*-values below 0.05 are filled with *orange* color. To increase clarity, the *y*-axis was limited from −3 to 3 excluding 176 nonsignificant results out of 1237 successful association results (minimum and maximum effect sizes were −174.1 and 188.7, respectively) (reproduced from Ref. 15 with permission from Creative Commons Attribution (CC BY) license)

Acknowledgments

This work was funded by the European Commission's Seventh Framework Programme (FP7/2007-2013, HEALTH-F2-2011, grant agreement No. 277984, TIRCON), the BMBF (German Ministry of Education and Research) grants 03IS2061A, 03ZIK012, 01GS0834, 01GS0833, 01GS0831, 01KU0908A, 01KU0908B, 0315536F, and the National Genome Research Network NGFNplus Atherogenomics, the Federal State of Mecklenburg-West Pomerania, the Caché Campus program of the

InterSystems GmbH, the Helmholtz Zentrum München (German Research Center for Environmental Health), the State of Bavaria, the German Center for Diabetes Research (DZD e.V.), the State of North-Rhine-Westphalia, the Leibniz Association (WGL Pakt für Forschung und Innovation), the government of Rheinland-Pfalz ("Stiftung Rheinland Pfalz für Innovation", contract AZ 961-386261/733), the Johannes Gutenberg-University of Mainz, and its contract with Boehringer Ingelheim and PHILIPS Medical Systems, the Agence Nationale de la Recherche, France (contract ANR 09 GENO 106 01), the European Union (HEALTH-2011-278913), and the DZHK (German Centre for Cardiovascular Research).

References

1. Ramasamy A, Mondry A, Holmes CC et al (2008) Key issues in conducting a meta-analysis of gene expression microarray datasets. PLoS Med 5:e184
2. Heinig M, Petretto E, Wallace C et al (2010) A trans-acting locus regulates an anti-viral expression network and type 1 diabetes risk. Nature 467:460–464
3. Ein-Dor L, Kela I, Getz G et al (2005) Outcome signature genes in breast cancer: is there a unique set? Bioinformatics 21:171–178
4. Ntzani EE, Ioannidis JPA (2003) Predictive ability of DNA microarrays for cancer outcomes and correlates: an empirical assessment. Lancet 362:1439–1444
5. Eysenck HJ (1994) Meta-analysis and its problems. BMJ 309:789–792
6. Campain A, Yang YH (2010) Comparison study of microarray meta-analysis methods. BMC Bioinformatics 11:408
7. Repsilber D, Fink L, Jacobsen M et al (2005) Sample selection for microarray gene expression studies. Methods Inf Med 44:461–467
8. Du P, Kibbe WA, Lin SM (2008) lumi: a pipeline for processing Illumina microarray. Bioinformatics 24:1547–1548
9. Lin SM, Du P, Huber W et al (2008) Model-based variance-stabilizing transformation for Illumina microarray data. Nucleic Acids Res 36:e11
10. Fu J, Wolfs MGM, Deelen P et al (2012) Unraveling the regulatory mechanisms underlying tissue-dependent genetic variation of gene expression. PLoS Genet 8:e1002431
11. Fehrmann RSN, Jansen RC, Veldink JH et al (2011) Trans-eQTLs reveal that independent genetic variants associated with a complex phenotype converge on intermediate genes, with a major role for the HLA. PLoS Genet 7:e1002197
12. Völzke H, Alte D, Schmidt CO et al (2011) Cohort profile: the study of health in Pomerania. Int J Epidemiol 40:294–307
13. Holle R, Happich M, Löwel H et al (2005) KORA–a research platform for population based health research. Gesundheitswesen 67(Suppl 1):S19–S25
14. Zeller T, Wild P, Szymczak S et al (2010) Genetics and beyond–the transcriptome of human monocytes and disease susceptibility. PLoS ONE 5:e10693
15. Schurmann C, Heim K, Schillert A et al (2012) Analyzing Illumina gene expression microarray data from different tissues: methodological aspects of data analysis in the MetaXpress consortium. PLoS ONE 7:e50938
16. R Development Core Team (2006) R: A language and environment for statistical computing. R Foundation for Statistical Computing, Vienna, Austria, http://www.R-project.org. ISBN 3-900051-07-0
17. Smith ML, Baggerly KA, Bengtsson H et al. (2013) illuminaio: an open source idat parsing tool for Illumina microarrays. F1000Research 2
18. Dunning MJ, Smith ML, Ritchie ME et al (2007) beadarray: R classes and methods for Illumina bead-based data. Bioinformatics 23:2183–2184
19. Du P, Kibbe WA, Lin SM (2007) nuID: a universal naming scheme of oligonucleotides for Illumina, Affymetrix, and other microarrays. Biol Direct 2: 16
20. Chen LS, Storey JD (2008) Eigen-r2 for dissecting variation in high-dimensional studies. Bioinformatics 24:2260–2262
21. Westra HJ, Peters MJ, Esko T et al (2013) Systematic identification of trans eQTLs as putative drivers of known disease associations. Nat Genet 45:1238–1243

Chapter 8

Microarray Technology and Its Applications for Detecting Plasma microRNA Biomarkers in Digestive Tract Cancers

Hirotaka Konishi, Daisuke Ichikawa, Tomohiro Arita, and Eigo Otsuji

Abstract

Many cancers are known to be regulated by microRNAs (miRNAs), and the relationships between tissue miRNA expression levels and the amounts of miRNA circulating in the plasma (or plasma miRNA) have been examined in many types of cancers, including digestive tract cancers. The role of plasma miRNAs has yet to be elucidated in detail; therefore a comprehensive analysis of plasma miRNAs using microarrays should assist in establishing the utility of liquid biopsy or companion diagnosis. We here described the 3D-Gene® miRNA microarray (TORAY) currently used in our laboratory and introduced a trial application in digestive tract cancer diagnosis.

Key words miRNA microarray, Plasma biomarker, Digestive tract cancer, Liquid biopsy

1 Introduction

A study by Mitchell et al. [1] has reported detection of circulating microRNAs (or miRNAs), in the plasma of patients suffering from many diseases, including cancer. The high stability of plasma miRNAs has been attributed to their inclusion in small vesicles, such as exosomes [2], and their binding to proteins, such as Argonaute2 (Ago2) or high-density lipoprotein (HDL) [3, 4]. However, the origin or role of plasma miRNAs remains unclear. Previous studies have assessed plasma miRNA as the biomarkers based on PCR or microarray methods, and so the use of plasma miRNAs as biomarkers for the diagnosis or evaluation of progression of cancer is expected.

In order to comprehensively analyze plasma miRNAs, we have used the 3D-Gene® human microRNA microarray platform "3D-Gene®" with the Human miRNA 2- or 4-array chip series. This is an oligonucleotide microarray chip that has been designed for use in the expression profiling of miRNAs, which are registered at the miRNA database called miRBase 19.0. This microarray platform has some unique characteristics: (a) a chip substrate with a

Paul C.H. Li et al. (eds.), *Microarray Technology: Methods and Applications*, Methods in Molecular Biology, vol. 1368, DOI 10.1007/978-1-4939-3136-1_8, © Springer Science+Business Media New York 2016

characteristic morphology, the so-called minute columnar structure, (b) an agitation method using microbeads during the reaction, as well as (c) a gene immobilization method by molecular control at the nano level [5, 6]. Moreover, the accompanying RNA extraction kit shows high efficacy in extracting small amounts of RNA. In addition to its high reproducibility and quantitability, the microarray chip provides ultra-high sensitivity at the level of 0.1 attomol (10^{-19} mol). Furthermore, the procedure is easier and quicker than those of the conventional array systems.

The microarray system has exhibited very high sensitivity and is considered convenient. Gene expression profiling can be performed with 0.1 μg of total RNA without any amplification, and is suitable for the examination of plasma with a minimal sample volume.

First, we detected candidate plasma miRNA biomarkers for gastric [7], esophageal [8], and pancreatic [9] cancer using PCR methods. These methods examined miRNAs, which were previously correlated with each type of cancer tissue [10]. We then compared pre- and postoperative plasma samples from gastric cancer patients using 3D-Gene® miRNA microarray, and identified two plasma miRNAs, miR-451 and miR-486, as novel plasma biomarkers for gastric cancer [10, 11]. This application of the miRNA microarray had some benefits for our research. First, we were able to perform the comprehensive analysis of plasma miRNA, which correlated with only gastric cancer, except for the individual variability. Second, these two miRNAs were reported to be tumor-suppressive miRNAs for gastric cancer [12, 13]. These miRNAs could not be detected by PCR methods based on the oncogenic miRNAs of gastric cancer. Third, the increase in tumor-suppressive miRNAs in the plasma of cancer patients may lead to a breakthrough in studying the origin of plasma miRNAs. For instance, some plasma miRNAs may also be derived from normal tissue, not only from cancerous tissue, in patients with cancer. [11, 14]. Therefore, a comprehensive analysis of plasma miRNAs in cancer patients using a miRNA microarray may lead to not only the detection of novel biomarkers in plasma, the so-called liquid biopsy [15, 16], but also the elucidation of the role or origin of plasma miRNAs.

There are other applications of the miRNA microarray. We also studied and compared the miRNAs in plasma samples obtained from healthy volunteers as well as from gastric cancer patients after undergoing distal and total gastrectomy (Fig. 1). We also analyzed and compared a series of plasma samples obtained from unresectable advanced gastric cancer patients who had received chemotherapy (Fig. 2). Alterations in plasma miRNAs between pre- and post-chemotherapy may provide an important insight when these miRNAs are used as markers for prediction of chemotherapeutic efficacy, the so-called companion diagnosis [17–19].

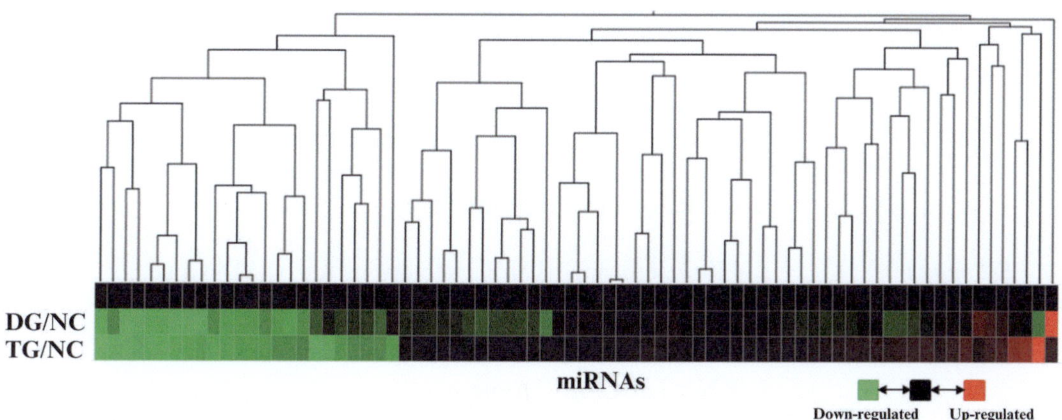

Fig. 1 Clustering analysis of plasma miRNA levels in patients who underwent distal gastrectomy (*DG*) or total gastrectomy (*TG*) and in healthy volunteers (*NC*). Plasma miRNA levels in DG or TG were compared with those in NC and alterations in miRNA level were indicated with a heat map. In this study, we assumed that miRNAs derived from normal gastric mucosa could be detected in plasma

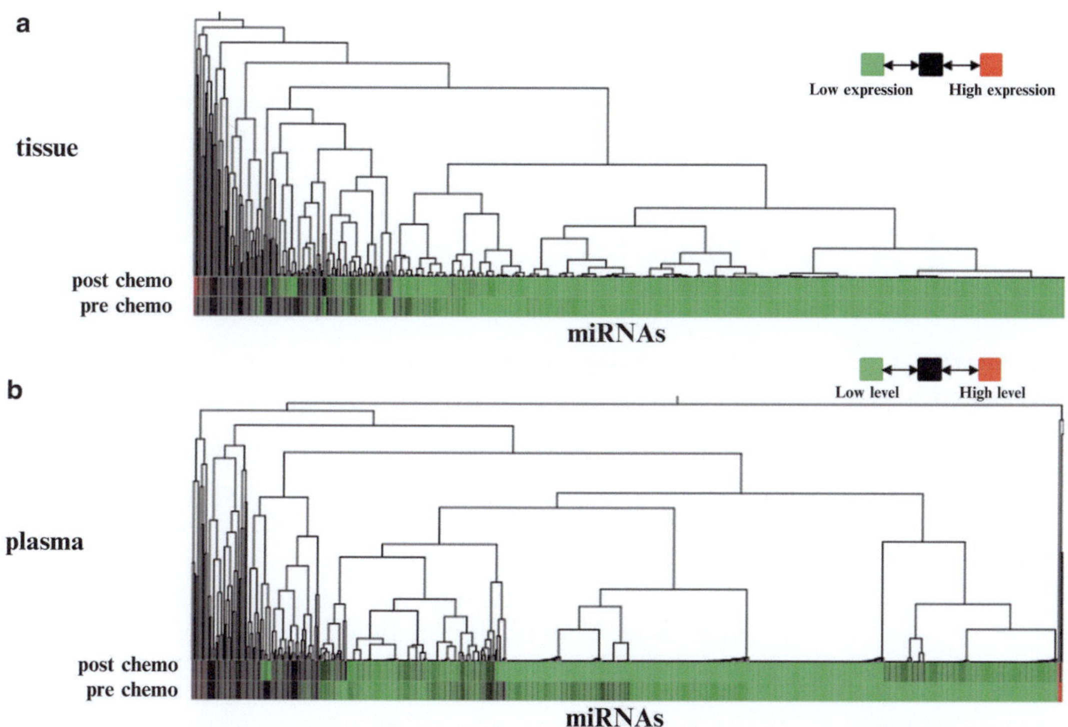

Fig. 2 Clustering analysis of the expression of (**a**) tissue miRNAs and (**b**) plasma miRNA levels. Tissue and plasma samples were collected from a patient with advanced gastric cancer before and after chemotherapy. The expression or levels of miRNAs were indicated with a *heat map*

2 Materials

Use nuclease-free distilled water and tips or tubes with low adsorption of RNAs; for example, siliconized products are recommended for all procedures. All reagents and products must be stored in the recommended manner until use.

2.1 Extraction of the RNA Sample

1. High efficacy RNA extraction kit ("3D-Gene®" TORAY, Kamakura, Japan).
2. Agilent 2100 Bioanalyzer (Agilent Technologies, Inc.).
3. RNA LabChip® (Agilent).
4. Micro-condensing centrifuge (PV-1200, Waken B Tech Co., Ltd.).
5. Small-scale spectrophotometer (NanoDrop Technologies Inc.).

2.2 Labeling of miRNA

1. microRNA Power labeling kit using the fluorescent label Hy5 (miRCURY LNA, 208030-A, Exiqon).
2. Spike control for enzymatic labeling of miRNA (TRT-XR304, cryopreserved at −80 °C) (*see* **Note 1**).
3. **Calf intestine phosphatase (CIP) master mix**. Add 0.5 μL of CIP, 0.5 μL of CIP buffer, and 1.0 μL of miRNA spike control (1×) into a reaction tube for each sample (*see* **Note 5**).
4. **Labeling master mix**. Add 3 μL of Labeling buffer, 1.5 μL of the fluorescent label (Hy5), 2 μL of DMSO, and 2 μL of the Labeling enzyme into a reaction tube for each sample.
5. **Heating block**
6. **Aspirator** (MDA-015, ULVAC, Inc.).

2.3 Preparation of Hybridization Solution

1. Vortex mixer.
2. Degasifier.
3. miRNA Hybridization Buffer Plus (refrigerated at 4 °C).
4. Hybridization buffer A (Block reagent) (refrigerated at 4 °C). These reagents are included in the microRNA Power labeling kit.
5. Block reagent master mix. Mix 2.9 μL of nuclease-free water and 0.6 μL of Hybridization buffer A (Block reagent) in a reaction tube for each sample.

2.4 Hybridization

1. Human miRNA chip ("3D-Gene®" TORAY, Kamakura, Japan). The protocol in this study is designed for the 2-array format of the chip. More details can be found at the "3D-Gene®" website (http://www.3d-gene.com).
2. Cover seal (included in the "3D-Gene®" kit).

3. Hybridization chamber (3D-Gene® Hybridization chamber, TRT-XN005: TORAY).
4. Shaker.
5. Thermostated oven.

2.5 Washing After Hybridization

1. 20 × SSC solution.
2. 10 % SDS solution.
3. Nuclease-free water.
4. First washing solution.
5. Second washing solution.
6. Third washing solution.
7. Staining rack.
8. Centrifuge for spin-drying the platform slides.

2.6 Measurement of Fluorescence Intensity

1. Microarray fluorescence scanner (e.g., a 3D-Gene® Scanner (TORAY Industries, Inc.) or other recommended scanners (ScanArray® Lite, Express, ExpressHT, PerkinElmer Japan Co., Ltd.)).
2. Analysis software: CRBIO IIe® (Hitachi Software Engineering Co., Ltd.) and GenePix® 4000B (Molecular Devices Corporation).

3 Methods

All the processes described in this section are performed in 2 days (Fig. 3). The processes described in Subheadings 3.1–3.4 are performed on day 1. The processes described in Subheadings 3.5 to 3.6 are then performed after almost 16 h on day 2.

All the methods described in this section comprise the protocol for detecting high amounts of total RNA extracted from tissue samples. An accompanying high efficacy RNA extraction kit (3D-Gene® RNA extraction reagent from a liquid sample kit) is suitable especially for the extraction of small amounts of RNA.

3.1 Extraction of RNA Samples

1. Extract total RNA from a 300 μL plasma sample and elute it to 25 μL of distilled water.
2. The amount of RNA extracted from plasma, serum, or the other body fluids should be adjusted to the appropriate concentration (*see* **Note 2**).
3. Check if the extracted RNA sample is of sufficient quality for the experiment by confirming a peak at a nucleotide size of 20–30 using capillary electrophoresis on the 2100 Bioanalyzer and RNA LabChip®, in accordance with the company instructions.

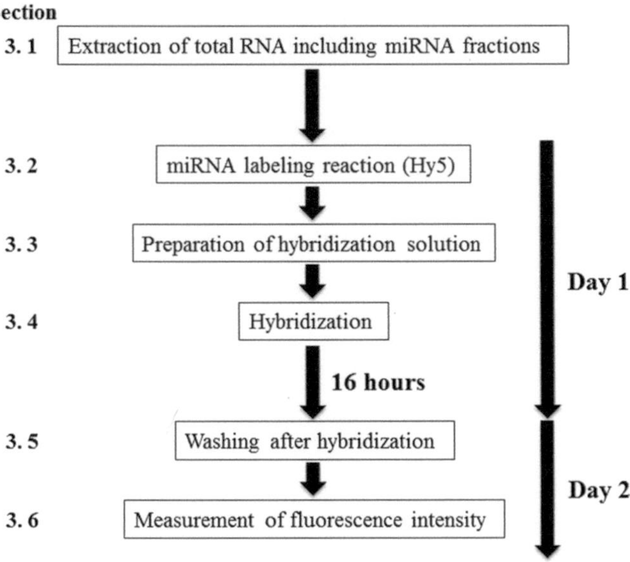

Fig. 3 The time course of the 3D-Gene® miRNA microarray experiment

3.2 Labeling of miRNA

1. Dissolve the Hy5 fluorescent label pellet in 29 μL of nuclease-free water.
2. Dissolve and dilute the spike control for enzymatic labeling in 110 μL of nuclease-free water (*see* **Note 3**). When used, a 5-μL aliquot of the spike control can be diluted.
3. Transfer total RNA to a 1.5-mL tube and adjust to 250 ng/μL with nuclease-free water (*see* **Note 2**).
4. Prepare the CIP master mix on ice to a total volume of 2 μL for each sample.
5. Mix 2 μL of adjusted total RNA (*see* **step 3**) and 2 μL of the CIP master mix in a new reaction tube.
6. Incubate this CIP reaction tube at 37 °C for 30 min (*see* **Note 4**).
7. Incubate the tube at 95 °C for 5 min to terminate the reaction (*see* **Note 4**).
8. Place the tube on ice (at 4 °C) for about 2 min.
9. Prepare the Labeling master mix on ice to a total volume of 8.5 μL for each sample (*see* **Note 5**).
10. Add this 8.5 μl of the Labeling master mix to the CIP reaction tube and the total volume is 12.5 μL.
11. Incubate this tube at 16 °C for 1 h (*see* **Note 6**).
12. Incubate this tube at 65 °C for 15 min to terminate the reaction (*see* **Note 6**).
13. Store the tube at 4 °C until hybridization.

3.3 Preparation of Hybridization Solution (*see* Note 7)

1. Bring the miRNA Hybridization Buffer Plus, Hybridization buffer A (Block reagent), and the 3D-Gene® chip to room temperature. Set the temperature of the heating block to 65 °C and thermostat of the shaker to 32 °C (*see* **Note 8**).
2. Prepare the Block reagent master mix to a total volume of 3.5 μL for each sample.
3. Mix the labeled miRNA sample (12.5 μL) and 3.5 μL of the Block reagent master mix to a total volume of 16 μL in a 1.5-mL tube at 4 °C (on ice) with light shielding.
4. Add 105 μL of miRNA Hybridization Buffer plus (65 °C) to the labeled miRNA sample in the tube. Avoid forming air bubbles during the mixing (*see* **Note 9**).
5. Subject the hybridization solution in the tube to degasification by a degasifier below 0.01Mpa at room temperature for 20 min with light shielding (*see* **Note 10**).
6. Incubate the degasified solution in the tube on the heating block at 65 °C for 3 min.
7. Centrifuge the tube at a maximum speed (15000× g) for 1 min at room temperature.
8. Incubate the solution at 32 °C for at least 5 min until hybridization (*see* **Note 11**).

3.4 Hybridization

1. Place the "3D-Gene®" Human miRNA chip on a flat bench and slowly remove the two protective sheets from the cover.
2. Add 110 μL (2-array format type) or 50 μL (4-array format type) of hybridization solution to the reaction chamber on the chip through the larger injection hole using a 200 μL micropipettor (*see* **Note 12**).
3. Wipe the surface with a cotton swab (*see* **Note 13**). Seal all injection holes with a cover seal and confirm that it has adhered well.
4. Mount the hybridization chamber on the shaker and put the miRNA array chips on the chamber.
5. Place 15 μL of nuclease-free water into the recesses on both sides of the chamber to maintain optimum humidity (*see* **Note 14**).
6. Seal the chamber and start shaking at 250 rpm, 32 °C for 16 h with light shielding (*see* **Note 15**).

3.5 Washing After Hybridization

1. Prepare two batches of first wash solutions, one batch of second wash solution, and two batches of third wash solutions (*see* **Note 16**).
2. Remove the miRNA array chip from the chamber. Put the chip in the first wash solution at room temperature and carefully take off the cover and adhesive using a flat forceps (*see* **Note 17**).

3. Promptly place the chip on the staining rack. Put the rack in the first wash solution at 30 °C (*see* **Note 16**).
4. Shake the staining rack in the first wash solution for approximately 5 times in a vertical direction and approximately three times in a horizontal direction at 1-min intervals. Continue this shaking operation for 5 min (i.e., five sets of shaking).
5. Remove the staining rack from the first wash solution and place it onto a paper towel to blot off the solution (*see* **Note 18**). Do not let the chip go dry (*see* **Note 19**).
6. Place the staining rack into the second wash solution at 30 °C and perform washing in the same shaking manner as in **step 4** for 10 min (i.e., ten sets of shaking).
7. Remove the staining rack from the second wash solution and place it onto a paper towel to blot off the solution (*see* **Note 18**). Do not let the chip go dry (*see* **Note 19**).
8. Place the staining rack into the third wash solution at room temperature and shake it in a horizontal direction approximately three times (*see* **Note 20**).
9. Promptly remove the staining rack from the third wash solution at room temperature and blot off the solution with a paper towel. Place the staining rack into another third wash solution at 30 °C to continue shaking for 5 min (i.e., five sets of shaking) in the same manner as in **step 4** (*see* **Note 20**).
10. Remove the chip from the staining rack and place it in a centrifuge with the spot surface facing down. Spin the chip for approximately 30 s to dry it.
11. Shield the chip from the light and store it in a case until scanning.

3.6 Measurement of Fluorescence Intensity

1. Perform measurements in accordance with the instruction manual provided with the DNA microarray scanner.
2. Ensure that no dust or material remains on the chip substrate before scanning. If some debris remains, remove it by snapping the reverse side of the chip or carefully blowing it with an air spray.
3. Chip scanning should be performed within 1 week of hybridization.

4 Notes

1. The solution of the spike control, which is the mixture at 500× concentration, is aliquoted into several tubes and cryopreserved at −80 °C until use.
2. If a small amount of total RNA is extracted, it is concentrated to a total volume of 5 μL using a condensing centrifuge. We

use 2 μL of this for the microarray analysis. If a high amount of total RNA is extracted from tissue samples, the concentration must be measured by using NanoDrop and adjusted to 250 ng/μL.

3. When we use the spike control, a 5-μL aliquot is added to a 1.5-mL tube followed by 495 μL of nuclease-free water to dilute the solution (the first diluted solution: 100× concentration). Thereafter, add 4 μL of the first diluted solution to a 1.5-mL tube followed by 16 μL of nuclease-free water (the second diluted solution: 1× concentration). This 1× concentration solution can be used for the microarray analysis.
4. The processes described in **steps 6** and 7 of Subheading 3.2 are performed using a heating block.
5. The preparation step for the labeling master mix should be performed as rapidly as possible in order to prevent the deactivation of fluorescent label, Hy5.
6. The processes described in **steps 11–13** of Subheading 3.2 are performed using a heating block with light shielding.
7. The preparation step for the hybridization solution should be conducted within 1–2 h.
8. Warm the miRNA Hybridization Buffer plus to 65 °C for 10 min with the cap loosened. Thoroughly mix the buffer by inverting it a few times until any solids disappear. Subject the tube containing hybridization buffer A (Block reagent) to flush centrifugation and collect the solution at the bottom of the tube. Place it on ice until use. The 3D-Gene® chip must be kept at room temperature for more than 30 min before use to prevent condensation in the reaction chamber.
9. Regarding the hybridization solution, tapping, vortex mixer, or similar instrument should not be used to prevent the formation of air bubbles. Thoroughly mix the solution by inverting it.
10. Gently centrifuge the 1.5-mL tube consisting of the hybridization solution. Set the tube in the tube rack with the cap open. Place the tube rack into the degasification container. Mount of the micropipettor perpendicularly against the opening of the injection hole. Slowly inject the hybridization solution, e.g., at a speed of injecting 110 μL in approximately 1 min. Do not apply pressure to the chip cover during injection.
13. Wipe the surface of the cover gently by a cotton swab. Do not touch the inside of the injection holes by the cotton swab to prevent the injected solution being absorbed.
14. The volume of nuclease-free water must be 15 μL to maintain optimum humidity within the hybridization chamber.
15. If the injection of hybridization solution will take a long time, thermal control is required and the completed chips should initially be set on the hybridization chamber at 32 °C.

16. Prepare two batches of the first wash solution including 10 mL of 20× SCC, 4 mL of 10 % SDS, and 386 mL of nuclease-free water, one batch at 30 °C and the other at room temperature; one batch of the second wash solution including 4 mL of 20× SSC, 4 mL of 10 % SDS, and 392 mL of nuclease-free water at 30 °C; and two batches of the third wash solution, including 1 mL of 20× SSC and 399 mL of nuclease-free water, one batch at 30 °C and the other at room temperature. Place each solution into the staining rack and warm the solution at the prescribed temperature. Up to 20 chips can be washed at one time per rack.
17. Check the bubble formation in the reaction chambers of chip for the confirmation of successful hybridization. Carefully insert the forceps between the cover and chip body when removing the cover to avoid damaging the spot area.
18. Blot off the remaining wash solution from the chip and staining rack. Do not carry the former wash solution to the next because the composition of the wash solutions will be changed.
19. Do not dry the surface of the chip in the following steps of the washing.
20. Washing in the third wash solution is important for removing the SDS contained in the first and second wash solution.

References

1. Mitchell PS, Parkin RK, Kroh EM et al (2008) Circulating microRNAs as stable blood-based markers for cancer detection. Proc Natl Acad Sci U S A 105:10513–10518
2. Kosaka N, Iguchi H, Yoshioka Y et al (2010) Secretory mechanisms and intercellular transfer of microRNAs in living cells. J Biol Chem 285:17442–17452
3. Arroyo JD, Chevillet JR, Kroh EM et al (2011) Argonaute2 complexes carry a population of circulating microRNAs independent of vesicles in human plasma. Proc Natl Acad Sci U S A 108:5003–5008
4. Vickers KC, Palmisano BT, Shoucri BM et al (2011) MicroRNAs are transported in plasma and delivered to recipient cells by high-density lipoproteins. Nat Cell Biol 13(4):423–433
5. Nagino K, Nomura O, Takii Y et al (2006) Ultrasensitive DNA chip: gene expression profile analysis without RNA amplification. J Biochem 139:697–703
6. Hisaoka M, Matsuyama A, Nagao Y et al (2011) Identification of altered microRNA expression patterns in synovial sarcoma. Genes Chromosomes Cancer 50:137–145
7. Tsujiura M, Ichikawa D, Komatsu S et al (2010) Circulating microRNAs in plasma of patients with gastric cancers. Br J Cancer 102:1174–1179
8. Komatsu S, Ichikawa D, Takeshita H et al (2011) Circulating microRNAs in plasma of patients with oesophageal squamous cell carcinoma. Br J Cancer 105(1):104–111
9. Morimura R, Komatsu S, Ichikawa D et al (2011) Novel diagnostic value of circulating miR-18a in plasma of patients with pancreatic cancer. Br J Cancer 105(11):1733–1740
10. Ichikawa D, Komatsu S, Konishi H et al (2012) Circulating microRNA in digestive tract cancers. Gastroenterology 142(5):1074–1078
11. Konishi H, Ichikawa D, Komatsu S et al (2012) Detection of gastric cancer-associated microRNAs on microRNA microarray comparing pre- and post-operative plasma. Br J Cancer 106(4):740–747
12. Bandres E, Bitarte N, Arias F et al (2009) microRNA-451 regulates macrophage migration inhibitory factor production and proliferation of gastrointestinal cancer cells. Clin Cancer Res 15(7):2281–2290

13. Oh HK, Tan AL, Das K et al (2011) Genomic loss of miR-486 regulates tumor progression and the OLFM4 antiapoptotic factor in gastric cancer. Clin Cancer Res 17(9):2657–2667
14. Pigati L, Yaddanapudi SC, Iyengar R, Kim DJ, Hearn SA, Danforth D, Hastings ML, Duelli DM (2010) Selective release of microRNA species from normal and malignant mammary epithelial cells. PLoS One 5(10):e13515
15. Page K, Guttery DS, Zahra N et al (2013) Influence of plasma processing on recovery and analysis of circulating nucleic acids. PLoS One 8(10):e77963. doi:10.1371/journal.pone.0077963
16. Esposito A, Bardelli A, Criscitiello C et al (2013) Monitoring tumor-derived cell-free DNA in patients with solid tumors: clinical perspectives and research opportunities. Cancer Treat Rev 40(5):648–55
17. Ma K, He Y, Zhang H et al (2012) DNA methylation-regulated miR-193a-3p dictates resistance of hepatocellular carcinoma to 5-fluorouracil via repression of SRSF2 expression. J Biol Chem 287(8):5639–5649
18. Neumeister VM, Anagnostou V, Siddiqui S et al (2012) Quantitative assessment of effect of preanalytic cold ischemic time on protein expression in breast cancer tissues. J Natl Cancer Inst 104(23):1815–1824
19. Gaultney JG, Sanhueza E, Janssen JJ et al (2011) Application of cost-effectiveness analysis to demonstrate the potential value of companion diagnostics in chronic myeloid leukemia. Pharmacogenomics 12(3):411–421

Chapter 9

Application of DNA Microarray to Clinical Diagnostics

Ankita Patel and Sau W. Cheung

Abstract

Microarray-based technology to conduct array comparative genomic hybridization (aCGH) has made a significant impact on the diagnosis of human genetic diseases. Such diagnoses, previously undetectable by traditional G-banding chromosome analysis, are now achieved by identifying genomic copy number variants (CNVs) using the microarray. Not only can hundreds of well-characterized genetic syndromes be detected in a single assay, but new genomic disorders and disease-causing genes can also be discovered through the utilization of aCGH technology. Although other platforms such as single nucleotide polymorphism (SNP) arrays can be used for detecting CNVs, in this chapter we focus on describing the methods for performing aCGH using Agilent oligonucleotide arrays for both prenatal (e.g., amniotic fluid and chorionic villus sample) and postnatal samples (e.g., blood).

Key words Array comparative genomic hybridization (aCGH), Chromosomal microarray analysis (CMA), Copy number variation (CNV), Prenatal diagnosis, Clinical utility

1 Introduction

Comparative genomic hybridization (CGH) was initially developed for cancer research. CGH was applied for analysis of chromosomal imbalances with an improved resolution of 2–3 megabases (Mb), as compared to 5–10 Mb offered by standard G-banding chromosome analysis [1]. A pivotal advance in CGH technology came with replacing metaphase chromosomes with BAC (bacterial artificial chromosome) or PAC (bacteriophage P1 artificial chromosome) clones. These clones are immobilized on a microarray glass slide for a hybridization reaction with the DNA targets. This advance led to array comparative genomic hybridization (aCGH) which enables the detection of copy number variants (CNV) throughout the genome with high resolution, and therefore, aCGH provided a basis for high-throughput analysis of genomic imbalances for clinical diagnostics [2, 3].

Our clinical laboratory initially offered Chromosomal Microarray Analysis (CMA) using BAC-based aCGH [4] for the

Paul C.H. Li et al. (eds.), *Microarray Technology: Methods and Applications*, Methods in Molecular Biology, vol. 1368, DOI 10.1007/978-1-4939-3136-1_9,

purpose of diagnosing genomic imbalances primarily in individuals with developmental delay, multiple congenital anomalies, and neuropsychiatric disorders. Included in the CMA array are probes for virtually all known microdeletion/duplication syndromes, pericentromeric and subtelomeric regions, as well as probes for some single gene disorders that may occur due to gain or loss of large DNA segments. Probes distributed randomly along all chromosome arms were also included to identify any full trisomies. Therefore, with a single test, CMA will detect almost all of the disorders detected by traditional cytogenetic analysis such as G-band chromosome analysis and fluorescent in situ hybridization (FISH). CMA provides a major advance in the diagnosis of patients in which a genetic cause of disability is strongly suspected but not detected by cytogenetic analysis. In an extensive review of 33 studies including 22,698 patients, the International Standard Cytogenomic Array Consortium found that CMA offered a diagnostic yield of 15–20 %, as compared to 3 % for G-banding chromosome analysis, in patients with intellectual disability or congenital anomalies [5]. Consequently, CMA is now recommended by the American College of Medical Genetics as the first-tier genetic test for the evaluation of individuals with multiple anomalies and nonsyndromic developmental delay/intellectual disability [6].

The use of chemically synthesized oligonucleotides (60 mer) as probes further increased the genomic resolution offered by CMA. The oligonucleotide probes are delivered with ink-jet printing technology to pre-determined locations on the glass chip surface [7, 8] allowing microarrays to be designed to target specific genomic regions of interest. A side-by-side comparison of the BAC and oligonucleotide array platforms demonstrated that the oligonucleotide array platform is far superior than the BAC platform in (1) providing a wider dynamic range which is extremely important in assessing the validity of experimentally detected CNV changes within regions covered by a single clone, (2) facilitating the detection of chromosomal mosaicism, and (3) the ability of oligonucleotide probes to examine CNV changes smaller than the average BAC size (~150 kb) [9].

With the rapid developments in array technology in recent years, the coverage of our array has approximately doubled every year. In June 2009, we launched our 180 K array (version 8) allowing interrogation of the exons of more than 4000 candidate and disease genes and the entire genome at an average resolution of 30 kb [10]. In the fall of 2010, we launched the 400 K array which includes the 180 K array with exon coverage and 60 K SNPs for comprehensive analysis of both copy number variants (CNVs) and clinically relevant copy number neutral loss of heterozygosity [11].

These different array versions can be customized for both prenatal and postnatal studies to detect chromosomal abnormalities ranging in size from whole chromosomes to a single exon [10, 12]. In this chapter, we describe the methods for performing microarray analysis using Agilent oligonucleotide arrays for prenatal [amniotic fluid (AF) and chorionic villus (CVS)] and postnatal samples (blood).

2 Materials

2.1 DNA Extraction and Measurement of Genomic DNA Concentration and Integrity

1. Puregene Blood Kit Plus (Qiagen): contains Red Cell Lysis Buffer, Proteinase K (20 mg/mL), Protein Precipitation Solution, Cell Lysis Solution/RNAase A (100 mg/mL), and glycogen.
2. Isopropanol.
3. DNA hydration buffer, DNA binding buffer, DNA wash buffer (Zymo Research).
4. Elution buffer AE (Qiagen).
5. Molecular-grade water.
6. DNA Clean Concentrator™-5 (Zymo-spin Columns and Collection tubes).
7. Screw-top tubes (1.5 mL).
8. Barrier pipet tips (ART 10 Reach, Sterile).
9. Alcohol-resistant marker.
10. Rocking incubator (Boekel).
11. NanoDrop spectrometer (ND-1000).

2.2 DNA Sample Digestion for Hybridization

1. Restriction enzyme RsaI kit (Promega R6371): contains RsaI (10.0 U/μL), 10× Buffer C, Acetylated BSA (10.0 μg/μL).
2. Restriction enzyme AluI kit (Promega R6281): contains AluI (10.0 U/μL), 10× Buffer B, Acetylated BSA (10.0 μg/μL).
3. Nuclease-free water.
4. Water Bath, 37 °C ± 2 °C and 65 °C ± 2 °C.
5. Floating Tube Racks.
6. Digestion master mix (made according to Table 1).

2.3 DNA Sample Labeling

1. Sure Tag labeling kit (Agilent Technologies, Cat # 5190-3400).
2. 1× Tris–EDTA (pH 8.0) (TE).
3. Centrifuge filters (Microcon YM-30).
4. PCR Cleanup Filter Plate (Millipore, MultiScreen PCRμ96).
5. Plate Tape.
6. Plate vortexer.
7. Vacuum manifold.
8. Dye Labeling Master Mix (made according to Table 2).

Table 1
Digestion master mix for each single-tube reaction

Reagent	Volume (μL)	Final concentration
1. Nuclease-free water	2.50	
2. 10× Buffer C	2.10	
3. Acetylated BSA (10.0 μg/μL)	0.40	0.67 μg/μL
4. AluI (10.0 U/μL)	0.50	0.83 U/μL
5. RsaI (10.0 U/μL)	0.50	0.83 U/μL
Total volume	6.00	

Table 2
Labeling master mix for single-tube reaction

Reagent	Volume (μL)
Nuclease-free water	2
5× Reaction buffer	10
10× dNTP	5
dCTP labeled dye (Cy3—1.0 mM or Cy5—1.0 mM)	3.0
Klenow (Exo-)	1.0
Total volume	21.0

2.4 DNA Hybridization of Patient Samples

1. Human Cot-1 DNA (1.0 mg/mL).
2. Oligo aCGHHybridization Kit: contains 10× Blocking Agent and 2× Hybridization Buffer (*see* **Note 1**).
3. Hybridization mix (made according to Table 3).
4. SureHyb chambers.
5. SurePrint G3 CGH+SNP Postnatal Research Microarray Kit, 2 × 400K (1860 genes with exon coverage and 60 K SNP).
6. SurePrint G3 CGH Postnatal Research Microarray Kit, 2×400K (1700 genes with exon coverage).
7. SurePrint G3 CGH+SNP Prenatal Research Microarray Kit, 4×180K (Prenatal array with enriched coverage for disease-associated regions).
8. SurePrint G3 CGH Postnatal Research Microarray Kit, 8×60K (Segmental disease tiling array with enriched density for disease-associated regions).

Table 3
Amounts of components in hybridization mix

Component		2 × 400K array (μL)	4 × 180K array (μL)	2 × 105K array (μL)	8 × 60K array (μL)
1	Cyanine 5-labeled sample	19.5	19.5	19.5	19.5
2	Cyanine 3-labeled sample	19.5	19.5	19.5	19.5
3	1× TE buffer	40.0	0.0	40.0	0.0
4	Cot-1 DNA (1.0 mg/mL)	25.0	5.0	25.0	2.5
5	Agilent 10× blocking agent	26.0	11.0	26.0	4.5
6	Agilent 2× hybridization buffer	130.0	55.0	130.0	22.5
Final hybridization sample volume		260.0	110.0	260.0	68.5
Pipetted volume for each microarray		245.0	100.0	245.0	40.0

9. Hybridization chamber gasket slides.
10. Hybridization oven and hybridization oven rotator.

2.5 Array Washing

1. Cover glass forceps.
2. Slide staining tray.
3. Stir bars.
4. Magnetic stir plate and magnetic heating stir plate.
5. *N*-lauroylsarcosine (an anion detergent that helps avoid excessive foaming during stirring).
6. Saline-sodium phosphate-EDTA (SSPE) hybridization buffer, 20×.
7. Oligo aCGH Wash Buffer 1 (1 L): add 25 mL of 20× SSPE and 0.25 mL of 20 % *N*-lauroylsarcosine (20 g of *N*-lauroylsarcosine dissolved in 100 mL of deionized water) to 975 mL of deionized water in a measuring cylinder. Shake well. For larger volume preparations, increase the volumes of the respective components accordingly.
8. Oligo aCGH Wash Buffer 2 (1 L): add 5 mL of 20× SSPE and 0.25 mL of 20 % *N*-lauroylsarcosine to 995 mL of deionized water in a measuring cylinder. Shake well. For larger volume preparations, increase the volumes of the respective components accordingly.

2.6 Slide Scanning

1. Agilent Laser Scanner (G2505C).
2. Agilent Feature Extraction Software (Version 10.10.0.23).

2.7 General Reagents and Apparatus

1. Phosphate buffered saline (PBS), 1×.
2. Ethanol (70 and 100 %).
3. 100 % Acetonitrile.
4. Bleach.
5. Kimwipes.
6. Parafilm.
7. Micropipettors (10, 20, 200, 1000 μL).
8. Transfer pipette.
9. Microfuge tubes (1.7 mL).
10. Minivortexer.
11. Eppendorf 5415D Microfuge and 5810 Centrifuge.

3 Methods

Wear disposable lab coat, nitrile gloves, and environmental safety-approved eye protection for all steps.

3.1 DNA Extraction from Whole Blood (See Note 2)

1. Mix the blood tubes by inverting them six times.
2. Attach the DNA number labels onto two 15-mL conical centrifuge tubes.
3. Pour 3 mL of blood into one labeled 15-mL tube (*see* **Note 3**).
4. Add 9 mL Red Cell Lysis Solution to the 3 mL of blood. Invert the tube 10 times to mix and incubate for 10 min at room temperature; invert at least once during incubation.
5. Centrifuge for 10 min at 2,465 ×g on the Eppendorf centrifuge 5810. Remove the supernatant by carefully decanting it into a waste bucket containing absorbent paper towel, leaving behind the visible white cell pellet and about 100–200 μL of the residual liquid. Visually inspect the size of each cell pellet to determine the amount of Cell Lysis solution to be used in **step 7**.
6. Vortex the tube vigorously at maximum speed to resuspend the cell pellet in the residual liquid; this greatly facilitates cell lysis in **step 7** below.
7. Add 3 mL of Cell Lysis +15 μL RNnase A Solution to the regularly sized resuspended pellet. *Gently* rock the tube up and down to ensure the nuclei are lysed. If necessary, pipet the solution *slowly* up and down using a 3-mL transfer pipette (*see* **Note 4**).
8. If cell clumps are visible, incubate at 37 °C (±1 °C) or room temperature until the solution is homogeneous. Allow samples to cool to room temperature.
9. Add 1 mL of Protein Precipitation Solution to the cell lysate.
10. Vortex *vigorously* at maximum speed for about 20 s to mix the Protein Precipitation Solution with the cell lysate. At this stage,

small clumps of proteins may appear throughout the lysate. This is a good indication of protein precipitation.

11. Centrifuge at 2,465 ×*g* for 10 min. The precipitated proteins will form a tight brown pellet; however, if the pellet does not appear to be tight repeat **steps 10** and **11**.
12. Pour the supernatant, which contains the DNA, into another labeled 15-mL conical centrifuge tube containing 3 mL of isopropanol.
13. Mix the content by inverting the tube gently for 50 times or until the threads of DNA form a visible white clump.
14. Centrifuge at 2,465 ×*g* for 3–5 min. The DNA will form a small white pellet.
15. Pour off the supernatant and add 1–3 mL of 70 % ethanol. Invert the tube several times in order to wash the DNA pellet.
16. Centrifuge at 2,465 ×*g* for 5 min. Carefully decant the tube on a clean absorbent paper to remove ethanol, and allow the DNA pellet in the tube to air-dry for 15 min.
17. Add 300 μL DNA hydration solution to the tube, and allow the DNA in the tube to rehydrate overnight with rocking at 37 °C (±1 °C) or 1 h at 55 °C (±1 °C) (*see* **Note 5**).
18. Mix the solution thoroughly by pipetting it *slowly* up and down using the 1-mL transfer pipette.
19. Quantitate DNA using the NanoDrop Spectrophotometer. Store DNA at 4 °C (±1 °C).

3.2 DNA Extraction from Amniotic Fluid (AF)

When performing microarray analysis on prenatal samples (e.g., AF and CVS), it is always recommended to aliquot sample perform maternal cell contamination studies (protocol not included: AmpfLSTR identifier PCR Amplification Kit from Life Technologies). Fetal gender must be determined by PCR of Amelogenin gene or FISH (protocol not included) so appropriate male/female control DNA can be used for the analysis.

1. Prepare the starting material by spinning the sample in a 15-mL tube in the Eppendorf 5810 microfuge at 2,465 ×*g* for 15 min at room temperature.
2. Remove the supernatant with a sterile transfer pipet and put it in a new 15-mL screw-top tube temporarily (*see* **Note 6**). This leaves the cell pellet with approximately 50 μL of media in the 15-mL tube.
3. Tap the 15-mL tube to resuspend the pellet in the remaining media. Ensure that the pellet is completely resuspended with no visible clumps.
4. Add 500 μL 1× PBS to the tube taking care to wash down the sides of the tube.

5. Transfer the cell sample using a Pasteur pipet to a newly labeled 1.5-mL screw-top tube. For a large cell sample, split it into two 1.5-mL screw-top tubes. For an extra large pellet, split it into three 1.5-mL screw-top tubes.
6. Add 500 μL of PBS to the empty 15-mL tube and transfer the solution to the 1.5-mL screw-top tube to transfer any remaining cells.
7. Centrifuge the tube at maximum speed in an Eppendorf 5415D benchtop microcentrifuge 16000 × *g* or 13200 rpm) for 30 s.
8. Remove the PBS and transfer it into a newly labeled 15-mL conical tube (*see* **Note 6**). This leaves ~20 μL in the tube. Vortex and resuspend the pellet.
9. Wash the cells with 1 mL of 1× PBS and centrifuge at 3,100 × *g* for 30 s. Remove the PBS with a Pasteur pipet, leaving ~20 μL. Repeat this step once.
10. Remove the PBS with a Pasteur pipet, leaving ~20 μL. Vortex the tube to resuspend the pellet.
11. Add 600 μL of Cell Lysis Solution to the resuspended pellet in the tube.
12. Add 100 μL of Proteinase K (20.0 mg/mL) to the tube.
13. Screw on the top of the tube, place Parafilm around the top, and invert the tube 25 times to mix.
14. Put the tube(s) in the oven at 37 °C ± 2 °C overnight with the "rocker" function turned on.
15. Remove the tube(s) from the oven (*see* **Note 7**).
16. Add 3.0 μL of RNase A solution (100.0 mg/mL) to the cell lysate. Screw on the top of the tube and place Parafilm around it.
17. Invert the tube 25 times and then incubate at 37 °C ± 2 °C in the oven for 15 min.
18. Remove the tubes from the oven. Briefly spin and place them on ice for 5 min.
19. Add 200 μL of Protein Precipitation Solution to the tube. Vortex vigorously for 20 s to uniformly mix the sample in the tube.
20. Centrifuge the tube at 16,000 × *g* for 5 min (*see* **Note 8**).
21. Pipet 600 μL of isopropanol into a newly labeled 2.0-mL tube. Place it on ice for at least 5 min to cool.
22. Transfer the supernatant to the tube containing the ice-cold isopropanol in order to precipitate the DNA.
23. Mix the sample in the tube by gently inverting it 50 times.
24. Centrifuge at 16,000 × *g* for 2 min.
25. Carefully pour off the isopropanol into a waste container, being careful to leave the DNA pellet undisturbed (*see* **Note 9**).

26. Centrifuge at 16,000 × *g* for an additional 30 s. Pipet to remove the residual isopropanol.
27. Add 600 μL of 70 % ethanol to the tube. Screw on the top of the tube and invert it several times to wash the DNA pellet.
28. Centrifuge at 16,000 × *g* for 1 min.
29. Carefully pour off the ethanol into a waste container. Be careful to leave the DNA pellet undisturbed.
30. Centrifuge at 16,000 × *g* for an additional 30 s. Pipet to remove the residual ethanol.
31. Invert the tube on a clean absorbent paper for 10–15 min. Make sure that all of the ethanol has evaporated before proceeding.
32. Add 50 μL of the DNA Hydration Buffer to the DNA pellet (*see* **Note 10**).
33. Incubate the tube at 55 °C ± 2 °C for 2 h to rehydrate the pellet, tapping the tube periodically.
34. If possible, leave the tube overnight at room temperature to thoroughly rehydrate the DNA.
35. Mix the sample by tapping the tube and briefly spin the sample.
36. Measure the DNA concentration using the NanoDrop spectrometer (*see* Measurement of Genomic DNA Concentration and Integrity).
37. Cleanup and concentrate the DNA (*see* Cleanup of Extracted DNA).

3.3 DNA Extraction from Chorionic Villus Sample (CVS)

When performing microarray analysis on prenatal samples (e.g., AF and CVS), it is always recommended to aliquot sample perform maternal cell contamination studies (protocol not included: AmpfLSTR identifier PCR Amplification Kit from Life Technologies). Fetal gender must be determined by PCR of Amelogenin gene or FISH (protocol not included) so appropriate male/female control DNA can be used for the analysis.

1. If the CVS (*see* **Note 11**) is received in a 15-mL tube, transfer it to a labeled 1.5-mL screw-top tube. Centrifuge the tube at 3100 × *g* for 2–5 s. Look for clumps of cells, and indicate on the benchsheet whether you have seen the clumps. Remove as much of the media as possible using a sterile transfer pipet (*see* **Note 6**).
2. Add 1.0 mL 1× PBS to the sample in the tube.
3. Centrifuge the tube at 16,000 × *g* for 15 s.
4. Remove the PBS from the tube, leaving ~20.0 μL.
5. Flick the tube gently to resuspend the pelleted sample and repeat **steps 2–4**.
6. Add 600 μL of the Cell Lysis Buffer and 200 μL of Proteinase K (20 mg/mL) to the tube. Parafilm the tube and invert it

25 times to mix. Place it on rocking platform of the oven at 37 °C ± 2 °C overnight.

7. Remove the sample from the rocking platform (*see* **Note 7**).
8. Add 3.0 μL of RNaseA (100.0 mg/mL) to the cell lysate. Invert the tube 25 times and incubate it in a 37 °C ± 2 °C incubator for 15 min.
9. Add 600 μL of isopropanol to a newly labeled 1.5-mL microfuge empty tube and place it on ice for at least 2 min before proceeding to **step 10**.
10. Quickly spin the tube and place it on ice for 5–10 min.
11. Add 200 μL of protein precipitation solution to the tube and vortex it vigorously for 20 s.
12. Spin at 16,000 × *g* for 5 min. If the protein pellet is not tight, place the tube on ice for another 5 min and centrifuge it again at 16,000 × *g* for 3 min.
13. Transfer the supernatant to the tube containing ice-cold isopropanol to precipitate DNA.
14. Mix the sample by gently inverting the tube 50 times.
15. Centrifuge at 16,000 × *g* for 2 min.
16. Pour off isopropanol from the tube into a clean labeled waste tube, being careful to leave the DNA pellet undisturbed. Spin for an additional 30 s and pipet out any residual isopropanol.
17. Add 600 μL of 70 % ethanol to the tube. Replace the top of the tube and invert it several times to wash the pellet.
18. Centrifuge at 16,000 × *g* for 1 min.
19. Pour off the ethanol into a labeled waste tube, being careful not to disturb the pellet. Spin for an additional 30 s and pipet out any residual isopropanol.
20. Leave the tube to air-dry for 10–15 min for the remaining ethanol to evaporate off.
21. Add 50 μL (or more for a large pellet) of the DNA Hydration Buffer to the DNA pellet.
22. Incubate the tube at 55 °C±2 °C for 2 h to rehydrate the pellet, tapping the tube periodically. If possible, leave the tube overnight at room temperature to thoroughly rehydrate the DNA.
23. Mix the sample by tapping the tube and briefly spin the sample. Measure the concentration using the Nanodrop spectrometer (*see* Measurement of Genomic DNA Concentration and Integrity) and then perform cleanup (*see* Cleanup of Extracted DNA) to produce a maximum of 5 μg of DNA on a Zymo column.

3.4 Cleanup of Extracted DNA

1. If the DNA concentration prior to cleanup is more than 500 ng/μL, parafilm the sample and let it sit overnight at room temperature.

2. Add 100 μL of Binding buffer to 5 μg of DNA sample. Vortex and let it sit for 5 min at room temperature.
3. Add the sample to the Zymo mini filter placed inside a collection tube. Spin for 30 s. Discard the waste.
4. Add 200 μL of wash buffer to the Zymo mini filter. Spin for 30 s. Discard the waste. Repeat once.
5. For elution A, transfer the mini filter to a newly labeled 1.7-mL tube. Have another technician check the labels on the tubes prior to transfer.
6. Add 18 μL of water to the Zymo mini filter. Let it sit for 10 min and spin for 30 s into the first labeled tube.
7. For elution B, transfer the mini column to another newly labeled 1.7-mL tube. Have another technician check the labels on the tubes prior to transfer.
8. Add 10 μL of water to the Zymo mini filter. Let it sit for 5 min. Spin for 30 s into a second labeled tube.
9. Determine the concentration of each elution using the Nanodrop spectrometer (*see* Measurement of Genomic DNA Concentration and Integrity).

3.5 Measurement of Genomic DNA Concentration and Integrity

1. Measure the DNA concentration using the Nanodrop UV spectrometer. Before using it, ensure that its daily maintenance has been carried out according to the Nanodrop Maintenance protocol.
2. Start the program "ND-1000 v3.30" on the desktop.
3. Select "Nucleic Acids."
4. Place 1.5 μL of sterile water on the pedestal. Lower the arm. Click OK. Wait for a reading.
5. Wipe off sample with a Kimwipe.
6. On the pedestal, place 1.5 μL of the liquid used to prepare the samples (Puregene Hydration Buffer for Puregene extracted DNA, AE buffer for Qiagen extracted DNA, Molecular-Grade water for Zymoed sample).
7. Click "Blank" or press F3. All readings should go to zero.
8. Wipe off the liquid with a Kimwipe and replace it with 2.0 μL of the same liquid.
9. Click "Measure" or press F1.
10. If the blanking procedure is performed correctly, the readings for "A-260 10 mm path" and of the "A-280 10 mm path" should both be between 0.00 and 0.05, the graph should be a fairly straight line, and the ng/μL reading should be between 0.0 and 1.5. If a sample is out of any of these ranges, repeat **steps 6–9**.

11. Spin all the samples at 16,000 × *g* for 2–10 s.
12. Manually enter the DNA number of the sample in the "Sample ID" window. Alternatively, the barcode on the sample tube is scanned into the instrument. Wipe dry the pedestal and place 1.5 μL of sample on the pedestal. Click "Measure" or press F1.
13. Repeat **steps 11–13** for each sample.
14. During sample measurement, if the program displays a "reblank" message, repeat **steps 5–9**. Place 1.5 μL of Hydration Buffer on the pedestal and measure it. If the measurement is within the limits described in **step 10**, proceed to **steps 11–13**. If it is not within those limits, repeat **steps 5–9** to reblank.
15. The ratio of absorbance at 260 and 280 nm is used to assess the integrity of the DNA. If the A260/A280 ratio is below 1.75 or above 2.00, the A260/A230 ratio is below 1.5, or the concentration is below 66 ng/uL, the integrity of the DNA sample is not adequate. Purify the sample using the Zymo column and re-measure the DNA concentration.

3.6 Sample Digestion for Hybridization (See Note 12)

1. Place labels on the top of closed 1.7-mL microfuge tubes for the samples of both the patients and the controls.
2. Add the appropriate amount of nuclease-free water to each labeled tube to dilute the genomic DNA to give a concentration of 1.0 μg DNA in 20.0 μL. Use long-reach barrier pipet tips whenever pipetting from genomic stock DNA tubes. Open only one genomic tube and one reaction tube at a time, and close both of them before proceeding to the next sample. Recheck the labels of the genomic tubes and the reaction tubes again after all of the DNA has been aliquoted.
3. Add 1.0 μg genomic gender-matched control DNA to a separate labeled 1.7-mL microfuge tube. If the control DNA used is at a higher concentration, add nuclease-free water to bring each tube to a final volume of 20.0 μL. The gender control should be thoroughly mixed by tapping or inverting before use. DNA should not be vortexed, as this will shear it.
4. Prepare the Digestion Master Mix according to Table 1. If you choose to make Mix for all reactions, multiply each number on Table 1 by the number of samples, and then multiply by 1.1 to allow for a 10 % pipetting overage.
5. Add 6.0 μL of the Digestion Master Mix to each reaction tube. Each tube should now have a total volume of 26.0 μL. Place the tubes on ice as soon as you have added the Digestion Master Mix. Mix well by flicking the bottom of the tubes and spin them at 3,100 × *g* for 2–10 s.
6. Place the sample tubes at 37 ± 2 °C (either in a floating rack in a water bath or in a standard rack in an oven) and incubate from 2 to 17 h (or overnight).

7. Transfer the sample tubes to a floating rack in a water bath at 65 ± 2 °C and incubate for 20 min to inactivate the enzymes. Place tubes on ice for at least 5 min.
8. Store the sample tubes at −20 °C until labeling (*see* Note 13).

3.7 Fluorescent Labeling of Genomic DNA (See Note 12)

1. If samples were stored at −20 ± 2 °C after enzyme digestion, thaw them at room temperature for 5 min.
2. Centrifuge the 1.7-mL tubes at 3,100 × *g* for 2–10 s.
3. Add 10.0 μL of 5× Random Primers to each reaction tube containing 26.0 μL of digested genomic DNA.
4. Flick the bottom of the tube to mix. Centrifuge the tubes at 3,100 × *g* for 2–10 s.
5. Denature the DNA in a heat block at 105 ± 2 °C for 3 min.
6. *Immediately* place the tubes on ice for 5 min, and then spin the tubes at 3,100 × *g* for 2–10 s. Return the tubes to ice.
7. Make a master mix for each dye according to Table 2 (Cy5-dCTP for the sample DNA and Cy3-dCTP for the control DNA). Add the components on ice in the order indicated. Mix by tapping and spin at maximum speed for 2–10 s.
8. Add 19.0 μL of Labeling Master Mix to each reaction tube. Flick the bottom of the tube to mix. Centrifuge the tubes at 3,100 × *g* for 2–10 s.
9. Place the sample tubes in a floating rack in a water bath at 37 ± 2 °C and incubate for 2 h.
10. Transfer the sample tubes to a water bath at 65 ± 2 °C and incubate for 10 min to inactivate the enzyme. Place the tubes on ice for at least 5 min.
11. The labeled DNA samples (50.0 μL) can be stored at −20 °C for a day before proceeding to the cleanup step.

3.8 Cleanup of Labeled Genomic DNA Using Millipore Filter Plates (See Note 12)

1. Add 50.0 μL of 1× TE (pH 8.0) to each reaction tube, bringing the total volume to 100.0 μL.
2. Add the contents of each tube to the appropriate well of the 96-well plate (*see* **Note 14**).
3. Place the plate on the vacuum manifold. Ensure that the valve between the collection flask and the manifold is closed. Ensure that the bleed valve is closed all the way (turn the silver knob clockwise). Turn on the house vacuum for drying the plate.
4. Slowly open the valve from the flask to the manifold while gently pressing down on the 96-well plate. Watch the pressure indicator and ensure that the pressure is below −20 in. of mercury.
5. Set a timer for 10 min. After 10 min, close the valve from the vacuum to the manifold, open the bleed valve, and remove the plate. Blot the bottom of the plate with a paper towel. Return

the plate to the manifold, close the bleed valve, and open the valve to the vacuum.

6. Set a timer for 10 min. After 10 min, remove the plate from the vacuum manifold (*see* **Note 15**). Close the valve between the vacuum flask and the manifold. Turn off the house vacuum. Release the pressure on the 96-well plate by slowly turning the silver knob on the bleed valve counterclockwise. Remove the plate when the vacuum has released it.
7. Add 25.0 μL 1× TE buffer (pH 8.0) to each well. Place a piece of plate tape over the sample wells and ensure that it is tightly adhered to the top of the wells. Place the 96-well plate in the plate vortexer and mix at a speed setting of "5" for 10 min.
8. Draw the contents of the first well into a pipet tip to measure its volume. Note the volume and transfer the sample to a properly labeled 1.7-mL tube.
9. Bring the total sample volume to 21.0 μL with 1× TE buffer (pH 8.0), if necessary.
10. Repeat **steps 8** and **9** for each well.
11. The labeled DNA can be stored overnight or over the weekend at −20 ± 2 °C, either before or after measuring the DNA concentration (*see* Quantitation of the Yield and Specific Activity of the Labeling).

3.9 Cleanup of Labeled Genomic DNA Using Millipore Microcon Columns (See Note 12)

1. Add 430.0 μL of 1× TE buffer (pH 8.0) to each reaction tube, bringing the total volume to 480.0 μL.
2. Place a Microcon filter into the labeled 1.7-mL microfuge tube and load the correct labeled DNA sample onto the filter. Spin for 10 min at 9,300 rpm (8,000 × *g*) in a microcentrifuge at room temperature. Discard the flow-through in a labeled Cy-3/Cy-5 waste container.
3. Add 480.0 μL of 1× TE buffer (pH 8.0) to each filter. Spin for 10 min at 8,000 × *g* at room temperature. Discard the flow-through in the labeled Cy-3/Cy-5 waste container.
4. Invert the filter and place it into a newly labeled 1.7-mL tube. Spin for 1 min at 8,000 × *g* at room temperature to collect the purified sample.
5. Measure and record the volume of each eluate in μL. If the sample volume is more than 21.0 μL, return the sample to its filter and spin for 1 min at 8,000 × *g* in a microcentrifuge at room temperature. Discard the flow-through in a labeled Cy-3/Cy-5 waste.
6. Repeat **steps 4** and **5** until each sample volume is less than, or equal to, 21.0 μL.
7. Bring the total sample volume to 21.0 μL with 1× TE buffer (pH 8.0) if necessary.

8. The labeled DNA can be stored overnight or over the weekend at −20 ± 2 °C, either before or after measuring the DNA concentration (*see* Quantitation of the Yield and Specific Activity of the Labeling).

3.10 Quantitation of the Yield and Specific Activity of the Labeling (See Note 12)

1. Start the program "ND-1000 v3.30" on the desktop of a computer attached to the NanoDrop Spectrophotometer.
2. Select "MicroArray" from the first screen.
3. Place 1.5 μL of Nuclease-Free water on the pedestal. Lower the arm. Click OK. Wait for a reading.
4. Select "DNA-50" for "Sample Type" from the next screen.
5. Wipe off the sample on the pedestal with a Kimwipe.
6. Place 1.5 μL of 1× TE buffer (pH 8.0) on the pedestal.
7. Click "Blank" or press F3. All readings should go to "0.0".
8. Wipe off the sample with a Kimwipe and replace with a fresh 1.5 μL of 1× TE buffer (pH 8.0).
9. Click "Measure" or press F1.
10. If the concentration reading is between 0.0 and 1.0 ng, proceed to the next step. Otherwise, repeat **steps 5–9** until the reading is in range. *Note:* If the reading is consistently negative, close the program, reopen it, and start again.
11. Enter the number of the first sample to be measured in the "Sample ID" field. Wipe off the pedestal and place 1.5 μL of the sample on the pedestal. Click "Measure" or press F1. Repeat for each sample and each control.
12. When all samples have been measured, click "Show Report," and then select "File, Print Window."
13. Match each patient sample with the gender-matched control that most closely matches its concentration.

3.11 Hybridization of Patient Samples (See Note 12)

1. According to the type of microarray used, the amounts of the components are different, as shown in Table 3. Add the components, in order, to a labeled nuclease-free tube.
2. Mix the sample by pipetting it up and down several times. Flick the bottom of the tube to mix. Centrifuge the tubes at 8,000 × *g* for 2–10 s.
3. Place sample tubes in a heat block at 105 ± 2 °C and incubate for 3 min.
4. Immediately transfer sample tubes to a water bath at 37 ± 2 °C and incubate for 30 min.
5. Remove sample tubes from the water bath. Spin for 1 min at 8,000 × *g* in a microcentrifuge.
6. Load a clean gasket slide into the Agilent SureHyb chamber base with the gasket label facing up and aligned with the

rectangular section of the chamber base. Ensure that the gasket slide is level and seated properly within the chamber base.

7. Check the Hybridization Worksheet to be sure of the destination for the first sample. Slowly dispense the hybridization sample mixture into the gasket well. Load all gasket wells, being extremely careful to dispense each sample into the correct well.
8. After checking the Hybridization Worksheet to be sure that you have the correct slide, place the slide microarray-side down onto the SureHyb gasket slide, so the numeric barcode side is facing up and the "Agilent"-labeled barcode is facing down. The label on the gasket slide should be lined up with the label on the array slide.
9. Gently place the SureHyb chamber cover onto the sandwiched slides and slide the clamp assembly onto both pieces.
10. Hand-tighten the clamp onto the chamber. Experience will show how to get the chambers tight enough, but not too tight to break the slide.
11. Vertically rotate the assembled chamber to wet the slides and assess the mobility of the bubbles. Tap the assembly on the palm of your hand if necessary to move bubbles.
12. Place the assembled slide chamber in the rotator rack in a hybridization oven set to 65 ± 2 °C. Be sure that the rotator is balanced both side-to-side and front-to-back. Set the hybridization rotator to 20 rpm.
13. Hybridize at 65 ± 2 °C for 20–68 h.
14. Proceed to Array Washing protocol.

3.12 Array Washing

1. Place at least 250 mL of Oligo aCGH Wash Buffer 2 in a 37 ± 2 °C water bath and a dish labeled "Wash BF2" in a 37 ± 2 °C incubator overnight, or at least 3 h before washing.
2. Fill a slide-staining dish labeled "Wash BF1" with approximately 350 mL of room-temperature Oligo aCGH Wash Buffer 1 and place it in the hood.
3. Place a slide rack into another slide-staining dish labeled "Wash BF1." Add a magnetic stir bar. Fill this dish with enough room-temperature Oligo aCGH Wash Buffer 1 to cover the slide rack (~250 mL). Place this dish on a magnetic stir plate in the hood.
4. Insdie the fume hood, fill labeled Acetonitrilewith approximately 300 mL of room-temperature acetonitrile (*see* **Note 16**).
5. The procedure of washing is conducted as depicted in Table 4.
6. Remove the hybridization chambers to be washed from the 65 ± 2 °C incubator.
7. Place the first slide into the wash buffer. Place the hybridization chamber assembly on a flat surface and disassemble it.

Loosen the thumbscrew, turning counterclockwise. Slide off the clamp assembly and remove the chamber cover. Remove the array-gasket sandwich from the chamber base by grabbing the slides from their ends.

8. Keep the microarray slide numeric barcode facing up and submerge the array-gasket sandwich into the first dish containing Oligo aCGH Wash Buffer 1. Do not let go of the slides.
9. With the sandwich completely submerged in Oligo aCGH Wash Buffer 1, pry the sandwich open from the barcode end. Insert one end of the plastic forceps between the slides and gently turn the forceps to separate the slides. Let the gasket slide drop to the bottom of the staining dish. Remove the microarray slide and place it into the slide rack in the second dish containing Oligo aCGH Wash Buffer 1, being very careful not to touch the array. Minimize exposure of the slide to air.
10. Repeat **steps** 7 through **9** for up to four additional slides, leaving a blank space between each pair of slides. No more than five slides can be washed in each wash batch. Larger dishes can be used for a maximum of ten slides.
11. When all slides in the batch are placed into the slide rack in the second dish, stir using a setting between 110 and 130 for 5 min (Table 4).
12. When there is 1 min left on the previous wash, remove a prewarmed dish and the Oligo aCGH Wash Buffer 2 from the 37 ± 2 °C incubator and water bath. Place the dish on a stir plate, add a magnetic stir bar, and fill the dish with Oligo aCGH Wash Buffer 2 to the top of the label.
13. When the time is up, transfer the slide rack to the dish containing the Oligo aCGH Wash Buffer 2 and stir on a setting between 110 and 130 for 1 min (Table 4).
14. Remove the slide rack from the dish and tilt the rack slightly to minimize wash buffer carry-over. Transfer the slide rack to the dish containing acetonitrile and leave for 1 min (Table 4).

Table 4
Procedure of microarray slide washing

Procedure	Wash buffer	Temperature	Time
Disassembly	Oligo aCGH wash buffer 1	Room temperature	Depends on the total number of slides
First wash	Oligo aCGH wash buffer 1	Room temperature	5 min
Second wash	Oligo aCGH wash buffer 2	37 ± 2 °C	1 min
Third wash	Acetonitrile	Room temperature	1 min

15. Remove the slide rack very slowly to minimize droplets on the slides. It should take about 10 s to remove the slide rack.
16. Scan slides immediately to minimize the impact of air oxidation on signal intensities, or store slides in slide boxes in a desiccator.
17. If necessary, repeat **steps 6–16** for the next group of five slides using fresh Oligo aCGH Wash Buffer 1 and pre-warmed Oligo aCGH Wash Buffer 2. The Oligo aCGH Wash Buffer 1 in the dish used to open the array-gasket assembly can be used throughout the day and should be discarded after all of the day's washing has been completed. The Oligo aCGH Wash Buffer 1 (on the stir plate) and Oligo aCGH Wash Buffer 2 can be used for up to five slides. If fewer than five slides were washed and more are to be washed later the same day, leave the Oligo aCGH Wash Buffer 1 in the hood and place the Oligo aCGH Wash Buffer 2 in its tray in the 37 ± 2 °C incubator until ready for use. Acetonitrile can be used for up to 20 slides. Leave them the Acetonitrile dish in the hood if they will be used later for another wash.
18. When all washing for the day has been completed, wash all of the dishes, slide rack, and stir bars. Pour used Oligo aCGH Wash Buffer 1 and Oligo aCGH Wash Buffer 2 down the lab sink. Rinse the dish and lid with tap water, and then fill the dish several times with Millipore water and empty. Rinse the lid with Millipore water. Air-dry the Oligo aCGH Wash Buffer 1 dishes. The Oligo aCGH Wash Buffer 2 dish can be placed directly in the incubator. Discard the acetonitrile in the acetonitrile waste container in the hood. Air-dry the dishes in the hood.

3.13 Scan Slides (see Note 17)

1. Scan the array slides according to instructions provided with the scanner used.
2. The Agilent software for extracting the data from the tiff file and analysis can be downloaded from their website (http://www.genomics.agilent.com).
3. Examples of the microarray principle and output are shown in Figs. 1 and 2.

4 Notes

1. The 2× Hybridization Buffer contains lithium chloride (LiCl), lithium dodecyl sulfate (LDS) and Triton. LiCl is hazardous and harmful by inhalation, contact with skin, and if swallowed. It may cause harm to breast-fed babies and impair fertility. The target organ is the central nervous system. LDS is harmful by inhalation and irritating to the eyes, respiratory system, and skin. Triton is harmful if swallowed and has the risk of serious damage to eyes.

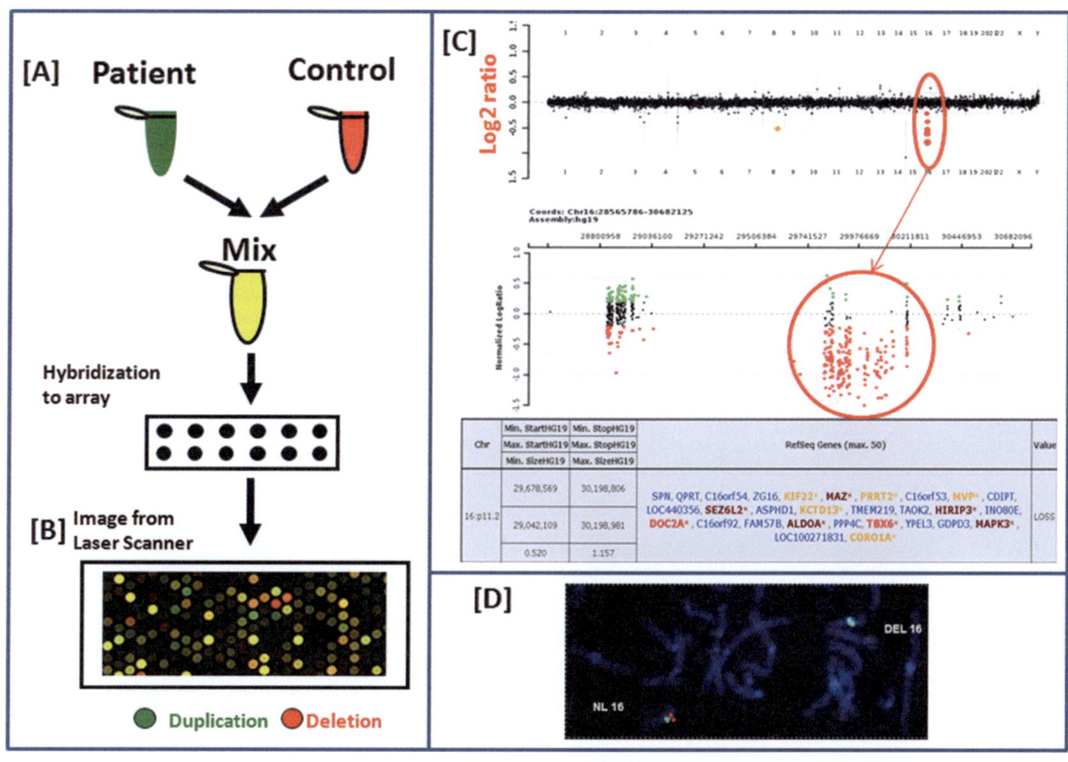

Fig. 1 Array comparative genomic hybridization (aCGH) showing the deletion of chromosome band 16p11.2. (**A**) A schematic of the process for conducting array CGH: Genomic DNA from the patient is labeled with a green fluorescent dye (Cy5) and genomic DNA from a control is labeled with a red fluorescent dye (Cy3). The two samples are mixed and co-hybridized to the array of DNA fragments on the slide. (**B**) A laser scanner reads the fluorescent signals on the slide, and the intensities of each color are quantified using specialized software. A copy number loss is indicated by a *red spot* (more control than patient DNA) and a copy number gain is indicated by a *green spot* (more patient than control DNA). If both the patient and control DNA are the same, the spot is *yellow* in color. (**C**) A typical result is presented in the Agilent software. The *top panel* depicts the whole genome view and the *middle panel* is the segmental view showing a deletion of chromosome band 16p11.2. The *bottom panel* is the data output from our customized software showing the minimum and maximum size of the deletion on chromosome 16 and genes involved in the region. (**D**) Confirmation of the deletion by fluorescent in situ hybridization (FISH), with the *red signal* representing the target probe and *green signal* the control probe, demonstrating the 16p11.2 deletion (reproduced from Ref. 4 with permission from Elsevier)

2. For best results, blood must not exceed 6 days old. DNA extraction for blood, amniotic fluid (AF), and chorionic villus sample (CVS) is adapted from published Puregene protocols (QIAmp DNA Blood Midikit Handbook protocols).
3. The DNA isolation protocol is designed for whole blood samples of 1–3 mL, with an expected yield of 50–150 μg of DNA. PCR is also performed on the negative control, which is included in the extraction batch, to ensure no contamination during extraction procedures.
4. In general, for a cell pellet of average size, use 3 mL of Cell Lysis/RNase solution. If the cell pellet is much smaller than

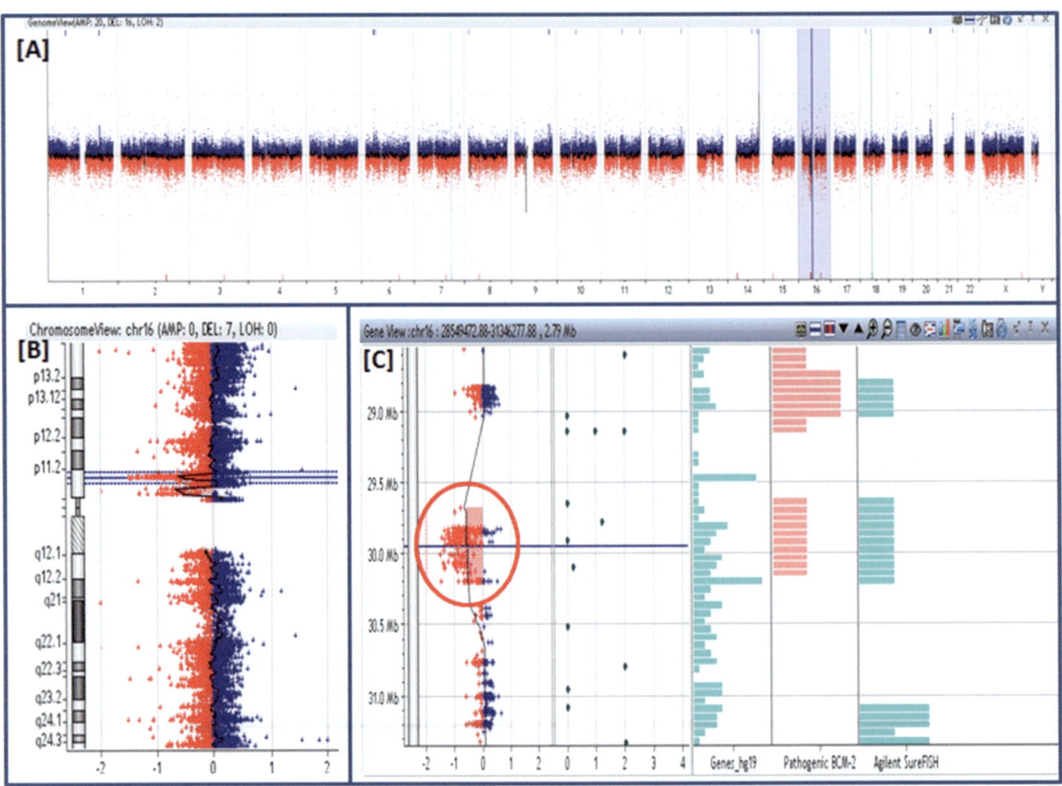

Fig. 2 The same deletion of 16p11.2 is presented in the Agilent software. (**A**) The *top panel* depicts the whole genome view. (**B**) The segmental view of the entire chromosome 16 showing the deletion of chromosome band 16p11.2. (**C**) The *red circle* magnifying the deleted region

average size, scale down the amount of Cell Lysis Solution used such as 1–2 mL. For very small pellets, use 300 μL for the Whole Blood Preparation procedure. Samples are stable in the cell lysis solution for at least 18 months at room temperature.

5. DNAs can be rehydrated at room temperature over the weekend; then they are placed at 55 ± 1 °C on the following Monday for 1 h.
6. The supernatant should be transferred to a newly labeled 1.5-mL screw-top tube. It should be kept until you are sure that sufficient DNA was isolated in this protocol. If there is sufficient DNA isolated, the supernatant can then be bleached and discarded.
7. The DNA isolation protocol can stop here and continue on the next day. Samples may also be left over the weekend with the protocol continued on Monday. Keep the cell lysate at room temperature for the extraction to be continued later.
8. After centrifugation, the precipitated proteins should form a tight white pellet at the bottom of the tube. If the pellet is not tight, place the vial on ice for 5 min and centrifuge it again for 3 min.

9. Pour the supernatant into a clean 1.7-mL tube temporarily so that if the pellet becomes dislodged, you will be able to retrieve the lost DNA from the supernatant. Therefore, you will not risk losing or contaminating the DNA.
10. The volume of hydration buffer used depends on the size of the DNA pellet. If the pellet does not go into solution with 50.0 μL of hydration buffer, more buffer may be added.
11. The chorionic villus sample (CVS) should be cleaned (i.e., free of any maternal decidua).
12. The hybridization protocol beginning with Sample Digestion for Hybridization to Hybridization of Patient Sample are adapted with modification from Agilent manual, "Agilent Oligonucleotide Array-Based CGH for Genomic DNA Analysis."
13. At this point, the DNA samples may be covered with aluminum foil and stored at -20 ± 2 °C for labeling the next day. Samples can also be left at -20 ± 2 °C over the weekend.
14. Each 96-well multiscreen plate can be used twice. For the second use of the plate, two columns of blank wells should be left between the used wells and the new ones.
15. If the samples still look wet, place the plate on the vacuum for 5 min.
16. Acetonitrile is used at 100 % and is also present in the Stabilization and Drying Solution. Acetonitrile is toxic and flammable and must be used in a suitable fume hood. It must be disposed of in a segregated collection bottle for disposal by Environmental Safety. Nitrile gloves are approved for only brief exposure (5 min or less) to acetonitrile. Change gloves if significant exposure to acetonitrile has occurred.
17. The Scan Slide step is adapted from Agilent manuals: Agilent G2565AA and Agilent G2565BA Microarray Scanner System and from Agilent G2567AA Feature Extraction Software (v.10.9).

References

1. Kallioniemi A, Kallioniemi OP, Sudar D et al (1992) Comparative genomic hybridization for molecular cytogenetic analysis of solid tumors. Science 258(5083):818–821
2. Solinas-Toldo S, Lampel S, Stilgenbauer S et al (1997) Matrix-based comparative genomic hybridization: biochips to screen for genomic imbalances. Genes Chromosomes Cancer 20(4):399–407
3. Snijders AM, Nowak N, Segraves R et al (2001) Assembly of microarrays for genome-wide measurement of DNA copy number. Nat Genet 29(3):263–264
4. Shinawi M, Cheung SW (2008) The array CGH and its clinical applications. Drug Discov Today 13(17):760–770
5. Miller DT, Adam MP, Aradhya S et al (2010) Consensus statement: chromosomal microarray is a first-tier clinical diagnostic test for individuals with developmental disabilities or congenital anomalies. Am J Hum Genet 86(5):749–764
6. Manning M, Hudgins L (2010) Array-based technology and recommendations for utilization in medical genetics practice for detection of chromosomal abnormalities. Genet Med 12(11):742–745

7. Hughes TR, Mao M, Jones AR et al (2001) Expression profiling using microarrays fabricated by an ink-jet oligonucleotide synthesizer. Nat Biotechnol 19(4):342–347
8. Cleary MA, Kilian K, Wang Y et al (2004) Production of complex nucleic acid libraries using highly parallel in situ oligonucleotide synthesis. Nat Methods 1(3):241–248
9. Ou Z, Kang SH, Shaw CA et al (2008) Bacterial artificial chromosome-emulation oligonucleotide arrays for targeted clinical array-comparative genomic hybridization analyses. Genet Med 10(4):278–289
10. Boone PM, Bacino CA, Shaw CA et al (2010) Detection of clinically relevant exonic copy-number changes by array CGH. Hum Mutat 31(12):1326–1342
11. Wiszniewska J, Bi W, Shaw C et al (2014) Combined array CGH plus SNP genome analyses in a single assay for optimized clinical testing. Eur J Hum Genet 22(1):79–87
12. Breman A, Pursley AN, Hixson P et al (2012) Prenatal chromosomal microarray analysis in a diagnostic laboratory; experience with >1000 cases and review of the literature. Prenat Diagn 32(4):351–361

Chapter 10

High-Throughput DNA Array for SNP Detection of KRAS Gene Using a Centrifugal Microfluidic Device

Abootaleb Sedighi and Paul C.H. Li

Abstract

Here, we describe detection of single nucleotide polymorphism (SNP) in genomic DNA samples using a NanoBioArray (NBA) chip. Fast DNA hybridization is achieved in the chip when target DNAs are introduced to the surface-arrayed probes using centrifugal force. Gold nanoparticles (AuNPs) are used to assist SNP detection at room temperature. The parallel setting of sample introduction in the spiral channels of the NBA chip enables multiple analyses on many samples, resulting in a technique appropriate for high-throughput SNP detection. The experimental procedure, including chip fabrication, probe array printing, DNA amplification, hybridization, signal detection, and data analysis, is described in detail.

Key words NanoBioArray (NBA), DNA microarray, Microfluidics, PDMS chip, Gold nanoparticles, Single nucleotide polymorphism (SNP)

1 Introduction

Variations in single nucleotide polymorphism (SNP) in the KRAS gene are important cancer biomarkers, and detection of these SNP variations is crucial for selection of the appropriate type of therapy for patients [1]. Currently, various methods are used for SNP detection and they can be categorized in three major groups based on DNA sequencing, real-time PCR and DNA hybridization [2]. Hybridization-based techniques, such as the DNA microarray, are simple and have the high sample-throughput potential. However, these techniques rely on diffusion-based mass transport to deliver target strands to the probe sites, and therefore a long hybridization time is required [3]. To address the issue, Wang et al. developed a microfluidic method which used the centrifugal force in a CD-like chip to induce a liquid flow to facilitate the mass transport and delivery of the target strands [4, 5]. In this method, the chip is made by sealing two PDMS slabs consisting of channels consecutively to the circular glass chip. First, the PDMS slab consisting of radial channels is sealed with the glass chip to create an array of

Paul C.H. Li et al. (eds.), *Microarray Technology: Methods and Applications*, Methods in Molecular Biology, vol. 1368,
DOI 10.1007/978-1-4939-3136-1_10, © Springer Science+Business Media New York 2016

radially patterned probe lines (Fig. 1). Second, after removing the first slab, the second PDMS slab consisting of spiral channels is sealed with the same glass chip, and the target strands are introduced through the channels (Fig. 2). DNA hybridization between the targets and the probes occurs at the intersections of spiral channels with the radial probe lines. Using this microfluidic method, multiple fungal pathogenic samples, of up to 100, were analyzed simultaneously [4]. Although this method is proved to be fast and efficient, SNP detection entails high-temperature hybridization which complicates the experimental procedure. To avoid the use of high temperature, SNP detection of target oligonucleotide and PCR products is conducted at room temperature using gold nanoparticles (AuNPs) [6–8]. In this AuNP method, the target strands are loaded on the surfaces of AuNPs prior to introduction to the channels of the NBA chip. Involvement of the target bases in binding with the surfaces of AuNPs changes the mechanism of DNA hybridization [9], which favors the target binding to the perfectly matched (PM) probe, but not the mismatched (MM) probe, and enables the SNP detection without the use of high temperature.

Herein, we report the detailed experimental procedure of SNP discrimination at room temperature in the CD-like NBA chip. In this method, the DNA probes are introduced into the radial

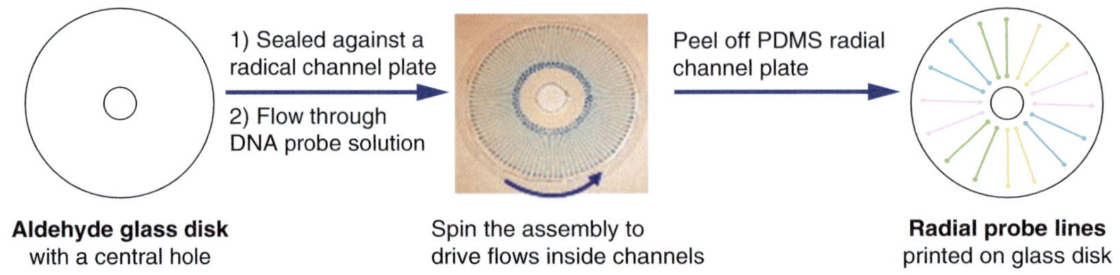

Fig. 1 Schematic diagram of probe array printing along radial channels in the CD-NBA chip (reproduced from ref. [4] with permission from Elsevier)

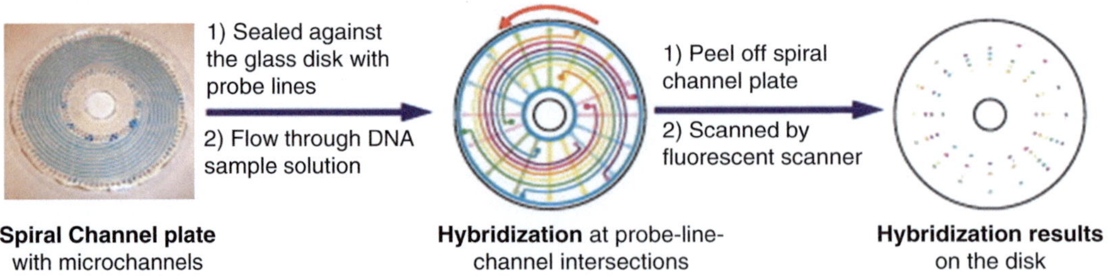

Fig. 2 Schematic diagram of DNA hybridization in the spiral channels in the CD-NBA chip (reproduced from ref. [4] with permission from Elsevier)

channels of the chip for probe printing. Then, the genomic DNAs are amplified using PCR, and purified. The DNAs are loaded on the surfaces of AuNPs, and introduced into the spiral channels of the chip for hybridization.

2 Materials

Prepare all aqueous solutions using ultrapure deionized (DI) water (resistance of 18 MΩ/cm at 25 °C).

2.1 PDMS Chip Fabrication

1. Negative photoresist SU-8 50 and its developer (MicroChem Corp, Newton, MA, USA).
2. Circular silicon wafers of 4-in. diameter (Cemat Silicon SA, Warszawa, Poland).
3. SYLGARD 184 silicon elastomer kit (Dow Corning Corp., Midland, MI, USA).
4. Pyrex crystallizing dish (500 mL).
5. Piranha solution: 70 % of sulfuric acid (98 %) and 30 % of hydrogen peroxide (30 %).
6. Photomask: design the channel pattern using a CAD software (Visual Basic or L-Edit) and print the photomask at high-resolution (>3000 dpi).
7. NOX solution: Dilute concentrated Liqui-Nox (Alconox, White Plains, NY, USA) in 10 volumes of DI water.
8. Silicone sealant 732 (Dow Corning Corp., Midland, MI, USA).
9. Dimethyldichlorosilane solution (repel silane), 2 % in octamethylcyclotetrasiloxane.
10. Spin-coater WS-400 (Laurell Technologies Corp., North Wales, PA, USA).
11. Near UV exposure system (Bachur & Associates, San Jose, CA, USA).

2.2 Surface Aldehyde Functionalization

1. Circular glass chips of 4-in. diameter (Precision Glass & Optics, Santa Ana, CA, USA).
2. Sparkleen detergent: Dissolve 10 % of Sparkleen powder in DI water in a wash bottle.
3. APTES solution: Prepare 2 % v/v of 3-aminopropyltriethoxysilane (APTES) in anhydrous ethanol.
4. PBS buffer 20×: Dissolve 60.00 g NaCl, 2.00 g KCl, 14.4 g Na_2HPO_4 and 2.4 g KH_2PO_4 in 500 mL of DI water and keep the solution at room temperature.
5. Glutaraldehyde solution: Mix 20 mL glutaraldehyde (25 %), 5 mL PBS 20×, and 75 mL of DI water (*see* **Note 1**).

2.3 Probe Array Printing

1. Probe stock solution: Prepare 500 μM DNA probes in DI water. *See* Table 1 for probe sequences.
2. Immobilization buffer: Prepare 0.15 M of sodium bicarbonate and 1.5 M of NaCl in DI water.
3. Probe solutions: Mix 2 μL of probe stock solution in 38 μL of immobilization buffer.
4. Reduction solution: Add 50 mg sodium borohydride, 0.75 mL PBS 20×, 0.1 mL Triton X 100 (10 %), 5 mL ethanol 95 % in a Falcon tube and dilute the mix with DI water to 20 mL.
5. Blocking solution: Dissolve 20 mg of bovine serum albumin in 20 mL DI water.
6. In-house built rotating platform.

2.4 Polymerase Chain Reaction

1. KRAS genomic DNA samples (Horizon Discovery Ltd., Cambridge, UK).
2. DNA amplification kit (Life Technologies, Burlington, ON, Canada): 10× amplification buffer, dNTP solution, $MgCl_2$, and Taq DNA polymerase.
3. Forward primer: Prepare 5 μM solution in DI water (*see* Table 1).
4. Reverse primer: Prepare 5 μM solution in DI water (*see* Table 1).
5. DNA amplification mixture: Mix 10 μL of 10× amplification buffer, 10 μL of dNTP solution, 10 μL of forward primer, 6 μL reverse primer, 5 μL $MgCl_2$, 0.5 μL Taq DNA polymerase, 5 μL DNA template, and 53.5 μL DI water in a 0.6-mL Eppendorf tube.
6. Thermocycler (Perkin Elmer, Waltham, MA, USA).

Table 1
The sequences of probe and primer oligonucleotides

	Name	Sequence
Probes	W	5′-/C12amine/CC TAC GCC ACC AGC TCC AAC-3′
	A	5′-/C12amine/CC TAC GCC AGC AGC TCC AAC-3′
	D	5′-/C12amine/CC TAC GCC ATC AGC TCC AAC-3′
	V	5′-/C12amine/CC TAC GCC AAC AGC TCC AAC-3′
Primers	Forward	5′-biotin-TGA CTG AAT ATA AAC TTG TGG TAG TTG GAG-3′
	Reverse	5′-ATG ATT CTG AAT TAG CTG TAT CGT CAA GGC-3′

7. QIAquick nucleotide removal kit: spin columns, buffer PNI, buffer PE, buffer EB, and collection tubes.
8. Tabletop microcentrifuge.

2.5 DNA Hybridization

1. 5-nm gold nanoparticle (AuNP) solution (Sigma Life Sciences, Oakville, ON, Canada).
2. Saline sodium citrate (SSC) 10×: Dissolve 44.3 g sodium citrate and 87.6 g of NaCl in 1000 mL of DI water and adjust the pH to 7 using 0.1 M HCl.
3. Sodium dodecyl sulfate (SDS) 1.5 %: Add 150 mg of SDS solid to 15 mL DI water.
4. Target solutions: Mix 4 μL of purified PCR products, 3.5 μL of AuNP solution, and 6.5 μL of DI water in a microcentrifuge tube. Place the tube in a water bath on a hot plate at 95 °C for 5 min. Cool down the solution by placing it in an ice bath for 5 min. Centrifuge the tube. Add 2 μL of SSC 10× and 4 μL of SDS 1.5 % and mix.
5. SA-Cy5 solution (50 μg/mL): Mix 5 μL of streptavidin-Cy5 stock solution (1 mg/L) with 10 μL of Tween 20 (1.5 %) and 85 μL of DI water.

2.6 Fluorescence Detection and Data Analysis

1. Confocal laser fluorescent scanner (Typhoon 9410, GE Healthcare).
2. IMAGEQUANT 5.2 software.

3 Methods

3.1 Polydimethylsiloxane (PDMS) Chip Fabrication

1. Place the silicon wafer inside a Pyrex dish and add 100 mL of piranha solution to the dish in the fume hood (*see* **Note 2**). Place the dish on a hot plate and incubate the solution for 15 min. at 80 °C. Swirl the dish once every minute.
2. Remove the wafer and rinse it with water, ethanol (95 %) and water. Dry the wafer by nitrogen gas.
3. Spin coating: Adjust the center of the silicon wafer on the stage of the spin coater and pour ~3 mL of SU-8 at the center of the wafer. Spin the wafer at a rate of 500 rpm for 5 s, and then of 3000 rpm for 30 s to create a SU-8 layer of 35 μm thick on the wafer.
4. Soft bake: bake the wafer at 65 °C for 5 min. to remove the SU-8 solvent.
5. Create the channel pattern on the coated wafer by covering it with the photomask and expose the uncovered SU-8 to UV radiation (270 mW/cm^2) for 5 s initiate polymer cross-linking.

6. Hard bake: Bake the wafer at 95 °C for 3 min. to complete the polymer cross-linking of the exposed SU-8.
7. Pattern development: Add 20 mL of SU-8 developer solution to the wafer in a Pyrex dish to dissolve the unexposed SU-8. Shake the solution for 10 min. at room temperature. The channel pattern should now show up on the wafer. Thereafter, dry the wafer by nitrogen gas (*see* **Note 3**). This is the master mold to be used later for PDMS casting.
8. Create a circular border on the wafer around the patterned region with silicone sealant 732 and leave the silicone to cure for 1 day at room temperature.
9. Prepare 10:1 mixture of PDMS elastomer base to curing agent, and leave the mixture at –20 °C for 1 h in order to remove the air bubbles introduced during mixing.
10. Treat the surface of master mold with the repel silane solution (a release agent) and leave the solution for 5–10 min to dry.
11. PDMS casting: pour the PDMS elastomer mixture on the master mold until a layer with 2 mm thickness is attained. Leave the elastomer to cure and harden at room temperature for 1 day.
12. Cut the edges of the PDMS slab using a blade and then gently demold and release the slab from the master mold surface.
13. Create the chip reservoirs using a sharpened hole punch (gauge 18 or 1.5-mm diameter). Insert the punch from the channel side of the chip. After punching all reservoirs, wash the chip with NOX solution and dry it by nitrogen gas.

3.2 Surface Aldehyde Functionalization

1. Wash the glass chip with Sparkleen solution and rinse it with water.
2. In the fume hood, place the chip inside a Pyrex dish and add 100 mL of piranha solution (*see* **Note 2**). Place the dish on a hot plate and heat the chip in the solution at 80 °C for 15 min. Swirl the dish once every minute.
3. Remove the chip and rinse it with water, ethanol (95 %) and water. Dry the chip by nitrogen gas (*see* **Note 3**).
4. In the fume hood, place the chip inside a Pyrex dish and add 100 mL of APTES solution. Purge the solution with nitrogen gas. Seal the dish with Parafilm and incubate at room temperature for 20 min. Swirl the dish once every minute.
5. Remove the chip from the Pyrex dish. Rinse the chip with 95 % ethanol (*see* **Note 4**). Dry the chip by nitrogen gas and incubate it in the oven for 1 h at 120 °C (*see* **Note 5**).
6. Place the glass chip in the Pyrex dish. Add 100 mL of glutaraldehyde solution. Put a lid on the dish and place it in the fridge for 1 h.
7. Remove the chip from the fridge, wash it with DI water, and dry with nitrogen gas.

3.3 Probe Array Printing

1. Seal the PDMS slab with radial channels on the aldehyde-functionalized glass chip.
2. Add 0.5 μL of probe solution to the inlet reservoirs and the CD-NBA chip and place the chip on the rotating platform. Fill the radial channels by spinning the platform at 400 rpm for 3 min, incubate the solutions inside the channels for 20 min. at room temperature, and then drive the solutions out by spinning the chip at 1800 rpm for 1 min.
3. Peel off the PDMS slab, wash the chip with DI water, and dry it by nitrogen gas.
4. Put the chip in a Pyrex dish. Add the reduction solution and incubate it at room temperature for 20 min. Swirl the dish once every minute. Remove the chip and wash it with DI water.
5. Add the blocking solution to the chip in a Pyrex dish and incubate it at room temperature for 15 min. Swirl the dish once every minute. Remove the chip, wash it with DI water, and dry it with nitrogen gas.

3.4 DNA Amplification and Purification

1. Prepare the DNA amplification mixture in an Eppendorf tube as given in Subheading 2.4, **item 5**. Add 50 μL of mineral oil to the top of the mixture and place the tube in the thermocycler.
2. Thermocycling protocol: initiate denaturation at 95 °C for 3 min, execute 30 cycles of denaturation (95 °C) 40 s, annealing (55 °C) for 30 s, and extension (72 °C) for 60 s, and complete the reaction at 72 °C (final extension) for 10 min.
3. Remove the tube from the thermocycler. Add 10 volumes of Buffer PNI to one volume of the PCR product solution and mix.
4. Put a QIAquick column in a tube. Add the PCR product solution to the column and flow the solution through by centrifuging the tube at 6000 rpm (3600 rcf) for 1 min.
5. Discard the flow-through and place the column back into the same tube. Add 750 μL of diluted Buffer PE (mix 6 mL Buffer PE with 24 mL anhydrous ethanol) to the column. Centrifuge the tube at 6000 rpm for 1 min.
6. Discard the flow-through and place the column back in the same tube. Centrifuge the tube at 13,000 rpm (17,900 rcf) for an additional 1 min (*see* **Note 6**). Place the QIAquick column in a clean 1.5 mL microcentrifuge tube. Add 40 μL of DI Water to the center of the QIAquick membrane, let the column stand for 1 min, and centrifuge the column for 1 min at 13,000 rpm (*see* **Note 7**).

3.5 DNA Hybridization

1. Seal the PDMS slab with spiral channels against the glass chip printed with radially patterned probe lines.

2. Add 1 μL of target solutions to the inlet reservoirs of the spiral channels and place the chip on the rotating platform. Spin the chip at 700 rpm for 10 min.
3. Add 0.5 μL of SA-Cy5 solution to the inlet reservoirs of the spiral channels and place the chip on the rotating platform. Spin the chip at 1500 rpm for 3 min.
4. Peel off the PDMS slab, wash the glass chip with DI water, and dry it with nitrogen gas.

3.6 Fluorescence Detection and Data Analysis

1. Put the glass chip on the Typhoon scanner. Adjust the excitation and emission wavelengths at 633 and 670 nm, respectively. Scan the chip at a resolution of 10 μm (Fig. 3).
2. Analyze the scanned image using IMAGEQUANT software. In order to obtain the signal intensity of the spots: draw a line across multiple spots, go to analysis window, and choose "create graph." The signal intensity of each spot is represented by a peak. Take the intensity of the baseline beside each peak as the background. The corrected signal intensity of each spot is the height of the corresponding peak minus the background.

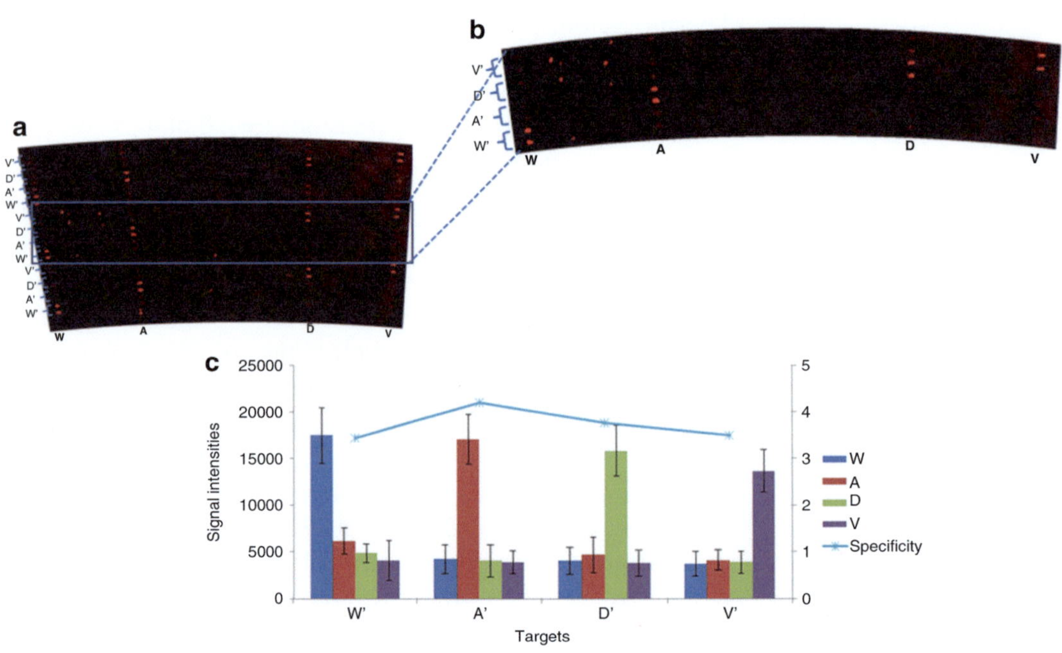

Fig. 3 DNA hybridization between PCR products amplified from four different KRAS samples. Original scanned image showing (**a**) 24 spiral × 4 radial channels of CD-NBA chip, (**b**) the inset inside the blue box and (**c**) the resulted histogram. DNA hybridization occurred between PCR products amplified from four different KRAS samples and the oligonucleotide probes immobilized on the surface of CD-NBA chip. One of the four samples was wild-type (W′) and the other three carried the mutations of G12A (A′), G12D (D′), and G12V (V′) at the KRAS gene codon 12

4 Notes

1. Glutaraldehyde is toxic and strongly irritant and it should be handled with care inside the fume hood. Glutaraldehyde degrades at room temperature and it should be stored at 4 °C.
2. Piranha solution is a hazardous and corrosive solution. Handle it with extreme caution. Use personal protective equipment such as full face shield and heavy-duty gloves. Always add hydrogen peroxide to the acid, not vice versa, and add the liquid very slowly.
3. Occasionally, some SU-8 residues remain, and they appear like dusts on the master mold surface. These residues can be removed by purging the surface with nitrogen gas, rinsing the wafer with NOX solution or performing the SU-8 development step for a second time.
4. Carefully inspect the chip for any dirt or stain on its surface. Repeat the wash and dry step until the surface is clean.
5. APTES-functionalized glass slides should not be in contact with water until after the 1-h incubation at 120 °C. Use dry glassware in the APTES-functionalization step and the subsequent incubation step.
6. Residual ethanol from the Buffer PE will not be completely removed unless the flow-through is discarded before this additional centrifugation.
7. Elution efficiency of the purified PCR product is dependent on pH. The maximum elution efficiency is achieved between pH 7.0 and 8.5.

References

1. Parsons BL, Myers MB (2013) Personalized cancer treatment and the myth of KRAS wild-type colon tumors. Discov Med 15:259–267
2. Domagała P, Hybiak J, Sulżyc-Bielicka V, Cybulski C, Ryś J, Domagała W (2012) KRAS mutation testing in colorectal cancer as an example of the pathologist's role in personalized targeted therapy: a practical approach. Pol J Pathol 3:145–164
3. Sedighi A, Li PC (2014) Challenges and future trends in DNA microarray analysis. Compr Anal Chem 63:25–46
4. Wang L, Li PC (2010) Optimization of a microfluidic microarray device for the fast discrimination of fungal pathogenic DNA. Anal Biochem 400:282–288
5. Wang L, Li PC, Yu H-Z, Parameswaran AM (2008) Fungal pathogenic nucleic acid detection achieved with a microfluidic microarray device. Anal Chim Acta 610:97–104
6. Wang L, Li PC (2010) Gold nanoparticle-assisted single base-pair mismatch discrimination on a microfluidic microarray device. Biomicrofluidics 4:032209
7. Sedighi A, Li PC (2014) Kras gene codon 12 mutation detection enabled by gold nanoparticles conducted in a nanobioarray chip. Anal Biochem 448:58–64
8. Sedighi A, Li PC (2013) Gold nanoparticle assists SNP detection at room temperature in the nanoBioArray chip. Int J Mat Sci Eng 1(1):45–49
9. Sedighi A, Li PC, Pekcevik IC, Gates BD (2014) A proposed mechanism of the influence of gold nanoparticles on DNA hybridization. ACS Nano 8(7):6765–6777

Chapter 11

Single Nucleotide Polymorphisms as Genomic Markers for High-Throughput Pharmacogenomic Studies

Annalisa Lonetti, Maria Chiara Fontana, Giovanni Martinelli, and Ilaria Iacobucci

Abstract

Genetic variations in patients have strong impact on their drug therapies and responses because the variations may contribute to the efficacy and/or produce undesirable side effects for any given drug. The Drug Metabolizing Enzymes and Transporters (DMET) assay is a high-throughput technology by Affymetrix that is able to simultaneously genotype variants in multiple genes involved in absorption, distribution, metabolism, and excretion of drugs for subsequent clinical applications, i.e., the assay allows for a precise genetic map that can guide therapeutic interventions and avoid side effects.

Key words Single nucleotide polymorphisms, Drug metabolizing enzymes and transporters

1 Introduction

Single Nucleotide Polymorphisms (SNPs) are single-base changes at specific genetic loci of a DNA sequence that distinguish between members of a species. SNPs in human occur in a significant proportion (more than 1 %) of a large population [1]. SNPs can be located in both coding and noncoding DNA sequences, and therefore for those located in noncoding regions, the majority of the SNPs are of no biological consequence, and only a fraction of the SNPs located in the coding regions have functional significance [2]. SNPs are common, great in number, and widely distributed throughout the entire human genome with an estimated frequency of about 1/1000 base pairs [3, 4]. All these properties of SNPs make them suitable genetic markers for high-throughput array-based approaches and support their clinical applications since SNPs might impact disease diagnosis and predisposition. More importantly, the presence of genetic variations strongly impact drug therapies and responses, by contributing to the efficacy and/or side effects for any given drug [5, 6]. Furthermore, it is important to

Paul C.H. Li et al. (eds.), *Microarray Technology: Methods and Applications*, Methods in Molecular Biology, vol. 1368, DOI 10.1007/978-1-4939-3136-1_11, © Springer Science+Business Media New York 2016

conduct the investigation of the genetic variations related to the absorption, distribution, metabolism, and excretion (ADME) of drugs, which is called pharmacogenomics. The main intent of pharmacogenomics is to identify personalized therapies in order to reduce the inter-individual variability and to improve efficacy and safety of drug therapy [5]. The Drug Metabolizing Enzymes and Transporters (DMET) assay by Affymetrix is an array-based genotyping platform that interrogates genetic variants in 231 ADME-related genes simultaneously. Among these genes, there are genes that are related to drug-metabolizing enzymes (thiopurine S-methyltransferase or *TPMT*, dihydropyrimidine dehydrogenase or *DYPD*, and members of the cytochrome P450 family), the UDP glucuronosyltransferases (primarily *UGT1A1* and *UGT1A9*), the ATP-binding cassette (ABC) transporters (*ABCB1* or P-glycoprotein, and *ABCG2* or breast cancer resistance protein), other novel drug transporters (e.g., sulfotransferase), and transcription regulators (e.g., *NR1I2*), and also genes known to induce other ADME genes (e.g., *PPARD*) [7]. The DMET genetic markers are collectively 1936, and they are classified into genotyping markers (biallelic SNPs, triallelic SNPs, and insertions/deletions of varying lengths) and five regions of copy number variation (CNV) [8]. The molecular inversion probes (MIP) amplification technology can genotype in a single reaction all the interrogated genomic sites [9]. Indeed, DMET array-based genotyping interrogates a sufficient number of polymorphisms related to the ADME of drugs, allowing the identification of clinically significant metabolic profiles for the patient-specific therapy. Moreover, the collected SNPs data are then converted into known haplotypes based on standardized nomenclature (i.e., star alleles nomenclature), routinely used in pharmacogenomic studies [10]. This conversion is essential for describing and interpreting known haplotypes. In recent times, several research groups have successfully used the DMET platform to identify genetic profiles that influence the sensitivity to specific therapies [6, 11–15].

2 Materials

2.1 Laboratory Configuration

During the DMET protocol it is essential to use two separate rooms to avoid contamination of the samples with amplified PCR products. The pre-Amp Lab is an area free of PCR products (can do PCR, but not cleanup of PCR products). In this Lab a fume hood is required to perform the multiplex PCR (mPCR), and this is called the mPCR Staging Area in the pre-Amp Lab. mPCR is a pre-amplification step to amplifiy some genetic markers before they join with other markers in the PCR amplification using the highly selective MIP technology. The post-Amp Lab is a separate room where the steps following PCR are performed. Each of these areas and Labs requires dedicated equipment, including pipettes and 96-well iceless coolers.

2.2 Affymetrix Reagents, Equipment, and Software

There are two types of DMET assay packs (*see* **Note 1**) and they contain the DMET Plus Reagent kit which consists of the Pre-Amp Kit, Labeling Kit, Hyb-Stain Kit, and Panel Kit.

1. Pre-Amp Kit consists of pre-amp water, Buffer A, Enzyme A, Gap Fill Mix 1 and 2, Exo Mix, Cleavage Enzyme, Universal Amp Mix and dNTP mix. Part of these reagents is used to make the Anneal Master Mix, *see* Table 1.
2. Labeling Kit consists of PCR cleanup mix, post-amp water, Frag buffer, Frag reagent, DNA labeling reagent, 5× TdT buffer, and TdT enzyme. Part of these reagents is used to make the Labeling Master Mix, *see* Table 2.
3. Hyb-Stain Kit consists of hybridization solution, oligo control reagent, stain buffer, and hold buffer.
4. Panel Kit consists of MIP panel, mPCR primer mix, 1× TE buffer, gDNA control 2, and PCR dilution buffer.
5. Wash Solution: Wash Solution A and Wash Solution B.
6. DMET Plus Array.
7. GeneChip® Fluidics Station 450.
8. GeneChip®Hybridization Oven 640.

Table 1
Preparation of the anneal master mix

Reagent	Volume of 1 reaction (μL)
Pre-amp water	16.6
Buffer A	5
Enzyme A	0.0625

Table 2
Preparation of the labeling master mix

Reagent	1 Reaction (μL)
Post-amp water	0.4
5× TdT buffer	7
DNA labeling reagent	0.9
TdT enzyme	1.7
Total	10

9. GeneChip® Scanner.
10. Affymetric GeneChip® Command Console (AGCC) Software.
11. DMET™ Console Analysis Software: provides a fast genetic analysis of the data obtained by the DMET™ Plus Reagent kit in order to discover the involvement of metabolic pathways in drug metabolism.

2.3 Other Reagents and Equipment

1. AccuGENE® Water (Lonza Group LTD).
2. Molecular biology grade water.
3. Strip tubes.
4. DNA Blood Mini Kit (Qiagen).
5. Quant-iT™ PicoGreen® dsDNA Assay Kit (Life Technologies), containing λ DNA (100 μg/mL). Several λ DNA standard solutions are made according to Table 3.
6. QIAGEN® Multiplex PCR Kit consists of multiplex PCR master mix, Q solution 5×, RNase-free water. The multiplex PCR master mix, which consists of HotStar Taq polymerase, dNTPs, $MgCl_2$, and multiplex PCR buffer, is used to make the mPCR Master Mix as shown in Table 4.
7. Streptavidin, R-phycoerythrin conjugate (SAPE).
8. Taq Polymerase (TITANIUM™, Clontech).
9. TE Buffer 1×, pH 8.0.

Table 3
Serial dilution of λ DNA standard solutions for quantitation of gDNA concentration

Serial dilution	Standard	TE 1× (μL)	Final dilution (ng/μL)	Final concentration (in assay) (ng/μL)
	30 μL of 100 μg/mL standard	1470	2	1
1:2	750 μL of 2 ng/μL standard	750	1	0.5
1:2	750 μL of 1 ng/μL standard	750	0.5	0.25
1:2	750 μL of 0.5 ng/μL standard	750	0.25	0.125

Table 4
Preparation of the mPCR mix

Reagent	Volume for 1 reaction (μL)
Multiplex PCR master mix	25
mPCR primer mix (3 μM)	5
5× Q-solution	5
RNase-free water	10

10. TECAN plate reader (Infinite 200 PRO).
11. Thermal cyclers.
12. 96-well iceless cooler.

3 Methods

3.1 Genomic DNA Preparation

To genotype patients affected by hematological malignancies, genomic DNAs (gDNAs) are required, and they will be obtained from remission samples (the "normal" counterpart). Blood samples are the major source of gDNA in pharmacogenomic studies. However, alternative and valuable gDNA sources might be saliva samples [16].

1. Extract gDNA from blood samples using the DNA Blood Mini Kit (Qiagen).
2. gDNA concentration should be determined using the Quant-iT™ PicoGreen® assay, which is an ultrasensitive fluorescent nucleic acid stain for the detection of double-stranded DNA (dsDNA). Prepare an aqueous working solution by making a 200-fold dilution of the concentrated PicoGreen® reagent in 1× TE buffer in a plastic container. For 6 mL of reagent solution, add 30 μL of PicoGreen® reagent to 5970 μL 1× TE buffer. Protect the solution from light.
3. Prepare the λ DNA standard solutions by adding 1470 μL of 1× TE buffer to 30 μL of the λ DNA standard to obtain a final concentration of 2 ng/μL. Proceed with serial 1:2 dilution in 1× TE buffer to obtain the λ DNA concentration of 1 ng/μL, 0.5 ng/μL, 0.25 ng/μL, *see* Table 3.
4. Dilute the extracted gDNA by ten times before quantification (i.e., 3 μL gDNA+27 μL 1× TE buffer). It is because of the high sensitivity of PicoGreen assay and the good yield of gDNA from the blood samples.
5. Prepare the microplate for gDNA quantification. For λ DNA standards, mix 50 μL of each standard solution with 50 μL of the PicoGreen working solution. The final λ DNA concentration in the assay will be 1 ng/μL, 0.5 ng/μL, 0.25 ng/μL, and 0.125 ng/μL (*see* Table 3). Ensure to insert the reagent blank, i.e., mixing 50 μL of 1× TE buffer to 50 μL of the PicoGreen working solution.
6. For each gDNA sample, mix 10 μL of diluted gDNA with 40 μL of 1× TE buffer and 50 μL of the PicoGreen working solution.
7. For both λ DNA standards and gDNA samples, they should be quantified at least in duplicate. Incubate for 2–5 min at room temperature. Protect the solutions from light.

8. Measure the fluorescence using a Tecan plate reader. Subtract the fluorescence value of the reagent blank from that of the λ DNA standard and gDNA samples. Plot the calibration curve using the λ DNA standards, and determine the gDNA concentration from the curve.
9. The gDNA samples should be normalized to a single concentration of 60 ng/μL. With the quantified gDNA concentrations, dilute the samples in 1× TE buffer to give a final concentration of 60 ng/μL.

3.2 Day 1 Operation: Pre-amplification and Annealing (Pre-Amp Lab)

1. Start the operation in the pre-Amp Lab at noon. Pay attention in working with enzymes because they are temperature-sensitive. Keep them at −20 °C until use. Spin down at 13,400 rcf for 30 s the enzyme solutions in the tubes to uniformly mix the content, and do not vortex. All other reagents should be vortexed after thawing and then spun down.
2. Thaw and then place on ice the reagents from the DMET Plus Reagent Kit, i.e., mPCR Primer Mix, gDNA Controls, 1× TE Buffer, and PCR Dilution Buffer.
3. Thaw and then place on ice the reagents from the QIAGEN Multiplex PCR Kit, i.e., Multiplex PCR Master Mix, Q-Solution, and RNase-free water. The Q-Solution contains some PCR additives that facilitate amplification of difficult templates.
4. In the pre-Amp Lab, prepare the Genomic Plate 1 (GP1 plate), which consists of 17 μL of each gDNA sample (at a concentration of 60 ng/μL) and 17 μL of each gDNA control. The gDNA controls in the Kit are already normalized to a working concentration of 60 ng/μL. The same sample scheme must be maintained in each step to facilitate the liquid transfer using the multichannel pipettes and to minimize pipetting errors. Seal the plate and spin it down at 800 rcf for 60 s. Maintain the plate at 4 °C using the 96-well iceless cooler.
5. Prepare the Genomic Plate 2 (GP2 plate). Using a multichannel P20 pipette, aliquot 10 μL of 1× TE Buffer and then transfer 2 μL of each sample or gDNA control from GP1 plate to the corresponding well of GP2 plate. Vortex and spin down GP2 plate. Seal GP1 plate and store it on ice or at 4 °C until use.
6. Prepare the mPCR Mix by adding reagents in the order shown in Table 4.
7. Transfer the mPCR Mix to a reagent reservoir. Aliquot 45 μL of the Mix to each well of a new 96-multiwall plate (mPCR plate) using a multichannel P200 pipette. Follow the same sample scheme of GP1/GP2 plates. Using a multichannel P20 pipette, transfer 5 μL of each sample/control from GP2 plate to the corresponding well of mPCR plate. Seal GP2 plate and keep it on ice until the pre-amplification stage is successfully

Table 5
Temperature program used for pre-amplification of some genetic markers

Temperature (°C)	Time	Cycle number
94	15 min	1
95 60 72	30 s 90 s 45 s	35
72	3 min	1
4	Infinity	1

completed. Seal, vortex, and spin down mPCR plate, then place it on a thermal cycler and run the temperature program shown in Table 5.

8. The mPCR products will be diluted twice. Two new plates are used for the dilution. After the temperature program is finished, spin down mPCR plate. In the mPCR Staging Area (i.e., fume hood), prepare Dilution Plate 1 (DP1 plate) and Dilution Plate 2 (DP2 plate). Using a multichannel P200 pipette, aliquot in DP1 and DP2 plates 153 μL PCR Dilution Buffer to the corresponding well of each sample/control in mPCR plate. Transfer 5 μL of each sample/control from mPCR plate to the corresponding well of DP1 plate and mix slowly by pipetting up and down ten times. Then, transfer 5 μL of each sample/control from DP1 plate to the corresponding well of DP2 plate, and mix slowly by pipetting up and down ten times (*see* **Note 2**).
9. In the Pre-Amp Lab, prepare the Anneal Master Mix by adding reagents from the Pre-Amp Kit in the order shown in Table 1. Pipette up and down the Anneal Mix five times using a P1000 pipette set to 900 μL and do not vortex.
10. Prepare the Anneal Plate (ANN plate) by adding 21.7 μL Anneal Master Mix per well. Transfer 13.4 μL of each sample from GP1 plate to ANN plate (final volume = 35.1 μL). Seal and transfer ANN plate to the mPCR Staging Area.
11. Add 5 μL diluted mPCR products from DP2 plate to the corresponding wells of ANN plate (final volume = 40.1 μL). Seal and transfer ANN plate to the Pre-Amp Lab, vortex, and spin. Put the ANN plate on the thermal cycler and start the temperature program shown in Table 6 (*see* **Note 3**):
12. At the end of the first 95 °C hold, press "pause" on the thermal cycler, remove the ANN plate and put it on an iceless cooler for 2 min. Then, spin the ANN plate, remove the seal and add 5 μL MIP Panel from the Panel Kit to each reaction well (final volume = 45.1 μL). Seal again the ANN plate, vortex, spin down and place it back on the thermal cycler and resume the temperature program (Table 6). Incubate the samples at 58 °C for 16–18 h.

Table 6
Temperature program for annealing

Temperature (°C)	Time
20	4 min
95	5 s
Add 5 μL MIP panel to each well	
95	5 s
58	Infinity

3.3 Day 2 Operation: Cleanup, Fragmentation, Labeling, Denaturation, and Hybridization

Start these steps in the morning, immediately after the Anneal stage (do not incubate samples for more than 18 h for annealing). The day 2 operation starts in the pre-Amp Lab. The first stage (Gap Fill Through Amplification) requires adding reagents during the thermal cycler reaction and stopping the program at specific time points. The following stages must be located in the post-Amp Lab. Before starting, thaw, spin down, and then place on ice until use the dNTP Mix and the Universal Amp Mix, whereas leave the Exo Mix, Cleavage Enzyme, Gap Fill Mixes 1 and 2, and the TITANIUM Taq Polymerase at −20 °C until ready to use.

1. In the pre-Amp Lab, prepare the Gap Fill Mix by mixing 190 μL Gap Fill Mix 2 and 10 μL Gap Fill Mix 1 in an Eppendorf tube. Mix well by pipetting slowly up and down and aliquot 14 μL of Gap Fill Mix to each 0.2 mL PCR tube of an 8-strip tube. Cap and spin down the strip tubes and place them in an iceless cooler.
2. Remove ANN plate from the thermal cycler, place the plate in an iceless cooler for 2 min and then spin down at 800 rcf for 60 s.
3. Using a multichannel P10 pipette, add 2.5 μL Gap Fill Mix to each reaction well of ANN plate. Seal, vortex, and spin.
4. Prepare the Assay plate (ASY plate) by transferring 12 μL each reaction from the ANN plate. Seal and spin the ASY plate.
5. Place the ASY plate in the thermal cycler and run the temperature program shown in Table 7 (*see* **Note 3**).
6. After the first 11 min at 58 °C, pause the thermal cycler, place the plate in an iceless cooler for 2 min, and spin down. Remove the seal and add 5 μL dNTP Mix to each reaction. Reseal the plate, vortex, spin down, and place the plate back on the thermal cycler. Resume the temperature program (Table 7).
7. When thermal cycler reaches 37 °C, pause the thermal cycler, place the plate in an iceless cooler for 2 min, and spin down.

Table 7
Temperature program for PCR

Temperature (°C)	Time	Cycles
58	11 min	1
Add 5 μL dNTP mix		
58	11 min	1
Add 5 μL exo mix		
37	15 min	1
95	5 min	1
Add 30 μL universal amp mix		
60	5 min	1
95	7 min	1
95 64 68	15 s 15 s 30 s	22
68	7 min	1
4	Infinity	

Remove the seal and add 5 μL Exo Mix to each reaction. Reseal the plate, vortex, spin down, and place the plate back on the thermal cycler. Resume the temperature program (Table 7).

8. In the pre-Amp Lab, prepare the Universal Amp Mix by directly adding 25 μL of Cleavage Enzyme and 70 μL of Taq Polymerase into the Universal Amp Mix Tube. Mix by pipetting up and down ten times.
9. When thermal cycler reaches 60 °C, pause the thermal cycler, place the plate in an iceless cooler for 2 min, and spin down. Remove the seal and add 30 μL Universal Amp Mix to each reaction. Reseal, vortex, spin down the plate, and place it back on the thermal cycler. Resume the temperature program (Table 7).
10. When the program has ended, transfer the sealed ASY Plate to the Post-Amp Lab and place it on ice.
11. In the post-Amp Lab, aliquot 2.5 μL PCR Cleanup Mix to each reaction of the ASY plate. Seal, vortex, and spin down the plate, then place it in a thermal cycler and run the temperature program shown in Table 8 (*see* **Note 4**).
12. Prepare the first QC gel (3 % agarose gel) to identify if the gDNA samples have amplified by PCR.

Table 8
Temperature program for PCR cleanup

Temperature (°C)	Time
37	15 min
80	15 min
4	Infinity

13. After the cleanup step, spin down the plate, remove the seal, and take 2 μL from each reaction well. Reseal the plate and place at 4 °C. For each reaction, prepare a loading sample by mixing 2 μL of PCR products, 2 μL of 2× loading buffer, and 8 μL water.
14. Load 10 μL of the loading sample onto the agarose gel and run gel electrophoresis at 120 V for 20 min. It should be detected for each sample a PCR product between 100 and 150 bp to ensure good-quality PCR.
15. In the post-Amp Lab, transfer 25 μL of each reaction to a new plate for fragmentation and labeling (Frag/Label plate).
16. Thaw and then place on ice the Post-Amp Water and the Fragmentation Buffer. Prepare the fragmentation mix by mixing 8.9 μL Post-Amp Water and 1 μL of Fragmentation Buffer per sample, and cool the mixture on ice for 5 min. Then, add 0.0675 μL/sample of Fragmentation Reagent to the mixture, then vortex, spin, and place on ice. Add 10 μL fragmentation mix to each reaction well (*see* **Note 5**). Seal, vortex, spin, and place the plate to thermal cycler to run the temperature program shown in Table 9 (*see* **Note 6**):
17. Prepare a second QC gel (3 % agarose gel) to check the fragmentation reaction to confirm acceptable fragment size. For each reaction well, take 10 μL of sample (from the Frag/Label plate) and mix it with 2 μL of 2× loading buffer. Load 10 μL of the loading sample onto the agarose gel and run gel electrophoresis at 120 V for 24 min. A gel that consists of fragments of less than 120 bp with the smear centered on around 50 bp shows the good quality in fragmentation.
18. In the post-Amp Lab, prepare the Labeling Master Mix by adding reagents in the order shown in Table 2.
19. Aliquot the Labeling Master Mix into strip tubes (45 μL/tube) and then add 10 μL of the mix to each sample of the Frag/Label plate by using a multichannel P20 pipette. Seal, vortex, and spin down the plate, then place it in a thermal cycler and run the temperature program shown in Table 10.

Table 9
Temperature program for fragmentation

Temperature (°C)	Time
37	15 min
95	15 min
4	Infinity

Table 10
Temperature program for labeling

Temperature (°C)	Time
37	60 min
95	15 min
4	Infinity

20. Before the hybridization step, preheat the Hybridization Oven to 49 °C with rotation on (35 rpm) (*see* **Note 7**).
21. Equilibrate the DMET Plus arrays to room temperature. Mark each array. For each array, insert a 200 μL pipette tip into the upper right septum.
22. Thaw the Hybridization Solution and Oligo Control Reagent, then prepare the Hybridization Master Mix by adding the Oligo Control Reagent directly to the Hybridization Solution tube. Pour the mix into a reagent reservoir on ice.
23. Prepare the Hybridization plate. With a multichannel (P200 and P20) pipette, aliquot 92 μL of Hybridization Master Mix and 8 μL of each sample from the Frag/Label plate (*see* **Note 8**). Seal, vortex, and spin the plate, then place the plate on a thermal cycler prewarmed at 95 °C. Leave the plate at 95 °C for 10 min to denature the samples before loading the arrays. Then, place the plate in an iceless cooler for 2 min and spin the plate.
24. Generate a sample batch registration file using the predefined template provided with the DMET Console (DMET. TEMPLATE) or using your own created template. The Batch Registration File could be a TSV or Excel file type and must contain sample information, such as sample file name, probe array type, sample name, source plate, source well, and sample type. Scan the array barcodes into this file.

25. Upload the Batch Registration File to Affymetrix GeneChip® Command Console™ (AGCC) software and click "save". The message "Batch Array Registration is complete" is displayed.
26. In each array, inject 95 μL of denatured sample via the pipette tip. Remove it from the upper right septum of each array and cover both the array septa with adhesive spots.
27. Place the array in the hybridization oven and allow to hybridize at 49 °C and 35 rpm for 16–18 h (*see* **Note 9**).
28. Thaw the Stain Buffer and Hold Buffer at 4 °C overnight.

3.4 Day 3 Operation: Stain, Wash and Scan

1. Prime the Fluidics Station, install the Wash Solution A and B bottles and fill the dH_2O container. Then, run the PRIME_450 script in the AGCC software.
2. Prepare SAPE Stain Solution by adding 90 μL of SAPE to the Stain Buffer tube. Mix by inverting the tube. Keep the mix at room temperature and protect it from light.
3. Setup the Software and the Fluidics Station according to the appropriate DMET protocol to wash and stain DMET Plus Arrays.
4. Wash no more than eight arrays in the Fluidics Station at any time, leave remaining arrays in the hybridization oven. Before washing, remove the adhesive spots from each array. Wash and stain arrays.
5. Warm up scanner for 10 min, then remove arrays from fluidics station and cover again both septa with adhesive spots (do not cover the window). Load arrays onto scanner and scan (*see* **Note 10**).

3.5 Analysis of DMET Data by the Affymetrix DMET™ Console Analysis Software

1. After scanning the DMET Plus Arrays, the files containing the data of hybridization intensity (CEL files) are generated by the AGCC software. Analysis files can thereafter be downloaded within DMET Console.
2. To start analyzing data in DMET™ Console, first create a workspace and add one or more datasets to the workspace. A dataset points to a collection of sample files (ARR), intensity files (CEL), and genotyping files (CHP). The CEL files are loaded in a new workspace and dataset are then converted to genotype calls (CHP files).
3. The CHP files get through the automatic Quality Control (QC), where the console divides the results in two groups: "In Bounds" and "Out of Bounds" depending on the default QC call rate threshold of 98 %. This means that samples with a QC call rate >98 % are "In Bounds" and samples with a call rate <98 % are "Out of Bounds."

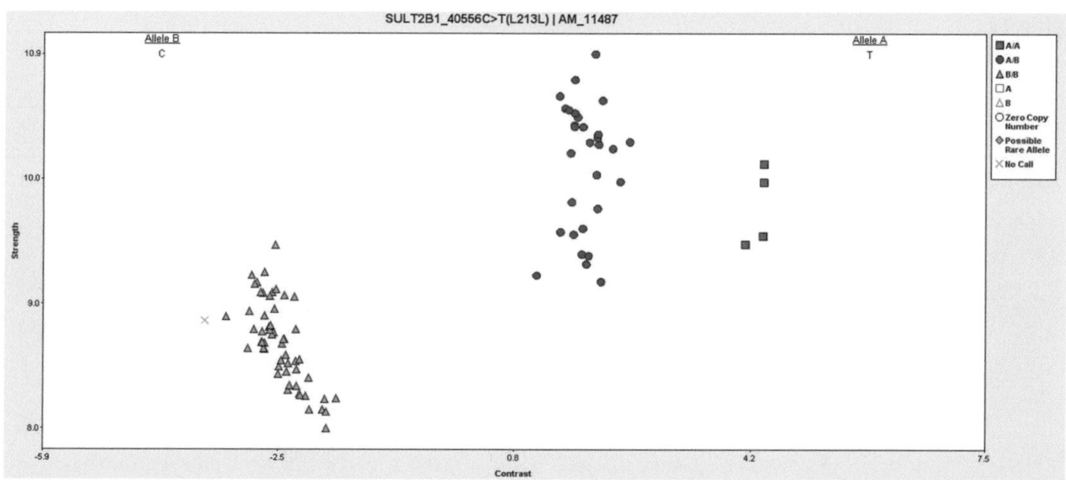

Fig. 1 Marker Summary cluster graph. The cluster graph displays the signal data for the marker selected by the user in the Marker Summary Table (*SULT2B1*_40556C>T). The genotype calls (AA, AB, BB, and noCall) are identified by color and shape (*see* Table 11). The graph legend uses abstract allele names, which is how the calls are stored in the genotype result files (CHP). In order to interpret these codes, the graph also displays the actual allele name: *SULT2B1*_40556C>T(L213L) on the reported strand (which is intended to be the gene coding strand)

4. Once the selected CHP files pass sample QC, the user can use a predefined marker list, or create one, to adapt the signal boundaries to the behavior of the samples.
5. The genotypes are translated into haplotypes alleles (i.e., SULT2B1_40556C>T) for many genes using the star allele nomenclature and then into a metabolizer condition which reveals the level of metabolic activity of the related enzyme.
6. The DMET™ Console software presents the genotyping results in the CHP Summary, Marker Summary, and Copy Number Summary tables [17]. The Marker Summary cluster graph is an X–Y scatter plot of the signal data for the selected marker across all CHP files in the user-selected results group (Fig. 1). Prior to genotyping, the DMET™ Plus array summarized signals are transformed using the "Minus vs. Average" signal transformation.
7. The marker summary cluster is summarized by the call codes listed in Table 11. The data in Fig. 1 include numerous AA, AB, BB calls, and one noCall, in which A represents the T allele and B represents the C allele of the *SULT2B1*_40556C>T mutation of the sulfotransferase *SULT2B1* gene.
8. A concordance check is used to compare the genotype results (CHP) with the reference data and help the user to evaluate data quality. The reference data can be text files (TXT) or

Table 11
Call codes used in the marker summary cluster graph

Call	Description
AA, BB, CC	Homozygous call for the respective allele, in which the algorithm applied a two copy number model
AB, AC, BC	Heterozygous call for the respective alleles, in which the algorithm applied a two copy number model
A, B	Hemizygous call for the respective allele, in which the algorithm applied a one copy number model For the DMET™ Plus array, these calls are only made for males on the X chromosome
Zero copy number	The algorithm determined that this marker is in a chromosome region that does not exist in this sample's DNA (zero copies of this marker)
PossibleRareAllele	The algorithm assigns this call to a genotype that was either never observed or very rarely observed in the training data that was used to derive the predefined genotype models A PossibleRareAllele call is made when the number of observations of the genotype in the training data is below the minimum number of prior observations specified in the analysis configuration used for genotyping
NoCall	The further a data point drifts from the expected cluster location, the less confidence the call is correct. If the confidence of a call is below the confidence threshold specified in the analysis configuration used for genotyping, the call is reset to NoCall

genotype results (CHP). The genotypes that are not found in both files, as well as genotypes reporting "NoCall" or "NotAvailable" for at least one of the two compared files, are excluded from the comparison.

9. The software shows the results in tables and graphs [17], so the user can:
   - Quickly identify possible rare alleles or missing data.
   - Identify haplotype and probe-level sequence variation in the test samples relative to a standard reference sequence.
   - Identify variation at other genes in the panel that impart functional or structural changes to the gene product.
   - Annotate the reported genotypes across probes to indicate genomic, mRNA, or peptide changes resulting from any observed variation in the analyzed samples.
   - Extract known functional or structural variants in a DMET profile to identify and summarize genes of potential altered activity.

- Predict general gene activity based on detected diplotypes.
- Integrate missing data into final study summary reports with the use of an Override Report.

3.6 Analysis of DMET Files by DMET Analyzer Software

This software was developed at the Department of Medical and Surgical Sciences, Magna Graecia University of Catanzaro, Italy. The software is freely available for academic purposes at https://sourceforge.net/projects/dmetanalyzer/files/. DMET-Analyzer lacks the quality control capabilities available in DMET Console and the possibility to manage Affymetrix binary files directly. The tool allows the following operations.

1. Automate the workflow of analysis of DMET-SNP data avoiding the use of multiple tools [18]. The user can prepare his/her own dataset in both excel file and delimited table files. The software is able to find the class-labels directly from the input files. DMET Analyzer automatically selects the relevant SNPs. In fact, the software includes several statistical software, such as the Fisher's test to check the association among the presence of SNP and the classes already determined, the Bonferroni test and False Discovery Rate to get better statistical significance of results, the Hardy-Weinberg calculator to analyze the linkage disequilibrium, and the Pearson's chi-square test to calculate the deviation from the Hardy-Weinberg equilibrium and to determine the significance of the deviation.
2. Automate the annotation of DMET-SNP data and to search in existing databases for SNPs (e.g., dbSNP) [19].
3. Associate SNP with pathways through the search in PharmaGKB, a major knowledgebase for pharmacogenomic studies.

4 Notes

1. There are two DMET assay packs and they are the DMET™ Plus Premier Pack and the DMET™ Plus Starter Pack. They contain the same types of reagents: the DMET™ Plus Starter Pack contains sufficient reagents for eight reactions (7 samples and 1 control), but the DMET™ Plus Premier Pack contains sufficient reagents for 48 reactions (45 samples and 3 controls).
2. Dilution Plate 1 (DP1) and Dilution Plate 2 (DP2) should be maintained at 4 °C in a 96-well iceless cooler.
3. Pay attention to the time when the program in the thermal cycler must be paused to add specific reagents.
4. The thermal cycler for the Cleanup Mix must be located in the post-Amp Lab. Do not return the PCR products to the

pre-Amp Lab to avoid contamination with mPCR products and samples.

5. During the fragmentation step, it is essential to work on ice to maintain the reagents, mix and samples (on Frag/Label plate) cool, and work quickly to avoid the reaction to start prematurely.
6. Place the Frag/Label plate onto the thermal cycler only when the temperature is exactly 37 °C.
7. Accurate hybridization temperature (49 ° C) is critical for this assay.
8. The Frag/Label plate can be stored at –20 °C for up to 1 week, if hybridizations need to be repeated.
9. When working with a large number of samples load multiple arrays and place them in the hybridization oven at the same time to reduce the effect of oven temperature changes on the hybridization results on different arrays.
10. Before scanning, inspect each array for the presence of bubbles and for blemishes/dirt/marks on the array window.

Acknowledgements

We are grateful to the financial support by European LeukemiaNet, Associazione Italiana control le Leucemie, AIRC, progetto Regione-Università 2010–2012 (L. Bolondi), and FP7 NGS-PTL project.

References

1. Brookes AJ (1999) The essence of SNPs. Gene 234(2):177–186
2. Collins FS, Guyer MS, Charkravarti A (1997) Variations on a theme: cataloging human DNA sequence variation. Science 278(5343): 1580–1581
3. Abecasis GR, Auton A, Brooks LD, DePristo MA, Durbin RM, Handsaker RE et al (2012) An integrated map of genetic variation from 1,092 human genomes. Nature 491(7422):56–65
4. Sachidanandam R, Weissman D, Schmidt SC, Kakol JM, Stein LD, Marth G et al (2001) A map of human genome sequence variation containing 1.42 million single nucleotide polymorphisms. Nature 409(6822):928–933
5. Scott SA (2011) Personalizing medicine with clinical pharmacogenetics. Genet Med 13(12): 987–995
6. Iacobucci I, Lonetti A, Candoni A, Sazzini M, Papayannidis C, Formica S et al (2013) Profiling of drug-metabolizing enzymes/transporters in CD33+ acute myeloid leukemia patients treated with Gemtuzumab-Ozogamicin and Fludarabine, Cytarabine and Idarubicin. Pharmacogenomics J 13(4):335–341
7. Affymetrix (2012) DMET plus allele translation reports: summary of comprehensive drug disposition genotyping into commonly recognized allele names. Affymetrix White Paper 1–19
8. Deeken J (2009) The Affymetrix DMET platform and pharmacogenetics in drug development. Curr Opin Mol Ther 11(3):260–268
9. Karlin-Neumann G et al (2007) Molecular inversion probes and universal tag arrays: application to highplex targeted SNP genotyping. In: Weiner MP, Gabriel SB, Stephens JC (eds) Genetic variation: a laboratory manual. Cold Spring Harbor Lab, Cold Spring Harbor, NY, pp 199–211

10. Robarge JD, Li L, Desta Z, Nguyen A, Flockhart DA (2007) The star-allele nomenclature: retooling for translational genomics. Clin Pharmacol Ther 82(3):244–248
11. Harris M, Bhuvaneshwar K, Natarajan T, Sheahan L, Wang D, Tadesse MG et al (2014) Pharmacogenomic characterization of gemcitabine response – a framework for data integration to enable personalized medicine. Pharmacogenet Genomics 24(2):81–93
12. Hertz DL, Roy S, Jack J, Motsinger-Reif AA, Drobish A, Clark LS et al (2014) Genetic heterogeneity beyond CYP2C8*3 does not explain differential sensitivity to paclitaxel-induced neuropathy. Breast Cancer Res Treat 145(1):245–254
13. Shiotani A, Murao T, Fujita Y, Fujimura Y, Sakakibara T, Nishio K et al (2014) Single nucleotide polymorphism markers for low-dose aspirin-associated peptic ulcer and ulcer bleeding. J Gastroenterol Hepatol 29(Suppl 4): 47–52
14. Bonifaz-Pena V, Contreras AV, Struchiner CJ, Roela RA, Furuya-Mazzotti TK, Chammas R et al (2014) Exploring the distribution of genetic markers of pharmacogenomics relevance in Brazilian and Mexican populations. PLoS One 9(11):e112640
15. Deeken JF, Cormier T, Price DK, Sissung TM, Steinberg SM, Tran K et al (2010) A pharmacogenetic study of docetaxel and thalidomide in patients with castration-resistant prostate cancer using the DMET genotyping platform. Pharmacogenomics J 10(3):191–199
16. Hu Y, Ehli EA, Nelson K, Bohlen K, Lynch C, Huizenga P et al (2012) Genotyping performance between saliva and blood-derived genomic DNAs on the DMET array: a comparison. PloS One 7(3):e33968
17. Affymetrix (2012) DMET™ console 1.3 user manual. DMET™ console 13 user manual
18. Guzzi PH, Agapito G, Di Martino MT, Arbitrio M, Tassone P, Tagliaferri P et al (2012) DMET-analyzer: automatic analysis of Affymetrix DMET data. BMC Bioinformatics 13:258
19. Day IN (2010) dbSNP in the detail and copy number complexities. Hum Mutat 31(1):2–4

Chapter 12

DNA Microarray-Based Diagnostics

Mahsa Gharibi Marzancola, Abootaleb Sedighi, and Paul C.H. Li

Abstract

The DNA microarray technology is currently a useful biomedical tool which has been developed for a variety of diagnostic applications. However, the development pathway has not been smooth and the technology has faced some challenges. The reliability of the microarray data and also the clinical utility of the results in the early days were criticized. These criticisms added to the severe competition from other techniques, such as next-generation sequencing (NGS), impacting the growth of microarray-based tests in the molecular diagnostic market.

Thanks to the advances in the underlying technologies as well as the tremendous effort offered by the research community and commercial vendors, these challenges have mostly been addressed. Nowadays, the microarray platform has achieved sufficient standardization and method validation as well as efficient probe printing, liquid handling and signal visualization. Integration of various steps of the microarray assay into a harmonized and miniaturized handheld lab-on-a-chip (LOC) device has been a goal for the microarray community. In this respect, notable progress has been achieved in coupling the DNA microarray with the liquid manipulation microsystem as well as the supporting subsystem that will generate the stand-alone LOC device.

In this chapter, we discuss the major challenges that microarray technology has faced in its almost two decades of development and also describe the solutions to overcome the challenges. In addition, we review the advancements of the technology, especially the progress toward developing the LOC devices for DNA diagnostic applications.

Key words DNA microarray, Stand-alone lab-on-a-chip (LOC) device, Diagnostic tool, Microfluidics, Label-free detection, Nanoarrays

1 Overview

The DNA microarray has achieved significant progress in both application and technology ever since it was first introduced in 1995 by Schena et al. [1]. This first microarray was fabricated by spotting or printing various complementary DNAs (cDNAs) on a glass microscope slide via a robotic printer and the microarray was used to monitor the differential expression of many genes in parallel. There are three major applications of the microarrays. First, the microarray platform, especially in the early years of developments,

Paul C.H. Li et al. (eds.), *Microarray Technology: Methods and Applications*, Methods in Molecular Biology, vol. 1368, DOI 10.1007/978-1-4939-3136-1_12, © Springer Science+Business Media New York 2016

has been used to obtain clinically relevant information from the gene expression levels [2]. For example, the microarray data were used to differentiate between cancer subtypes, to provide prognostic information (e.g., likelihood of recurrence or metastasis) and predictive information (e.g., efficacy of chemotherapy). Second, the microarrays have been developed for genotyping to characterize the DNA (or RNA) in order to detect human gene mutations (or characterize viral pathogens). While simple genotyping arrays consist of hundreds of features or spots, complex arrays utilize thousands of features to investigate mutations in many genes or to characterize multiple sequences in the pathogen genomes. Third, the microarrays are used to conduct the array-based comparative genomic hybridization (array-CGH) which provides a high-resolution tool for screening copy number variations (CNV) in the whole genome and offers several advantages over classical techniques [3]. In addition to the various novel applications, the microarray platforms experienced many technical advances after the commercial vendors took over the developments. For instance, the probes to be immobilized on the microarrays have shifted from cDNA to short oligonucleotides, and they were either pre-synthesized or synthesized in situ. These oligonucleotides demonstrated a higher specificity than cDNA probes. Glass is still the predominant substrate used in the microarray platform, but materials such as silicon and polymers have also been used as the microarray substrate. In terms of signal transduction to generate the microarray data, label-free detection techniques have also been developed [4].

Although the microarray technology should have a high potential for clinical applications, it has not experienced a smooth path of development. There are numerous challenges, more on biostatistics than on technical issues. One challenge is that the microarray data have notoriously been considered as being "noisy" [5]. The reproducibility of the data and the validity of the data interpretation reported by prominent microarray studies have been criticized because of a lack of appropriate standardization, adequate quality control measures and reliable data processing [6]. The uncertainty about the validity of the microarray data interpretation hindered the approval of array-based clinical tests by regulatory organizations as well as the subsequent adoption of the tests by clinical communities. The concerns about the validity of the microarray data interpretation is more serious in applications such as the introduction of new biomarkers (e.g., expression profiling), rather than in applications such as genotyping which are dealing with preexisting biomarkers.

Another challenge of microarray-based tests is the advent of the competitive PCR-based and sequencing-based tests. For instance, simple microarray tests, when only a few genes are being monitored or a limited number of mutations are being interrogated, have to compete with the well-known PCR-based tests. On the other hand, complex microarray tests, which provide large amounts of information unattainable by PCR-based techniques, are facing a strong competition with the newly emerged next-generation sequencing (NGS) techniques. They generate detailed information about the whole genome with the prices that have been tremendously lowered in recent years [7].

This chapter is dedicated to discuss the major challenges that the microarray technology has faced in the pathway of its growth since its inception. We also highlight the progress that has been achieved by the research and commercial communities to overcome the obstacles.

2 Reliability of Microarray Data

Since the advent of the DNA microarray technology, some concerns have been raised regarding the reliability and reproducibility of the microarray data [6, 8]. A meta-analysis was performed on the reproducibility of the data of seven large scale studies on cancer prognosis that used microarray-based expression profiling [8]. Surprisingly, in five of these studies the reported data were not reproducible. The analysis of the other two studies provided much weaker prognostic information than given by the original data [8]. Following the awareness about the shortage of standardization measures, the scientific community put much effort in preparing appropriate standards, controls and tools [9, 10]. Aiming to provide a basis for reporting the microarray results, the standard called MIAME (minimum information about a microarray experiment) was proposed [9], which ensured that the microarray data can be easily interpreted and independently verified. Commercial vendors of the microarray platforms improved their technologies over the years, and they also set up a series of quality control measures to enhance the reproducibility and accuracy of the data produced by their products. Together with the regulatory agencies, the vendors started the MAQC (Microarray quality control) project [10], which established thresholds and metrics for inter-platform comparison of microarray data.

In microarray analysis, especially in gene expression profiling where a massive amount of information is commonly produced, it is critical to provide the biological interpretation with statistical significance. In order to decide if a gene is outcome-related, the expression level of the gene in the patient sample is usually compared with the one from a normal sample, and a fold-change and clustering analysis are used to provide the biological interpretations, e.g., class comparison or prediction of the disease in cancer patients [6]. However, the expression levels of the genes naturally vary between different individuals and between different samples from the same individual [8]. Hence, a simple fold-change statistic does not account for the variability across specimens, leading to false positive outcomes. For instance, in 2007, Dupuy and Simon who reviewed 23 studies that reported results of outcome-related gene-finding analyses found out false-positive results in nine of them [5]. Because of the recent advances in bioinformatics, valid data analyses are currently available. The so-called supervised data interpretation methods are able to make distinctions among the specimens based on predefined information and to create valid information for clinical decision-making [6].

3 Microarrays Integrated with the LOC Devices

Integration of various steps of microarray assay in a miniaturized, portable and stand-alone lab-on-a-chip (LOC) device is a crucial requirement for a variety of applications, especially point-of-care (POC) diagnostics [11]. Current microarray technologies use separate instruments for sample preparation, DNA hybridization, signal visualization and data interpretation. Moreover, some of these components such as the fluorescent scanners used for signal visualization are bulky instrument that are only available in well-equipped laboratories. The development in sampling and detection technologies certainly accelerates the process of integration. For instance, some of the sample preparation steps for sample labeling can be avoided when label-free detection approaches are used [12–14]. Miniaturization of the microarray spots also alleviates the need of fluorescent scanning and so no bulky fluorescent scanners are required. More importantly, with the aid of microfluidic network, all steps of the microarray test can be integrated in a single miniaturized device. In the following section, we will present the advances and challenges in developing the technologies required for the microarray tests to be integrated in a stand-alone LOC device.

3.1 DNA Microarrays Combined with Microfluidic Networks

The developments in the microlithography techniques have enabled microfluidics which is used to create LOC devices [15, 16]. Coupling the microfluidic operations with microarray assays potentially adds precious value to them. One obvious benefit of microfluidic liquid-handling is the reduced sample and reagent consumption due to small micrometer-sized channels. In addition, the highly efficient and controllable pressure-driven flow for liquid handling and delivery in these microchannels allows for integration of different steps of microarray assays that are essential to implement the portable LOC devices [17]. More importantly, the target molecules in the samples are delivered to the probe spots in the microarrays using the convective flow, in additional to diffusion, and thereby reducing the hybridization times from hours to minutes. Furthermore, the benefit of the microfluidic microarray LOC device is in its potential on high sample throughput as well as sufficient number of probes. Conventional microarray experiments usually allow 1–10 samples to be applied on one chip [18], and so replicate analysis will require multiple chips. As discussed in **Subheading 2**, one of the challenges of DNA microarrays is the inevitable variations among samples [6], and these variations make replicate analysis of several samples necessary, and therefore the multi-sample analysis capability of the microfluidic microarray chips is highly valuable.

There are two ways to conduct the microarray analysis in the microfluidic chips, either in the microfluidic chambers or in the microfluidic channels [19–21]. First, large microfluidic chambers are used to enclose the area that is pin-spotted with arrayed probes, where the sample DNA molecules are hybridized with the probe molecules [22]. These chambers are compatible with both low-density and high-density microarrays, but it is always a challenge to design how the liquid will flow uniformly over the large chamber in such a way to achieve an equally distributed liquid movement across the arrays. Second, the microfluidic channels are used and they provided a better flow control of target solutions over the probe arrays. Various microfluidic chips containing straight and serpentine microchannels have been implemented, mainly for low-density microarrays [23]. In these cases, the pin-spotted probe regions are usually enclosed along the channel length of the microchips. Especially in the second way of conducting the microarray experiments, the microfluidic operations benefit the effective interactions between the target molecules in the sample solutions and the probe molecules immobilized on the channel surfaces due to the use of a dynamic flow.

In addition to benefit effective sample delivery, the dynamic microfluidic flow is used to facilitate the probe printing on the microarray surface in a uniform manner. The performance of hybridization in the microarray assay is heavily influenced by the morphology of the printed spots on the chip surface. However, since the probe solutions are exposed to air in the conventional probe-spotting method using pins, the solutions are subject to various problems, such as splashing, uneven evaporation and cross contamination [24], leading to unacceptable spot morphology. Even worse, during the blocking and cleaning procedures after probe-spotting, the unreacted probe molecules could diffuse away from the spot locations and smear to form comet-like spots [16, 25]. Furthermore, since the dynamic-flow hybridization is to be used with the spotted microarray, additional apparatus such as steel clamps must be used to ensure that the entire hybridization microchannel is well sealed and aligned to the probe rows [16].

These problems would be resolved by using the microchannel network for microprinting the probes, resulting in high homogeneity of the probe regions printed on the microfluidic microarrays. For instance, Wang et al. used a network of microchannels in two steps: first for probe printing and second for DNA hybridization, producing the 2D microarrays [20]. In the first step of this method, called the 2-step intersection approach, the probe solutions flowed in the horizontal microchannels, in the first polydimethylsiloxane (PDMS) slab sealed with a glass chip, in order to print an array of horizontal probe lines on a glass chip surface. In the second step, the target solutions flowed in the vertical channels, in the second PDMS slab sealed with the same glass chip, in order to hybridize with the spotted probes at the intersections between the vertical microchannels and the horizontal probe lines. The 2D microfluidic microarray is well suited for parallel sample hybridizations and, unlike the low-density DNA microarray spots printed by pins, the use of long and narrow probe line in microfluidic microarrays alleviates the need of time-consuming alignment between the hybridization channels and the printed probes.

The 2D microarrays can be used for many diagnostic applications. Since in many of these applications that deal with many samples, once a relatively small number of gene mutations or single nucleotide polymorphisms (SNPs) are identified, low-density microarrays can be employed to screen these mutations across many patient samples. This low-density microarray approach has been demonstrated to be reliable, cost-effective, and fast in data analysis and interpretation [26–28]. In order to perform SNP analysis for the KRAS mutation on the chip, Sedighi et al. replaced the regular free DNA targets by the gold nanoparticle-loaded targets to enhance the specificity of DNA hybridization reactions on the chip surface [28, 29].

So far, the dynamic flow used in microfluidic microarray chips has been achieved by pressure. An alternative to the pressure-driven flow is to achieve liquid pumping by centrifugal forces. Centrifugal pumping used for dynamic liquid flow has several advantages such as easy implementation and insensitivity to the physiochemical properties of the liquid. Using centrifugal force, the liquid can be transferred in a parallel manner in multiple channels of a disk-like chip by spinning it. Furthermore, the implementation of centrifugal force by disk spinning is compatible with the CD/DVD technology and its related industries which have been well developed. In most of the reported applications that utilize the centrifugal platform, only the radial channels are used for liquid handling and delivery. For instance, Bin et al. reported a CD-like device capable of generating the flow of DNA samples within the twelve PDMS microchannels for DNA sample delivery to the 1D microarray (Fig. 1), with the sample hybridization time reduced to 15 min and the sample volume as low as 1.5 μL [30]. However, this format for

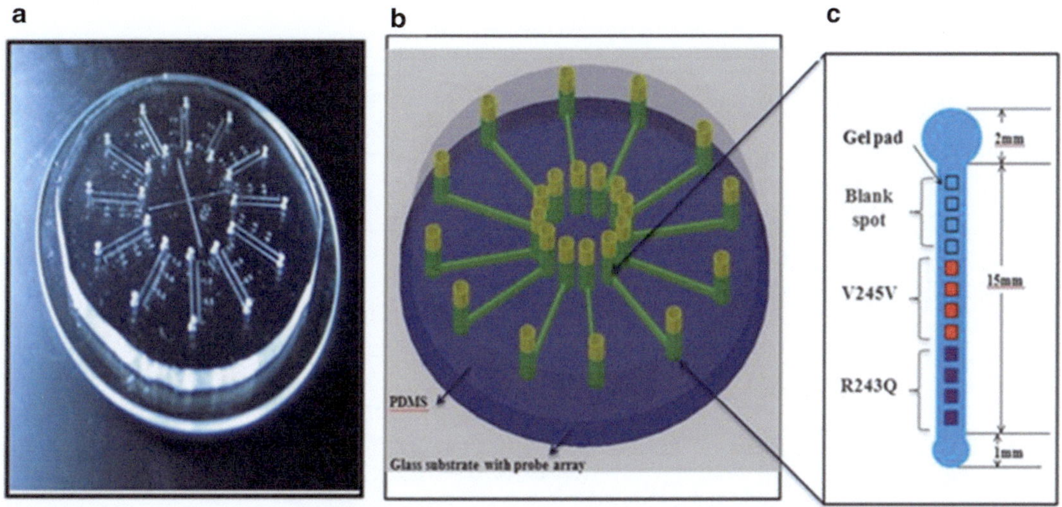

Fig. 1 (**a**) The photograph of the PDMS-glass CD-like chip. (**b**) The two-layer structure of the microfluidic chip including a top PDMS slab containing 12 hybridization microchannels, and a bottom glass disk with the immobilized 1D microarrays. (**c**) The DNA arrays for phenylketonuria (PKU) screening for R243Q and V245V mutations as well as the negative control probes (reproduced from ref [27] with permission from Elsevier

centrifugal liquid delivery has a design limitation because there is not enough space to accommodate the multiple fluid structures in the radial format [11]. For example, if the centrifugal platform is built on a 92-mm CD with a 15-mm center spindle hole, the maximum limit of the length of a radial microchannel is 38.5 mm. For such a short microchannel, the capillary effect may dominate the liquid flow and the flow velocity cannot be easily controlled. More importantly, by using centrifugal pumping only once in the radial direction, the intersection method cannot be applied to generate the 2D microfluidic microarray.

To integrate centrifugal pumping with the 2D microarray, Peng et al. exploited the centrifugal force twice based on the sequential use of two chips with specially designed channels in order to create a 2D microarray [31–33]. As shown in Fig. 2, in addition to the radial microchannels, which were used for probe printing, spiral microchannels were employed to implement target hybridization by the intersection method. In this method, a PDMS slab containing radial microchannels was first sealed against the glass disk and used for printing the radial probe line arrays on the disk. After the first slab was removed, a second PDMS slab that consisted of the spiral microchannels was sealed against the same disk, and DNA hybridization occurred at the intersections between the spiral microchannels and the radial probe lines. Dynamic target delivery facilitated by the centrifugal force can be conveniently controlled and synchronized [31–33]. The 2D microarrays generated using CD microfluidics also demonstrated a high sensitivity

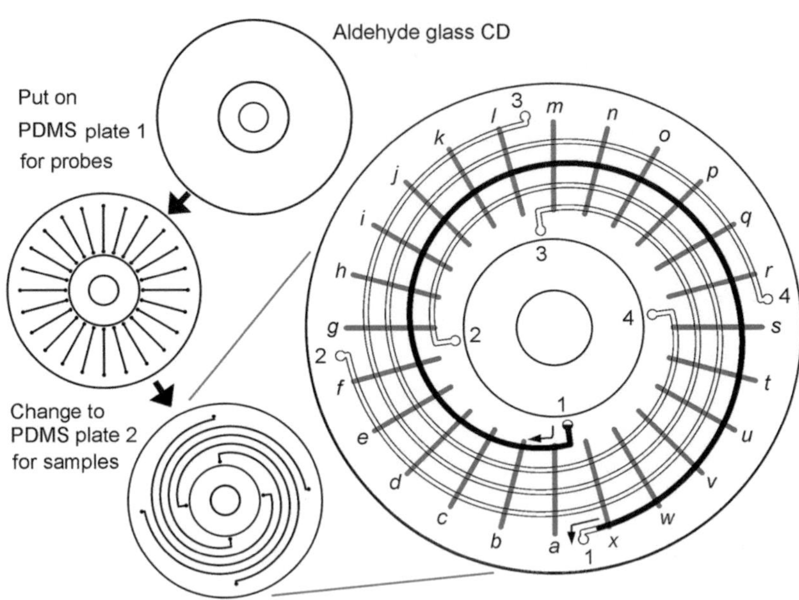

Fig. 2 2D microfluidic microarray analysis using the intersection method. Printing of DNA probes using the radial channel chip 1, with 24 DNA probe lines printed (*solid radial lines* marked by a–x). DNA Hybridization using the spiral channel chip 2, with four samples flowing through the four spiral channels (hollow spiral channels marked by 1–4). Liquid flow in spiral channel 1 was indicated by *black* color and the *two arrows*. Hybridizations occur at the intersections of the spiral channels and radial probe lines (reproduced from ref. [28] with permission from Elsevier)

and specificity for DNA analysis [34, 35]. So far, a higher spot density (384 × 384) has been achieved for the high-throughput microarray analysis on a 92-mm CD-like glass disk [36].

3.2 Advances in Detection Techniques

Fluorescence detection is commonly used for DNA microarray assays, in which the sensitivity and stability of the detection method have vastly improved over the years due to the discovery of new fluorescent dyes and more effective labeling techniques [37]. However, bulky fluorescent scanners are still required to achieve a high sensitivity and resolution, which is a limitation to the miniaturization requirement for LOC device developments. The efficiency of target labeling and fluorescence quenching of the dye also affects the reproducibility of the results [38]. In addition, fluorescent labeling of the target molecules adds complications and cost to the assays, and so several approaches have recently been developed in order to avoid target labeling [38].

One novel approach is the use of molecular beacon (MB) that takes advantage of both the sensitivity of fluorescence detection and convenience of no target labeling. However, the MB probes are still labeled [39], and they are single-stranded nucleic acids that retain a stem-and-loop structure and keep a pair of fluorophore-quencher at both ends of the stem strand in close proximity and thus the fluorescence emission of the fluorophore is quenched. In the presence of a non-labeled target molecule, the loop-region of MB probe hybridizes to the target, while the stem opens up and fluorescence of the fluorophore occurs when the quencher moves away [39].

Other novel detection techniques, based on the optical, electrochemical and microwave properties of the target molecules, have also been developed in order to alleviate the need of target labeling [38–43]. Surface plasmon resonance imaging (SPRi) is an attractive detection approach for microarrays because none of the targets or probes requires labeling, and thus this is a true label-free detection [40]. Moreover, Özkumur et al. developed the spectral reflectance imaging biosensor (SRIB) for high-throughput analysis of SNPs on a glass microarray chip [43]. This technique, which is based on optical interferometry, has allowed the single-nucleotide mismatched target oligonucleotides to be distinguished from perfectly matched ones, through dynamic data acquisition during the washing step using a low ionic concentration buffer [43].

Furthermore, an electrochemical technique using multiwalled carbon nanotubes (MWNT) nanoelectrode arrays was developed by Koehny et al. to detect unlabeled PCR amplicons [41]. In this technique, with the aid of $Ru(bpy)_3^{2+}$, the guanine bases in the DNA targets serve as the signal transduction moieties, providing an amplified anodic current associated with the oxidation of guanine groups at the nanoelectrode surface. The abundance of guanine bases in the target strands led to a high sensitivity and low detection limit, i.e., less than ~1000 target amplicons on a microspot are detectable. Another label-free detection is near-field scanning microwave microscopy (NSMM), which has been used by Lee et al. for detection of both DNA and RNA molecules [42]. NSMM monitors the microwave reflectance, which is dependent on the length and surface coverage of the nucleic acid strands, as well as on the hybridization state of the molecules (e.g., unhybridized single-stranded probe vs. hybridized double-stranded). The NSMM technique has demonstrated an acceptable resolution (potentially less than 50 μm) and a sensitivity comparable to fluorescent detection [42].

3.3 Miniaturization of Microarray Features

A major challenge in developing portable microarray devices is the need of large-format fluorescent scan. It is because the fluorescence detector, which utilizes a high numerical-aperture (NA) microscopic objective in order to detect the weakly fluorescent microarray spots, has a narrow field of view and so covers only a few microarray spots. Miniaturization of the microarray features in such a way that the whole array would be visible in the field of view of the high NA objectives would render the scanner unnecessary. Other than alleviating the need for a scanner, miniaturized arrays also favor fast mass and heat transport, and therefore reduce the assay time [44].

Creating an array with sub-micrometer features requires very accurate probe printing techniques with nanometer resolutions. Among different nanoprinting methods, the scanning probe microscopy (SPM)–based technique received a high level of attention in this regard. The SPM-based technique produces probe printing in a high positioning precision and also its non-vacuum operational condition is compatible with biomolecules [45]. As an example of this technique, Demers et al. used dip-pen nanolithography (DPN) to directly couple hexanethiol-modified oligonucleotides on the gold surface, and acrylamide-modified oligonucleotides on silica substrates [46]. They managed to reduce the size of the microarray features to 100 nm which allows an array containing ~100,000 features to be generated in an area compatible with the size of a typical AFM scanner [46]. Despite the extremely high resolution of DPN, this technique is intrinsically serial, and thus a significant amount of time must be allocated in order to generate the microarrays [47]. Nanoimprint lithography (NIL) is another approach employed for the reduction of feature dimensions in microarrays. In NIL, probe oligonucleotides were either synthesized in-situ on the chemically modified nanostructures created on a polymer surface [48], or physically tethered to the surface of nanostructures before they were delivered (or stamped) on the substrate [47]. For instance, two research groups independently developed a technique to replicate the whole DNA array in a single cycle [49, 50]. In this technique, a first master substrate made of oligonucleotide probe features was immersed in a solution containing the complementary target DNAs. Afterwards, a second substrate was brought into contact with the first substrate to adsorb the target DNAs, thus replicating the features on the second substrate. Such an effective nanostamping method was able to reproduce DNA arrays with features as small as 14 nm, with the spacing of 77 nm [51]. Advances in miniaturization of the size of the printed probes, in the nanometer scale, are significant steps moving toward the development of portable microarray platforms. It also makes the microarray test cheaper by avoiding the scanner and by reducing the amount of sample biomolecules.

3.4 Advances in Integration of LOC Devices

Many LOC microdevices have been developed to perform individual steps for DNA microarray assays. However, integration of these devices to give a POC diagnostic system in an efficient manner remains a challenge. All steps in an integrated system such as liquid handling, reagent metering, thermal and pressure control and signal transduction must be compatible with each other. Despite these challenges, great effort has been made by the researchers to develop the system that is able to perform the many steps of the microarray assay [52–58]. Anderson et al. reported one such system, which was

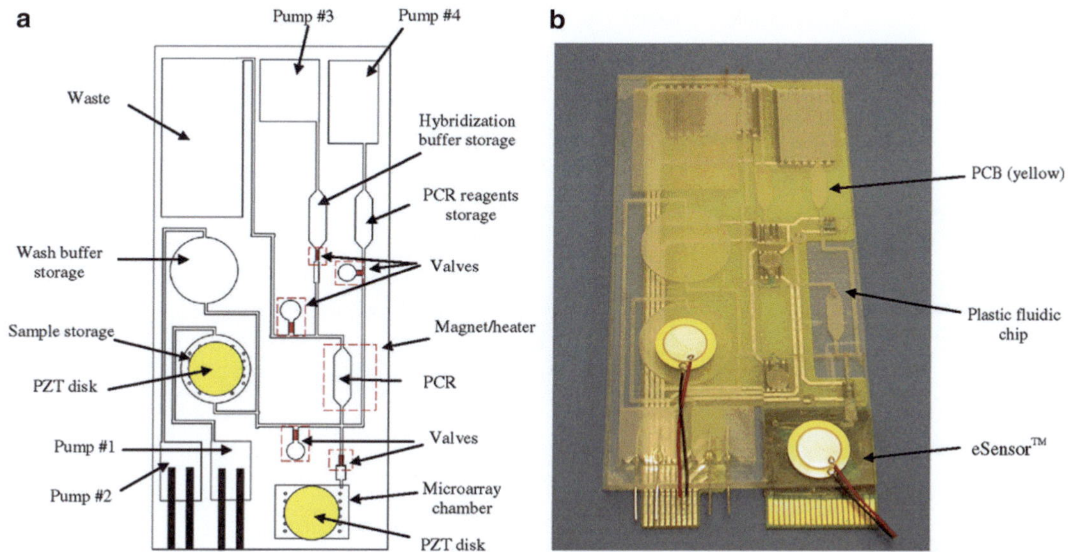

Fig. 3 The integrated microchip for DNA microarray detection. (**a**) Schematic and (**b**) photograph of the integrated device, which consists of a plastic fluidic chip, a printed circuit board (PCB), and an eSensor located in the microarray chamber. The two PZT disks are piezoelectric devices that assist in reagent mixing (reproduced from ref. [49] with permission from American Chemical Society)

capable of extracting and concentrating nucleic acids from aqueous samples, performing chemical amplification and serial enzymatic reactions (fragmentation, dephosphorylation, and labeling), metering and mixing, and microarray-based nucleic acid hybridization, for the detection of mutations in the HIV genome [52].

Liu et al. developed an integrated microchip for DNA microarray detection of bacterial pathogens in blood [53]. Their device consisted of a plastic chip, a printed circuit board (PCB), and an eSensor. The plastic chip included a mixing unit for cell capture, cell preconcentration, purification and lysis, a PCR unit for nucleic acid amplification, and a DNA microarray chamber for hybridization. Thiol-terminated DNA oligonucleotides were immobilized on the eSensor for electrochemical detection of hybridized target DNA [53] (Fig. 3). Liu et al. also integrated a DNA microarray platform, containing 12,000 features within a microfluidic cartridge, in order to automate the fluidic handling steps required for gene expression profiling assay [57]. Microarray hybridization and subsequent washing and labeling steps were all performed at the self-contained device [57].

Yeung et al. also developed an integrated microchip that was based on a multi-chamber design for multiplex pathogen identification [56] (Fig. 4). In their silicon-glass chip, the oligonucleotide probes were individually positioned at each indium tin oxide (ITO) electrode within the microfluidic chamber. Several microfluidic-controlled steps, which included thermal lysis, magnetic particle-based

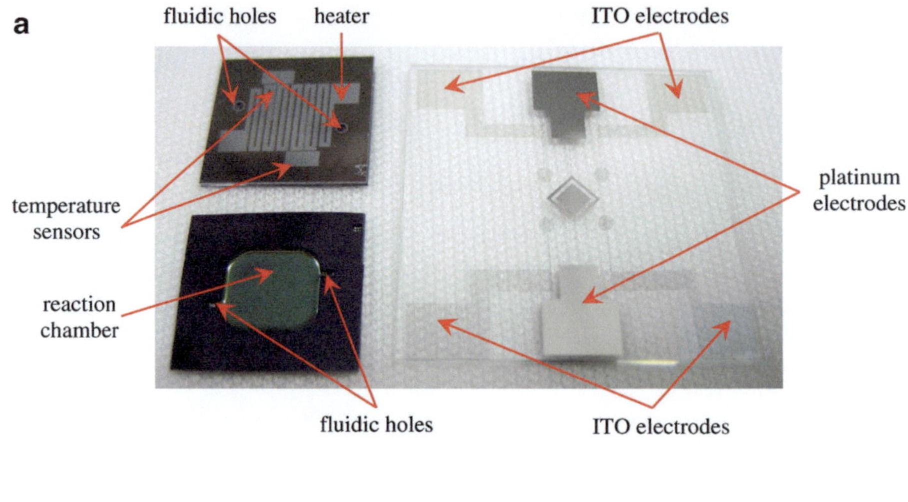

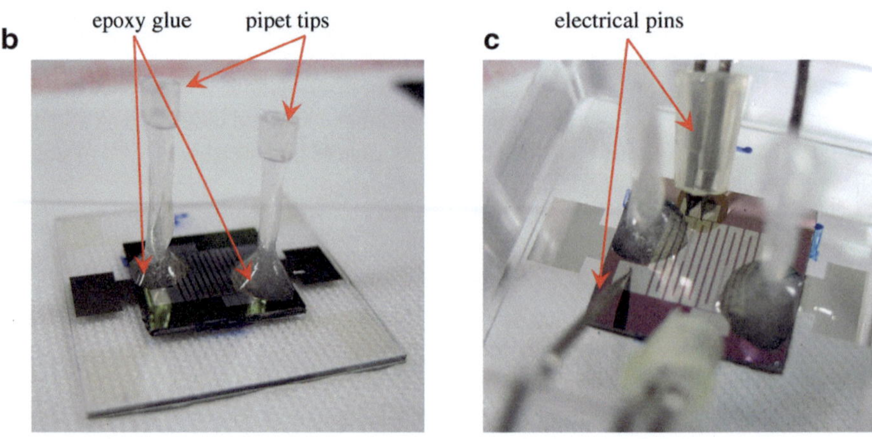

Fig. 4 The silicon-glass microchip for multiplex pathogen detection. (**a**) *Upper left*: top view of the silicon chip showing the fluidic holes along with thin-film platinum heater and temperature sensors; *lower left*: bottom view of the silicon chip showing the 8 mL reaction chamber and the through-hole for sample introduction; *right*: glass chip with patterned indium tin oxide (ITO) electrodes. (**b**) The assembled silicon-glass microchip, on which pipet tips were glued to the fluidic holes to form solution reservoirs. (**c**) Electrical connection of the contact pins to the ITO electrodes in the chip housed in a Plexiglas holder

isolation of the target genomes, asymmetric PCR, and electrochemical detection using silver-enhanced gold nanoparticles, were performed in the integrated device [56]. Liu et al. also integrated a DNA microarray platform, containing 12,000 features in with a microfluidic cartridge, in order to automate the fluidic handling steps required for gene expression profiling assay [57]. Microarray hybridization and subsequent washing and labeling steps were all performed at the self-contained device [57]. Choi et al. integrated an allele-specific PCR unit with a disposable DNA microarray chip for multiplex SNP detection. Convective flows, created by pneumatic micropumps, were used in this integrated system to accelerate the hybridization process, resulting in the whole assay completed on-site in 100 min. In this assay, a miniaturized fluorescent scanner, instead of a conventional bulky one, was conducted for hybridization detection [58]. More work still needs to be achieved to further develop these LOC devices to become stand-alone (sample-in-answer-out) devices.

4 Commercialization and Clinical Use

The microarray technology has been slow to penetrate the molecular diagnostic market, i.e., only 10 % in 2010 [7]. The poor reproducibility of the microarray data, which is due to technical limitations or natural variations between different samples, has prevented the microarray tests to have sufficient robustness required for a diagnostic test. In order to receive regulatory approval and even clinical acceptance for expression profiling tests, they are required to demonstrate the result reliability and the correlation of their results to the clinical outcomes [59]. Gaining regulatory approval is difficult for gene expression profiling because it is commonly based on new research studies that correlate the clinical outcomes to the levels of expression of new genes, which are not predefined and well-known biomarkers. Furthermore, these tests face the clinical utility question: how do the microarray data improve the outcome of the patient? For example, what is the suitable type of treatment, and how is over-treatment in chemotherapy prevented [59]? MammaPrint (Agendia, Netherlands) was the first expression profiling test that received the FDA approval in 2007 for prognosis application in breast cancer [60]. MammaPrint, based on a research reported by Vijver et al. assesses the expression profile of a set of 70 cancer-related genes. However, the test suffers from a strong competition from the PCR-based Oncotype test (Genomic Health, USA), primarily because the latter is able to analyze the widely used FFPE (formalin-fixed paraffin-embedded) samples. To address this sample need, a new microarray-based test (Tissue of origin, Pathwork Diagnostics, USA), with the capability to operate on FFPE samples, has recently entered the market. Another competition is from next-generation sequencing techniques, which are well developed and their prices are no longer prohibitive [61]. These techniques (e.g., RNA-Seq) are more reliable and informative since they provide the sequence information without prior knowledge. Moreover, these sequencing techniques are convenient since they provide digital, instead of analog, data.

Unlike gene expression arrays, genotyping arrays have fewer obstacles to overcome in gaining regulatory approval as well as clinical acceptance. Genotyping, which aims to characterize previously established sequence variations among the genome, does not have to prove its clinical correlation and utility. Nevertheless, the genotyping arrays are still required to be technically reliable and competitive in price. High-throughput genotyping arrays contain a significantly high numbers of features to investigate mutations in several genes or mixtures of pathogens and thus provide a huge amount of information. As an example, AmpliChip CYP450 obtained FDA approval in 2004 (the first microarray-based clinical test). The test uses 15,000 features on an Affymetrix platform to assess several types of variations in two genes, involved in the metabolism of many psychoactive drugs. On the other hand, low-throughput genotyping arrays use a fairly small number of features (up to few hundreds on a chip) for characterization of a pathogen or investigation of several SNP sites in a particular gene. The PapilloCheck test provided by Greiner Bio-One Company (Frickenhausen, Germany) is a fairly successful low-throughput genotyping test, which obtained the US regulatory approval in 2009. PapilloCheck utilizes an array of 140 oligonucleotides to determine the HPV (human papillomavirus) subtypes in cervical smear samples.

5 Future Perspectives of Microarray-Based Diagnostics

Recently, a number of LOC devices have been developed to integrate several steps of the microarray assay on a miniaturized platform. However, these devices still need to use the conventional methods for the signal detection [52]. The use of nanoarrays helps replace the bulky fluorescent scanner with the miniaturized fluorescence detector because the whole array can fit in the field of view of the detector. Other than fluorescence detection, label-free detection techniques will further simplify the future LOC devices. Different steps of the microarray assays are integrated in stand-alone LOC devices in a harmonious way, which make them capable of performing the sample-in-answer-out assays. These LOC devices exploit the microfluidic networks not only to connect the different compartments in the devices, but also to make the device faster, smaller and conveniently controlled.

DNA microarrays have faced fierce competition from next-generation sequencing (NGS) at the high end of throughput and from PCR-based techniques at the low end in the molecular diagnostic market. However, the cost of NGS assays, although not prohibitive anymore, are still more expensive than the ones offered by the microarray vendors. The presence of some unresolved difficulties in the sequencing techniques, like the necessity of multiplexing in RNA-sequencing as well as their extensive sample preparation and data interpretation needed for the techniques [62], will be in favor of the microarrays in the competition. On the other hand, since the microarray technique is more flexible in the sample matrix that it can process it is more familiar to the clinicians; the microarrays are in a better situation in competition with PCR-based techniques. Because of the wealth of knowledge offered by the human genome, the size of the molecular diagnostic market has largely expanded in the past decade and is predicted to be doubled in 2017 [63]. More microarray-based diagnostic tests are currently gaining regulatory approvals and entering the market. According to a recent report, the microarray-based tests share the largest portion of the molecular diagnostic market with the PCR-based tests [63], and the market of the former type of test is expected to grow in the coming years [64].

6 Summary

Owing to technical quality issues, the microarray techniques face serious concerns and criticisms about the reproducibility of the data they provided as well as the reliability of the inferred biological interpretations. These shortcomings in the microarray techniques themselves as well as the harsh competitions from other molecular diagnostic techniques deter the microarrays to be as successful in the clinical market as they were in the research counterparts.

However, DNA microarray vendors and researchers managed to resolve many of those issues, which led to regaining the product confidence in the market. This is demonstrated as the growth of the number of microarray-based tests that receive regulatory approvals as well as positive foresights by business reports.

Meanwhile, much advancement has been achieved in the prerequisite technologies for developing portable and stand-alone LOC devices. Future DNA microarrays are expected to be faster, smaller and more accurate. Supporting systems, well-matched with the microarray platforms, will be developed and integrated into the complex stand-alone LOC devices. These LOC devices play a central role in personalized medicine in the future. Technical issues, observed in the early microarray platforms, are mostly resolved in the current platforms. New discoveries about the human genome, thereby increasing the depth of our knowledge about it, will initiate new clinical utilities for the DNA microarray. Enjoying all of these developments, DNA microarray will maintain a reasonable share in the fast growing molecular diagnostic market.

References

1. Schena M, Shalon D, Davis RW, Brown PO (1995) Quantitative monitoring of gene expression patterns with a complementary DNA microarray. Science 270:467–470
2. Alizadeh AA et al (2000) Distinct types of diffuse large B-cell lymphoma identified by gene expression profiling. Nature 403:503–511
3. Shaw-Smith C (2004) Microarray based comparative genomic hybridisation (array-CGH) detects submicroscopic chromosomal deletions and duplications in patients with learning disability/mental retardation and dysmorphic features. J Med Genet 41: 241–248

4. Storhoff JJ et al (2004) Gold nanoparticle-based detection of genomic DNA targets on microarrays using a novel optical detection system. Biosens Bioelectron 19:875–883
5. Dupuy A, Simon RM (2007) Critical review of published microarray studies for cancer outcome and guidelines on statistical analysis and reporting. J Natl Cancer Inst 99:147–157
6. Simon R, Radmacher MD, Dobbin K, McShane LM (2003) Pitfalls in the use of DNA microarray data for diagnostic and prognostic classification. J Natl Cancer Inst 95:14–18
7. Jordan BR (2010) Is there a niche for DNA microarrays in molecular diagnostics? Expert Rev Mol Diagn 10:875–882
8. Michiels S, Koscielny S, Hill C (2005) Prediction of cancer outcome with microarrays: a multiple random validation strategy. Lancet 365:488–492
9. Brazma A et al (2001) Minimum information about a microarray experiment (MIAME) – toward standards for microarray data. Nat Genet 29:365–371
10. Chen JJ, Hsueh H-M, Delongchamp RR, Lin C-J, Tsai C-A (2007) Reproducibility of microarray data: a further analysis of microarray quality control (MAQC) data. BMC Bioinformatics 8:412
11. Wang L, Li PC (2011) Microfluidic DNA microarray analysis: a review. Anal Chim Acta 687:12–27
12. Vollmer F, Arnold S (2008) Whispering-gallery-mode biosensing: label-free detection down to single molecules. Nat Methods 5:591–596
13. Wang WU, Chen C, Lin K, Fang Y, Lieber CM (2005) Label-free detection of small-molecule–protein interactions by using nanowire nanosensors. Proc Natl Acad Sci U S A 102: 3208–3212
14. Crespi A et al (2010) Three-dimensional Mach-Zehnder interferometer in a microfluidic chip for spatially-resolved label-free detection. Lab Chip 10:1167–1173
15. Stone HA, Stroock AD, Ajdari A (2004) Engineering flows in small devices: microfluidics toward a lab-on-a-chip. Annu Rev Fluid Mech 36:381–411
16. Sedighi A, Wang L, Li PCH (2013) 2D nanofluidic bioarray for nucleic acid analysis. In: Iniewski K, Selimovic S (eds) Nanopatterning and nanoscale devices for biological applications. Taylor & Francis, CRC press, Boca Raton, pp 183–205
17. Hong J, Edel JB, deMello AJ (2009) Micro- and nanofluidic systems for high-throughput biological screening. Drug Discov Today 14:134–146
18. Lagarde AE (2003) DNA microarrays: a molecular cloning manual. Am J Hum Genet 73:218
19. Liu J, Williams BA, Gwirtz RM, Wold BJ, Quake S (2006) Enhanced signals and fast nucleic acid hybridization by microfluidic chaotic mixing. Angew Chem Int Ed 45:3618–3623
20. Peytavi R (2005) Microfluidic device for rapid (<15 min) automated microarray hybridization. Clin Chem 51:1836–1844
21. Campàs M, Katakis I (2004) DNA biochip arraying, detection and amplification strategies. Trends Anal Chem 23:49–62
22. Lee HJ, Goodrich TT, Corn RM (2001) SPR imaging measurements of 1-D and 2-D DNA microarrays created from microfluidic channels on gold thin films. Anal Chem 73: 5525–5531
23. Situma C et al (2005) Fabrication of DNA microarrays onto poly(methyl methacrylate) with ultraviolet patterning and microfluidics for the detection of low-abundant point mutations. Anal Biochem 340:123–135
24. Wang L, Li PCH (2007) Flexible microarray construction and fast DNA hybridization conducted on a microfluidic chip for greenhouse plant fungal pathogen detection. J Agric Food Chem 55:10509–10516
25. Sedighi A, Li PC (2013) Gold nanoparticle assists SNP detection at room temperature in the nanoBioArray chip. Int J Mat Sci Eng 1(1): 45–49
26. Bouchie A (2002) Organic farmers sue GMO producers. Nat Biotechnol 20:210
27. Meneses-Lorente G et al (2003) An evaluation of a low-density DNA microarray using cytochrome P450 inducers. Chem Res Toxicol 16:1070–1077
28. Sedighi A, Li PCH (2014) Kras gene codon 12 mutation detection enabled by gold nanoparticles conducted in a nanobioarray chip. Anal Biochem 448:58–64
29. Sedighi A, Li PCH, Pekcevik IC, Gates BD (2014) A proposed mechanism of the influence of gold nanoparticles on DNA hybridization. ACS Nano 8:6765–6777
30. Chen B et al (2011) Rapid screening of phenylketonuria using a CD microfluidic device. J Chromatogr A 1218:1907–1912
31. Peng XY (Larry), Li PCH, Yu H-Z, Parameswaran M (Ash), Chou WL (Jacky) (2007) Spiral microchannels on a CD for DNA hybridizations. Sens Actuators B Chem 128: 64–69
32. Peng XY, Li PCH (2008) Centrifugal pumping in the equiforce spiral microchannel. Can J Pure App Sci 2:551–556

33. Wang L, Kropinski M-C, Li PCH (2011) Analysis and modeling of flow in rotating spiral microchannels: towards math-aided design of microfluidic systems using centrifugal pumping. Lab Chip 11:2097
34. Wang L, Li PCH, Yu H-Z, Parameswaran AM (2008) Fungal pathogenic nucleic acid detection achieved with a microfluidic microarray device. Anal Chim Acta 610:97–104
35. Wang L, Li PCH (2010) Optimization of a microfluidic microarray device for the fast discrimination of fungal pathogenic DNA. Anal Biochem 400:282–288
36. Chen H, Wang L, Li PCH (2008) Nucleic acid microarrays created in the double-spiral format on a circular microfluidic disk. Lab Chip 8:826
37. Epstein JR, Biran I, Walt DR (2002) Fluorescence-based nucleic acid detection and microarrays. Anal Chim Acta 469:3–36
38. Sassolas A, Leca-Bouvier BD, Blum LJ (2008) DNA biosensors and microarrays. Chem Rev 108:109–139
39. Fang X, Liu X, Schuster S, Tan W (1999) Designing a novel molecular beacon for surface-immobilized DNA hybridization studies. J Am Chem Soc 121:2921–2922
40. Nelson BP, Grimsrud TE, Liles MR, Goodman RM, Corn RM (2001) Surface plasmon resonance imaging measurements of DNA and RNA hybridization adsorption onto DNA microarrays. Anal Chem 73:1–7
41. Koehne J et al (2003) Ultrasensitive label-free DNA analysis using an electronic chip based on carbon nanotube nanoelectrode arrays. Nanotechnology 14:1239
42. Lee K et al (2013) Label-free DNA microarray bioassays using a near-field scanning microwavemicroscope. Biosens Bioelectron 42: 326–331
43. Özkumur E et al (2010) Label-free microarray imaging for direct detection of DNA hybridization and single-nucleotide mismatches. Biosens Bioelectron 25:1789–1795
44. Tsarfati-BarAd I, Sauer U, Preininger C, Gheber LA (2011) Miniaturized protein arrays: model and experiment. Biosens Bioelectron 26:3774–3781
45. Xu S, Miller S, Laibinis PE, Liu G (1999) Fabrication of nanometer scale patterns within self-assembled monolayers by nanografting. Langmuir 15:7244–7251
46. Demers LM et al (2002) Direct patterning of modified oligonucleotides on metals and insulators by dip-pen nanolithography. Science 296:1836–1838
47. Truskett VN, Watts MPC (2006) Trends in imprint lithography for biological applications. Trends Biotechnol 24:312–317
48. Moorcroft MJ et al (2005) In situ oligonucleotide synthesis on poly(dimethylsiloxane): a flexible substrate for microarray fabrication. Nucleic Acids Res 33:e75
49. Yu AA et al (2005) Supramolecular nanostamping: using DNA as movable type. Nano Lett 5:1061–1064
50. Lin H, Sun L, Crooks RM (2005) Replication of a DNA microarray. J Am Chem Soc 127:11210–11211
51. Akbulut O et al (2007) Application of supramolecular nanostamping to the replication of DNA nanoarrays. Nano Lett 7:3493–3498
52. Anderson RC, Su X, Bogdan GJ, Fenton J (2000) A miniature integrated device for automated multistep genetic assays. Nucleic Acids Res 28:e60
53. Liu RH, Yang J, Lenigk R, Bonanno J, Grodzinski P (2004) Self-contained, fully integrated biochip for sample preparation, polymerase chain reaction amplification, and DNA microarray detection. Anal Chem 76: 1824–1831
54. Trau D et al (2002) Nanoencapsulated microcrystalline particles for superamplified biochemical assays. Anal Chem 74:5480–5486
55. Lee TM-H, Carles MC, Hsing I-M (2003) Microfabricated PCR-electrochemical device for simultaneous DNA amplification and detection. Lab Chip 3:100–105
56. Yeung S-W, Lee TM-H, Cai H, Hsing I-M (2006) A DNA biochip for on-the-spot multiplexed pathogen identification. Nucleic Acids Res 34:e118
57. Liu RH et al (2006) Fully integrated miniature device for automated gene expression DNA microarray processing. Anal Chem 78: 1980–1986
58. Choi JY et al (2012) An integrated allele-specific polymerase chain reaction-microarray chip for multiplex single nucleotide polymorphism typing. Lab Chip 12:5146–5154
59. Simon R (2008) Lost in translation: problems and pitfalls in translating laboratory observations to clinical utility. Eur J Cancer 44: 2707–2713
60. Van De Vijver MJ et al (2002) A gene-expression signature as a predictor of survival in breast cancer. N Engl J Med 347:1999–2009
61. Metzker ML (2010) Sequencing technologies – the next generation. Nat Rev Genet 11:31–46

62. Morozova O, Hirst M, Marra MA (2009) Applications of new sequencing technologies for transcriptome analysis. Annu Rev Genomics Hum Genet 10:135–151
63. Molecular Diagnostics Market & Forecast (By Application, Technology, Countries, Companies & Clinical Trials) to 2017: Global Analysis, ReportLinker (2013) http://www.reportlinker.com/p01158111-summary/Molecular-Diagnostics-Market-Forecast-By-Application-Technology-Countries-Companies-Clinical-Trials-to-Global-Analysis.html
64. Ledford H (2008) The death of microarrays? Nat News 455:847

Part IV

Applications of Microarray Technology to Other Biological Assays

Chapter 13

High-Throughput Screening of Substrate Specificity for Protein Tyrosine Phosphatases (PTPs) on Phosphopeptide Microarrays

Liqian Gao, Su Seong Lee, Jun Chen, Hongyan Sun, Yuliang Zhao, Zhifang Chai, and Yi Hu

Abstract

Phosphatases are a family of enzymes responsible for the dephosphorylation of biomolecules. Phosphatases play essential roles in cell cycle regulation, signal transduction, and cellular communication. In recent years, one type of phosphatases, protein tyrosine phosphatases (PTPs), emerges as important therapeutic targets for complex and devastating diseases. Nevertheless, the physiological roles, substrate specificity, and downstream targets for PTPs remain largely unknown. To demonstrate how microarrays can be applied to characterizing PTPs, we describe here a phosphopeptide microarray strategy for activity-based high-throughput screening of PTPs substrate specificity. This is followed by a kinetic microarray assay and microplate assay to determine the rate constants of dephosphorylation by PTPs. This microarray strategy has been successfully applied to identifying several potent and selective substrates against different PTPs. These substrates could be used to design potent and selective PTPs inhibitors in the future.

Key words High-throughput screening, Protein tyrosine phosphatases, Phosphopeptide microarrays, Substrate specificity

1 Introduction

Reversible phosphorylation of tyrosine residues in proteins plays essential roles in modulating cellular events including cell growth, cell differentiation, cell-cycle regulation, and immune response [1, 2]. Tyrosine phosphorylation by kinases accounts for less than 1 % of all cellular phosphorylation events in normal tissues, and can be significantly increased upon oncogenic transformation or growth factor stimulation [3–5]. On the other hand, dephosphorylation is a key process involved in cell signaling via removal of a phosphate group from a protein or other small molecules by a phosphatase. Two opposing enzyme superfamilies, protein tyrosine kinase (PTK) family and protein tyrosine phosphatase (PTP) family,

Paul C.H. Li et al. (eds.), *Microarray Technology: Methods and Applications*, Methods in Molecular Biology, vol. 1368, DOI 10.1007/978-1-4939-3136-1_13, © Springer Science+Business Media New York 2016

control the physiological balance of tyrosine phosphorylation inside the cells [5]. It has been reported that both genetic mutation and overexpression of PTPs could be implicated in human diseases such as diabetes and cancer [6]. Therefore, PTPs have been recognized as important therapeutic targets.

Since different PTPs are highly homologous in their active sites, they present a daunting challenge to develop selective and potent small molecule PTP inhibitors. The successful development of PTP inhibitors relies on detailed understanding of the molecular mechanism by which a given PTP carries out its enzymatic reaction, as well as on the substrate specificity of PTPs. Compared with many well-studied kinases [7], phosphatases have not been extensively characterized and their cellular partners/substrate specificities remain to be further explored [6, 8–10]. Conventional strategies for determining substrate specificity of PTPs, such as combinatorial peptide libraries [11], phage display [12], and SPOT synthesis (a technique for parallel synthesis of peptide libraries on the membrane) [13], have some limitations in terms of the throughput and sensitivity of the assays. In recent years, Waldmann group and Yao group have independently developed peptide microarrays for large-scale profiling of phosphatase substrate specificities [1, 4, 8]. With these high-throughput screening approaches, the enzymatic activity of different classes of protein phosphatases, including PTPs and Ser/Thr phosphatases, could be thoroughly examined, both qualitatively and quantitatively, against a library of phosphopeptides.

The microarray is a well-established high-throughput screening platform with a diverse spectrum of biological applications across different biomolecule types [14]. Microarrays could comprise hundreds or even thousands of molecules that are immobilized as micrometer-sized spots on a planar surface. This compact format enables simultaneous processing of samples on a large scale and is therefore amenable to high-throughput screening [15–17]. More specifically, peptide microarray offers a unique and versatile platform for determining the peptide substrate fingerprints of enzymes. In this study, we have successfully extended the previous phosphopeptide microarray approach for an activity-based high-throughput study of the substrate specificity of PTPs. In Fig. 1, the biotin-containing peptides were synthesized, immobilized onto avidin-coated arrays and screened against PTPs of interest. There were 144 peptides extrapolated from the putative PTP protein substrates. Each PTP will dephosphorylate its own peptide substrate. After specific fluorescent staining of the phosphate group in the phosphopeptide, microarray scanning, and data analysis (Fig. 2), the preferred peptide substrates for each PTP could be readily identified. One of the most classical PTPs, T-cell protein

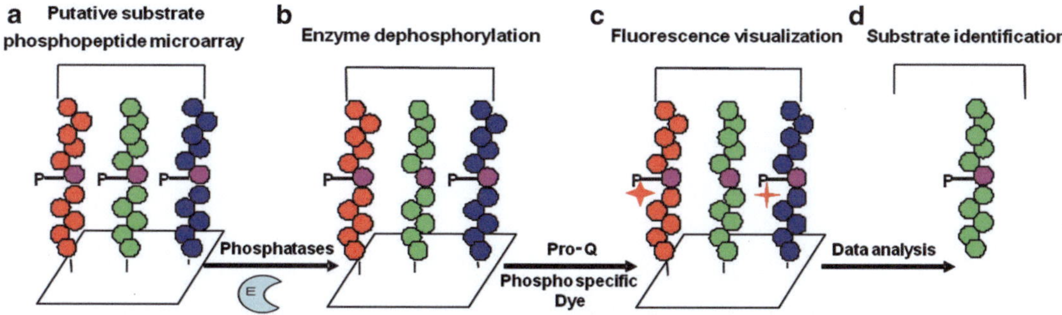

Fig. 1 The workflow of screening strategy: (**a**) Prepare a phosphopeptide microarray (P stands for phosphate). (**b**) Perform dephosphorylation assay. When incubated with PTPs, the immobilized phosphopeptides will be dephosphorylated based on PTP substrate specificity. (**c**) Perform Pro-Q dye assay, followed by fluorescence scanning with a microarray scanner to identify the phosphopeptides that are dephosphorylated by PTPs. (**d**) Carry out data analysis to identify the preferred peptide substrates for PTPs

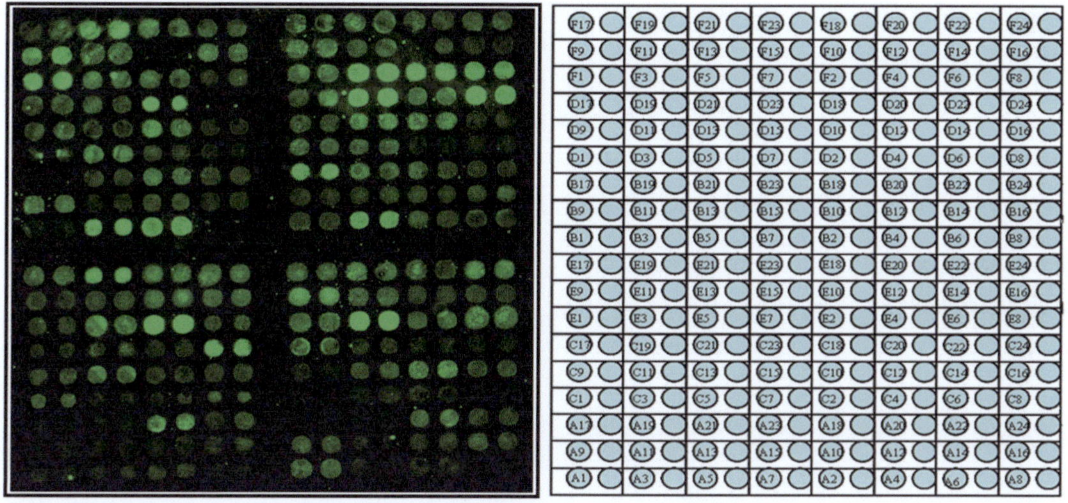

Fig. 2 Single-time-point (30 min) fingerprint results of the phosphopeptide microarrays treated with TCPTP (reproduced in part from ref. [4] with permission from John Wiley & Sons Inc.)

tyrosine phosphatase (TCPTP) was chosen in this study. Compared with the well-studied protein tyrosine phosphatase-1B (PTP1B), TCPTP is highly homologous to PTP1B and they share more than 70 % identity in protein sequence within the catalytic domain [18]. Through the example of TCPTP (Fig. 2), we will showcase the advantages of this phosphopeptide microarray approach in: (1) generating unique peptide substrate fingerprints of PTPs; (2) carrying out kinetic measurements and obtaining the kinetic constants of multiple peptides against PTPs; (3) identifying potent peptide substrates for PTPs.

2 Materials

2.1 Expression of TCPTP

1. Competent *E. coli* BL21 (DE3).
2. Construct of *TCPTP.*
3. Luria–Bertani (LB) media.
4. Kanamycin (antibiotic).
5. Isopropyl β-D-1-thiogalactopyranoside (IPTG).
6. Circular shaker (Heidolph).

2.2 Purification of TCPTP

1. Chromatographic column (1.5 cm diameter × 10 cm length).
2. Nickel-nitrilotriacetic acid (Ni-NTA) agarose resin (Qiagen, Cat. No. 1018244).
3. His-tag purification lysis buffer (50 mM Tris base, 300 mM NaCl, pH 8.0).
4. His-tag purification wash buffer (50 mM Tris base, 300 mM NaCl, 20 mM imidazole, pH 8.0).
5. His-tag purification elution buffer (50 mM Tris, 300 mM NaCl, 200 mM imidazole, pH 8.0).
6. SDS gel (10–12 %).
7. Protein marker.

2.3 Preparation of Avidin Slides

1. Piranha solution (H_2SO_4:H_2O_2 = 7:3).
2. APTES solution (400 mL): 12 mL aminopropyltriethoxisilane dissolved in 380 mL 100 % ethanol and 8 mL H_2O.
3. Avidin solution (for 30 slides): 1.5 mg avidin, 1.47 mL of Milli-Q water, 30 μL of 0.5 M $NaHCO_3$ (pH 9.0).
4. Carboxylic acid activation solution: 400 mL of DMF (HPLC grade), 7.4 g succinic anhydride, and 18 mL of 1 M $Na_2B_4O_7$ (pH 9.0).
5. NHS activation solution: 15 mL of DMF, 565 mg O-benzotriazole-*N,N,N′,N′*-tetramethyl-uronium-hexafluoro-phosphate (HBTU), 173 mg *N*-hydroxysuccinimide (HOSu), and 550 μL of *N,N*-diisopropylethylamine (DIEA). This will give 100 mM of HBTU, 100 mM of HOSu, and 200 mM of DIEA in 15 mL of DMF.
6. Quenching solution: 2 mM aspartic acid in 0.5 M $NaHCO_3$ (pH 9.0).
7. 150 °C oven.
8. Slide dish.
9. Metallic tray.
10. Magnetic stirrer.

2.4 Microarray Spotting

1. Phosphopeptide library stock solutions.
2. Phosphate buffered saline (PBS) (pH 7.4).
3. 384-well polypropylene microarray plate.
4. Stealth Micro Spotting pins (Telechem International, cat. ID. SMP7B).
5. Arrayer (ESI SMA™, Ontario, Canada).

2.5 Pro-Q™ Assay

1. Pro-Q™ Diamond dye (Invitrogen, cat. No. P33300). This dye binds to phosphate on the phosphopeptide to produce a baseline fluorescent signal.
2. Pro-Q destaining buffer (25 mL): 20 % acetonitrile in sodium acetate (50 mM, pH 4).

2.6 TCPTP Phosphatase Activity Assay on the Microarray Slide

1. Blocking solution: 1 % BSA in TBS.
2. TCPTP activity buffer: 50 mM HEPES, 100 mM NaCl, 2 mM EDTA, 0.01 % Brij 35, 1 mM DTT, pH 7.0.
3. TBS buffer.
4. TBST buffer: TBS buffer containing 0.05 % of Tween 20.

2.7 Solution-Based Microplate Assay

1. Malachite green assay kits (i-DNA Biotechnology Pte Ltd, Cat. No. POMG-25H).
2. Microplate shaker.
3. Tecan microplate reader (Tecan Group Ltd, Switzerland).
4. Greiner 384-well transparent plates (Practical Mediscience Pte Ltd).
5. Multichannel pipette (Practical Mediscience Pte Ltd).

2.8 Software

1. Array-Pro-Analyzer software (*Tecan* Trading *AG*, Switzerland).
2. GraphPad Prism v4.03 software (GraphPad, San Diego, USA).
3. Microsoft Excel.

2.9 General Apparatus

1. Adhesive film (ABgene, Cat. No. AB-0558).
2. Centrifuge (Eppendorf, 5415 R and 5810 R).
3. 15 and 50 mL centrifuge tubes.
4. Centrifuge filters (Microcon YM-3, cutoff 3 kDa).
5. Microscope glass slides: 75 mm × 25 mm × 1 mm.
6. Coverslips: 22 mm × 60 mm and 22 mm × 22 mm coverslips.
7. Marker pen.
8. Dessicator/dry storage box.
9. Ice box.
10. Humidity incubation chamber.

11. Powder-free gloves.
12. 1.5 mL reaction tubes and 14 mL sterile PP tube.
13. Slide-A-Lyzer dialysis cassettes 7K MWCO, 0.1–0.5 mL capacity (Thermo Scientific, Cat. No. 0066375).
14. Slide staining rack and slide staining jar.
15. Sonics Vibra-Cell sonicator (ITS Science & Medical Pte. Ltd.).
16. Microarray *Scanner* (Launch LS Reloaded, *Tecan* Trading *AG,* Switzerland).
17. UV/Visible spectrophotometer (Amersham Biosciences, Ultrospec 2100 Pro).
18. 96-well polypropylene stock plates: solid U-bottom and solid F-bottom (flat).

3 Methods

3.1 Expression of TCPTP

1. Transform competent *E. coli* BL21(DE3) with *TCPTP* construct and grow the single bacterial colony on LB-agar plates with kanamycin.
2. Inoculate a single bacterial colony from the agar plate into a 14 mL PP sterile tube containing 3–5 mL LB-kanamycin media. Grow the culture overnight at 37 °C in an orbital shaker with constant shaking at 200–250 rpm.
3. Dilute 100 times of the overnight culture into 200 mL fresh LB-kanamycin (50 μg/mL) media in a conical flask. Incubate the diluted LB culture at 37 °C with constant shaking at 200–230 rpm till OD_{600} reaches about 0.6–0.8 (it will take approximately 2–3 h).
4. Pipette a 10 mL culture of uninduced *E. coli* cells to be kept as a negative control for TCPTP expression.
5. The optimum absorbance (OD_{600}) for induction is 0.6–0.8. Check the absorbance periodically. If the cells are overgrown, repeat **steps 3** and **4**. Add IPTG (a final concentration of 0.1 mM) to induce TCPTP expression. Incubate the LB culture overnight (12–20 h) at 16 °C with constant shaking at 230 rpm.
6. Collect a 100 mL of induced LB culture, centrifuge, discard the supernatant, and store the cell pellets at −20 °C before use. Cell pellets can be stored at −20 °C for up to 1 month without any significant degradation of proteins. However, long-term storage (more than 3 months) of pellets may result in a decrease in protein activity.
7. Resuspend the induced bacterial cell pellets in 5–10 mL ice-cooled (4–8 °C) His-tag lysis buffer, and vortex if necessary.

3.2 Purification of TCPTP

1. Lyse the induced bacterial cells by sonicating them on ice at 30 % amplitude in a sonicator (5 s × 5 bursts, 10 s × 1 interval between each burst).
2. Centrifuge the bacterial lysate at 16,100 × *g* for 30 min at 4 °C. Transfer the supernatant to a fresh 1.5 mL tube on ice.
3. Pipette suitable amount of Ni-NTA resin beads into a chromatographic column (according to the expression levels, 100 mL of bacterial culture may need 150 μL of bead volume). Rinse the Ni-NTA resin with lysis buffer for 3–5 times (5–10 mL each time) before incubation with lysate.
4. Load the cell lysate and incubate it with the resin for about 45 min to 1 h with gentle shaking at 4 °C (*see* **Note 1**). The target protein TCPTP will bind to the Ni-NTA resin.
5. After incubation, let the lysate flow through the column slowly, and wash the column with His-tag wash buffer for 3–5 times (5–10 mL each time) (*see* **Note 2**).
6. Elute the target protein TCPTP with 0.3–1 mL of His-tag elution buffer for 2–4 times. When necessary, incubate the elution buffer with the resin for 2–5 min, and then collect all fractions of the eluent.
7. Measure the protein concentration by Bradford assay and combine the fractions that contain the highest amounts of TCPTP proteins (usually the first few fractions).
8. Run SDS gel electrophoresis and perform Western blotting to confirm the successful purification of TCPTP (Fig. 3).

3.3 Functionalization of Avidin onto the Surface of Microarray Slides

1. Carefully prepare piranha solution by slowly adding H_2O_2 to H_2SO_4 (*see* **Note 3**).
2. Place the slides in a tray and soak them in the piranha solution, and then occasionally shake the slides up and down.
3. Soak the slides for at least 4 h, and then take out the slides.
4. Carefully rinse the slides, first with H_2O and then with ethanol. Then dry the slides by flushing nitrogen gas or putting them in the fumehood for about 10 min till the slides are dry.
5. Prepare the APTES solution and pour it into a glass jar.
6. Put the glass slides into the glass jar with a stir bar, and then keep stirring the solution for 2 h.
7. After stirring, take out the slides and carefully wash them with 95 % ethanol for three times.
8. After washing, put the slides in a glass slide dish (with the lid) and then place it in a 150 °C oven for at least 2 h or overnight (*see* **Note 4**).
9. Take the slides out, and let them cool to room temperature.

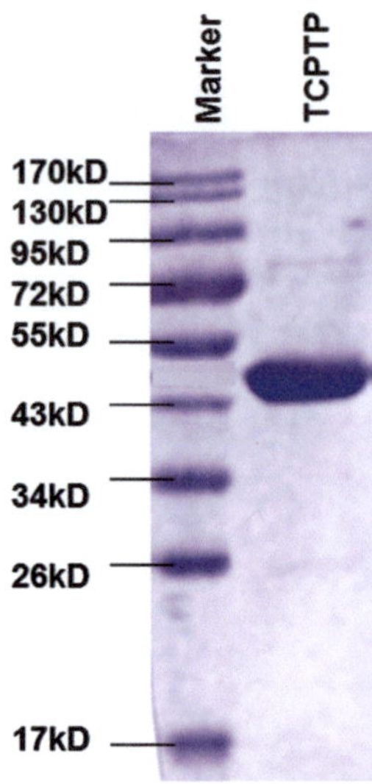

Fig. 3 Purification of TCPTP. This is a Coomassie blue-stained gel of the purified TCPTP protein

10. Wash the slides with 95 % ethanol three to five times (ethanol wash bottle) and dry the slides as described above.
11. Fill the slide tray with 400 mL of DMF and add 7.4 g of succinic anhydride. Put in a stir bar to stir the solution. After the solid is completely dissolved, add 18 mL of 1 M $Na_2B_4O_7$ (boric acid + NaOH, pH 9), and subsequently soak the slides into the solution (*see* **Note 5**). This step is for carboxylic acid activation.
12. Shake the slide tray once in a while to avoid small air bubbles adhering on the slides.
13. Meanwhile, start heating up 2 L of deionized water (use a stir bar) until it reaches about 95 °C.
14. Soak the slides in the hot water for 20 min with constant stirring (*see* **Note 6**).
15. Take out the slides and rinse them three to five times with ethanol using a wash bottle. Then briefly dry them with nitrogen gas or place the slides in the fumehood for about 10 min till the slides are dry.
16. Prepare the NHS activation solution and add 0.75 mL of the solution onto each slide.
17. Place the large coverslips (22 mm × 60 mm) on the slides, close the lid of the glass slide dish, and incubate for at least 3 h.
18. Transfer the slides to a metallic tray.
19. Wash the slides three to five times with 95 % ethanol using a wash bottle. This step takes about 3–5 min.

20. Prepare 1 mg/mL of avidin solution in 10 mM $NaHCO_3$.
21. Apply about 49 μL of the avidin solution onto each slide, and then incubate for 30 min under a coverslip (22 mm × 60 mm).
22. After the avidin immobilization reaction, quench the unreacted NHS groups with the quenching solution for 30 min and rinse the slides thoroughly with water.
23. Dry the slides using a stream of nitrogen gas, and store them at 4 °C before use.

3.4 Phosphopeptide Spotting and Quality Control

1. Prepare 16 μL of each biotinylated peptide solution in PBS/DMSO (1:1) to reach a final concentration of approximately 1 mM. This concentration would completely saturate the avidin group on the slides and it ensures that an adequate amount of peptides are immobilized (*see* **Note 7**).
2. Prepare the biotinylated peptide library in a 384-well polypropylene microarray plate for spotting, which is compatible with the microarray spotter.
3. Seal the spotting plate with adhesive film and store it at −20 °C. Thaw the plate when needed.
4. Before spotting, briefly wash the avidin-coated slides with deionized H_2O, and then quickly dry them with nitrogen gas or put them in the fumehood for about 10 min till the slides are dry.
5. Spot the biotinylated peptide library to the avidin-coated slides using the arrayer. After the spotting completes, the slides are allowed to stay inside the spotter for another 2–4 h. This will provide sufficient time for biotin/avidin interaction.
6. Put the slides into a slide dish containing 20 mL of deionized water and shake for at least 10 min to remove unbound peptides. Rinse the slides with distilled water. After the slides are briefly dried, store them at 4 °C before use.
7. To test the spotted slides and ensure the good quality of any given batch (i.e., the immobilization of phosphopeptides onto the surface of the avidin-coated slides is consistent between different batches), perform the Pro-Q™ assay as follows.
8. Dry the spotted slides by putting them in the fumehood for 10 min or using a stream of nitrogen gas.
9. Pipette 1 mL of Pro-Q™ Diamond dye and apply it onto the surface of the spotted slides for 1 h in a humidified incubation chamber under the coverslips (22 mm × 60 mm) at room temperature.
10. Wash the spotted slides with the Pro-Q™ destaining solution for 0.5–1 h. After washing, scan the slides with the microarray scanner. A typical image of Pro-Q™ stained slide is shown in Fig. 4.

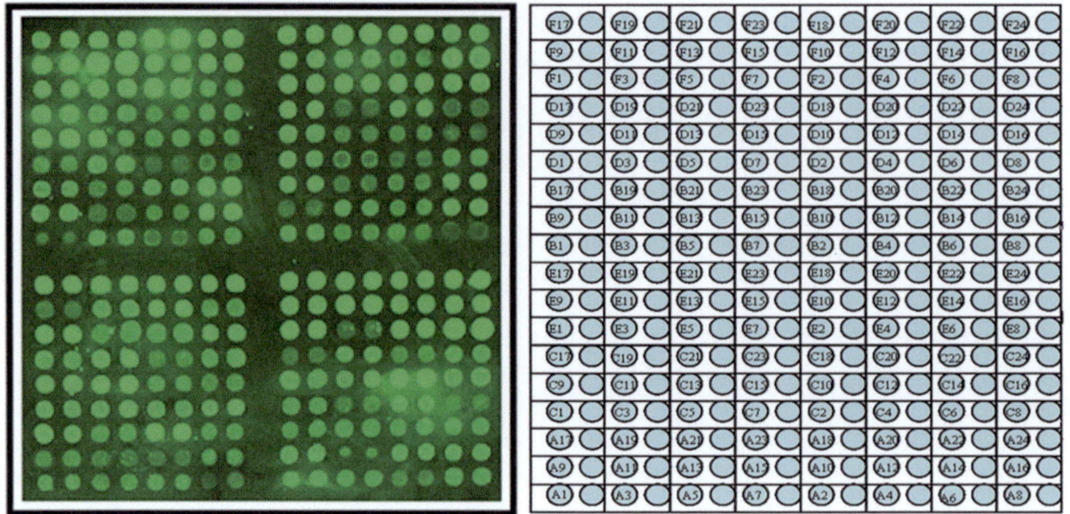

Fig. 4 Pro-Q staining image (*left*) and microarray spotting format (*right*). Pro-Q staining is used to confirm the good quality of spotted slides. The slide comprises a 144-phosphopeptide library printed in duplicate (reproduced in part from ref. [4] with permission from John Wiley & Sons Inc.)

3.5 Screening of TCPTP Substrates Spotted on the Microarray Slide

1. Bring the spotted slides to room temperature. Rinse them with distilled H_2O and shake for about 10 min to remove the unbound peptides.
2. Apply the blocking solution to the spotted slide for 1 h.
3. Prepare the TCPTP activity buffer.
4. As negative controls, use denatured TCPTP (heating at 95 °C for 10 min followed by incubation on ice) or activity buffer alone (*see* **Note 8**).
5. Perform the single-time-point microarray experiments to delineate the substrate specificity of TCPTP. Typically, the incubation time is 0.5 h and the volume of TCPTP applied to the slide is 75 μL. The enzyme can be applied directly onto the grid which is marked using a permanent marker pen to form rectangular frames. No coverslip is used.
6. After incubation, wash the slides with deionized H_2O to remove the TCPTP solution.
7. Gently rinse slides with deionized H_2O, and then quickly dry the slides with nitrogen gas or in the fumehood.
8. Perform the Pro-Q assay, which has been described in Subheading 3.4, **steps 7** and **10**.
9. After the Pro-Q assay, wash the slides with deionized H_2O to remove the Pro-Q reagent. When necessary, wash with TBST buffer to reduce the background so as to increase the signal-to-noise ratio (*see* **Note 9**).

10. Dry the slides with nitrogen gas or in the fumehood, and then scan them using microarray scanner to obtain the microarray image (Fig. 2).
11. Analyze the microarray images and perform quantitative data analysis. Confirm the well-known TCPTP peptide substrate (D14: TDKEYpYTVKDD) [19]. Identify unknown substrates (e.g., A06, D02) of TCPTP. In addition, the Top-10 peptide substrates identified from the single-time-point experiments for TCPTP are listed in Table 1.
12. To guide the design of potent inhibitors for PTPs in the future, perform further data analysis of the amino acid preference of the phosphopeptides at each position (namely from −5 to +5 positions, or from P−5 to P+5, counting from N- to C-terminus, with the location of phosphotyrosine (pY) defined as P) with the Top-10 peptides for TCPTP and the results are graphically presented in Fig. 5.
13. To better understand the kinetic information of TCPTP, time-course TCPTP activity assays are carried out for the Top-5 peptide substrates identified from the single-time-point microarray experiments. Different subgrids of peptide array were incubated with TCPTP solution in a humidified chamber for different time ranging from 5 min to 2 h (*see* **Note 10**).

Table 1
Top-10 peptide substrates identified from the single-time-point microarray experiments for TCPTP

Ranking	TCPTP
1	DSGGF*pY*TSRT **(A10:SRC1)**
2	EEEPV*pY*EAEPE **(F18:HS1)**
3	REGLN*pY*MVLAT (D02:ROS1)
4	DEKVD*pY*VQVDK **(A06:GAB2)**
5	DEELH*pY*ASLNF **(A04:CD33)**
6	VYESP*pY*SDPEE **(D22:ZAP70)**
7	TNDIT*pY*ADLNL **(D16:SIRP)**
8	SDDVR*pY*VNAFK **(D04:VEGF)**
9	VSSTH*pY*YLLPE **(F23:CDC42)**
10	MTGDT*pY*TAHAG **(C05:ABL)**

pY represents phosphotyrosine. The note underneath the peptide sequence indicates the peptide ID, and the original protein source (reproduced in part from ref. [4] with permission from John Wiley & Sons Inc.)

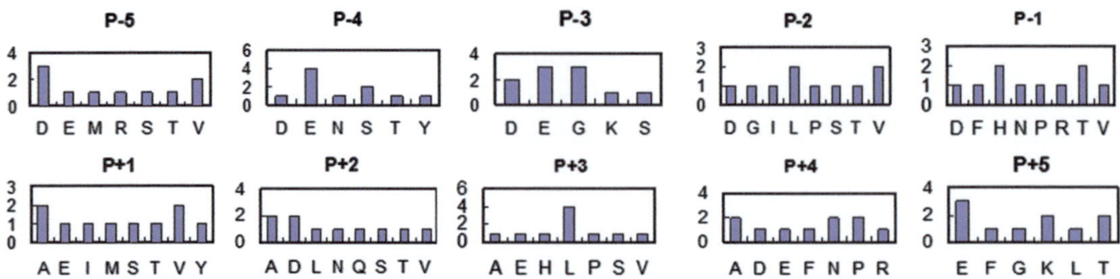

Fig. 5 Summary of amino acid preference of the phosphopeptides from their −5 to +5 positions (e.g., P+5 is 5 amino acid positions toward the C-terminus with phosphotyrosine located at the P position). The data are extrapolated from the Top-10 peptide substrates (*see* Table 1) identified from the single-time-point experiments using TCPTP

14. After the TCPTP incubation, stain the slides by repeating the same **steps 6–10**. The time-dependent microarray images are shown in Fig. 6a.
15. Analyze the array images and perform quantitative data analysis. Calculate the corresponding k_{obs} value using the GraphPad software (Fig. 6b and Table 2). The larger the k_{obs} value is, the faster TCPTP will dephosphorylate the corresponding phosphopeptide.

3.6 Validation of the Identified Hits Using the Solution-Based Microplate Assay

1. Transfer 20 μL of TCPTP activity buffer containing 20 μM of the peptides into each well of a transparent 384-well microplate.
2. Pipette the TCPTP solution (20 μL) into each well at different time points (0, 5, 15, 30, 60, and 120 min). The kinetic data for dephosphorylation of each phosphopeptide at each time point are determined in duplicates.
3. Stop the dephosphorylation reaction by addition of malachite green solution which contains sulfuric acid to denature the TCPTP. If the reaction solution contains free phosphate groups, which are generated through dephosphorylation of the phosphopeptide substrates by TCPTP, the reaction solution will then become green in color.
4. When the green color of the reaction solution become stable (around 10 min), measure the absorbance of the malachite green-phosphate complex at 650 nm using a microplate reader.
5. Perform background subtraction ($t=0$ min) and data normalization.
6. Process microplate data with the software GraphPad and fit the data with the following equation to plot the kinetic curves and derive the kinetic parameter k_{obs} (Fig. 6c and Table 2):

$$\mathrm{Abs}_{obs} = \mathrm{Abs}_{max} \times \left(1 - \exp\left(-k_{obs} \times \mathrm{t}\right)\right).$$

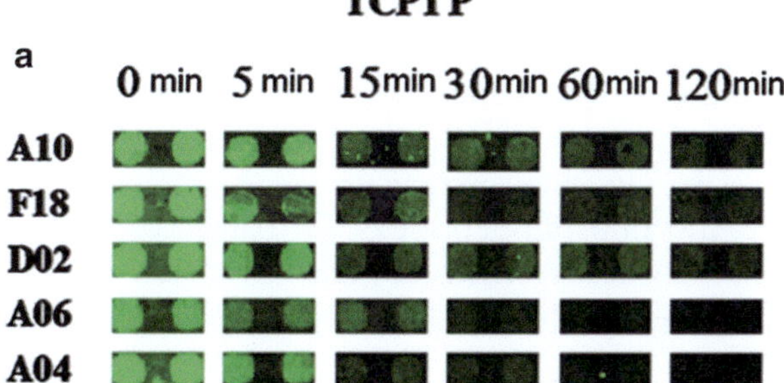

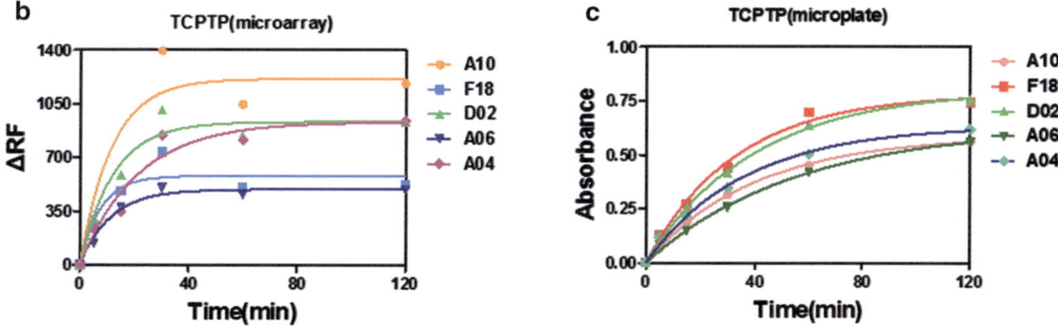

Fig. 6 Time-dependent, semi-quantitative dephosphorylation experiments of selected top five peptides with TCPTP. (**a**) Microarray images of selected spots from the time-dependent kinetic experiment of top five peptides (*see* Table 1) against TCPTP; (**b**) Plots of extrapolated microarray data and k_{obs} fitted curves; (**c**) Microplate-based time-dependent experiments (k_{obs} fitted curves) of the same five peptides against TCPTP (reproduced in part from ref. [4] with permission from John Wiley & Sons Inc.)

Table 2
Summary of kinetic constants (k_{obs}) determined from the microarray and microplate experiments for Top-5 peptide substrates (*see* Table 1) against TCPTP (reproduced in part from ref. [4] with permission from John Wiley & Sons Inc.)

TCPTP	k_{obs} (array, min^{-1})	k_{obs} (plate, min^{-1})
A10	0.086	0.025
F18	0.132	0.03
D02	0.082	0.025
A06	0.089	0.018
A04	0.047	0.027

where Abs_{obs} represents the absorbance reading at time x; Abs_{max} represents the fitted constant which corresponds to the maximum absorbance obtained when the enzymatic reaction is complete; t is the time.

Good agreement in the k_{obs} values is obtained for most of the phosphopeptide/PTP pairs between the time-dependent microarray-based experiments and the standard microplate assays (Table 2).

In summary, a phosphopeptide microarray-based approach has been described for high-throughput screening of substrate specificity of TCPTP. By this approach, we have successfully obtained substrate fingerprints for TCPTP, amino acid preference of TCPTP at each position, as well as the kinetic constants of peptides substrates for TCPTP. The high-throughput substrate fingerprint strategy can also be applied for studying other PTPs or Ser/Thr phosphatases.

4 Notes

1. To avoid the decrease in the enzymatic activity of TCPTP, carry out this step of incubation at 4 °C.
2. The columns are designed for low-pressure applications (less than 1 atm or 14 p.s.i.). When a pump is used for the liquid flow, any disruption in the flow must be avoided since this disruption may cause a rapid increase in back pressure. If this increase does happen, immediately switch off the pump and check the gel bed for liquid leakage.
3. When preparing the piranha solution, pour H_2SO_4 in the beaker first and then add H_2O_2. Make sure the temperature of the solution does not increase too much since it can get extremely hot. Be careful when handling the piranha solution since this hot acid solution is extremely corrosive.
4. To prepare the slides with high-quality surfaces, avoid sticking the slides to each other when putting the slide rack in the 150 °C oven.
5. Always immerse the slides in the reaction solution. Otherwise, the slide will be unevenly cleaned, and this will lead to uneven surfaces for the subsequent preparation of avidin slides, which will greatly compromise the quality of microarray.
6. Similarly, keep the slides immersed in water at all times. Otherwise, the surfaces of the slides may not be uniformly cleaned, which would greatly compromise data quality of the subsequent microarray experiments.
7. To prepare the spotted microarray slides with good quality, prepare the peptide solution at a high concentration (i.e., not

less than 1 mM) for spotting. Otherwise, the amount of peptides may not be sufficient to saturate the slide surface during immobilization.

8. There should be no (or negligible) signal difference between slides treated with negative controls (denatured TCPTP or activity buffer solution alone) and untreated slides. A decrease in fluorescent signals (stained using the Pro-Q dye) will be observed if the TCPTP is active and the phosphopeptides are the substrates of TCPTP so that the phosphate groups are removed from the peptides.
9. The frequency and duration of slide washing could be optimized as required. We typically use one to three times of 5-min washes which would produce very good results.
10. When the coverslips are put on the slides, care must be taken that no air bubbles are trapped between the coverslips and the slides. Otherwise, no dephosphorylation reaction would occur and this would generate false negatives. Besides, to acquire good and consistent fingerprint results with low backgrounds, carefully apply the TCPTP solution uniformly across the surface of the slides.

Acknowledgment

The authors acknowledge funding support from National Basic Research Program of China (No. 2011CB933101), Hundred Talents Program of the Chinese Academy of Sciences, and the City University of Hong Kong (Grant No. 7004025, 9667091), and also the funding from the Institute of Bioengineering and Nanotechnology (Biomedical Research Council, Agency for Science, Technology and Research, Singapore).

References

1. Kohn M, Gutierrez-Rodriguez M, Jonkheijm P et al (2007) A microarray strategy for mapping the substrate specificity of protein tyrosine phosphatase. Angew Chem Int Ed 46: 7700–7703
2. Alonso A, Sasin J, Bottini N et al (2004) Protein tyrosine phosphatases in the human genome. Cell 117:699–711
3. Sefton BM, Hunter T, Ball EH et al (1981) Vinculin: a cytoskeletal target of the transforming protein of Rous sarcoma virus. Cell 24:165–174
4. Gao L, Sun H, Yao SQ (2010) Activity-based high-throughput determination of PTPs substrate specificity using a phosphopeptide microarray. Biopolymers 94:810–819
5. Simoncic PD, McGlade CJ, Tremblay ML (2006) PTP1B and TC-PTP: novel roles in immune-cell signaling. Can J Physiol Pharmacol 84:667–675

6. Bialy L, Waldmann H (2005) Inhibitors of protein tyrosine phosphatases: next-generation drugs? Angew Chem Int Ed 44:3814–3839
7. Pawson T, Scott JD (1997) Signaling through scaffold, anchoring, and adaptor proteins. Science 278:2075–2080
8. Sun H, Lu CH, Uttamchandani M et al (2008) Peptide microarray for high-throughput determination of phosphatase specificity and biology. Angew Chem Int Ed 47:1698–1702
9. Liang F, Kumar S, Zhang ZY (2007) Proteomic approaches to studying protein tyrosine phosphatases. Mol Biosyst 3:308–316
10. Zhang ZY (2003) Mechanistic studies on protein tyrosine phosphatases. Prog Nucleic Acid Res Mol Biol 73:171–220
11. Garaud M, Pei D (2007) Substrate profiling of protein tyrosine phosphatase PTP1B by screening a combinatorial peptide library. J Am Chem Soc 129:5366–5367
12. Walchli S, Espanel X, Harrenga A et al (2004) Probing protein-tyrosine phosphatase substrate specificity using a phosphotyrosine-containing phage library. J Biol Chem 279:311–318
13. Espanel X, Walchli S, Ruckle T et al (2003) Mapping of synergistic components of weakly interacting protein-protein motifs using arrays of paired peptides. J Biol Chem 278:15162–15167
14. Hu Y, Uttamchandani M, Yao SQ (2006) Microarray: a versatile platform for high-throughput functional proteomics. Comb Chem High Throughput Screen 9:203–212
15. Foong YM, Fu J, Yao SQ et al (2012) Current advances in peptide and small molecule microarray technologies. Curr Opin Chem Biol 16:234–242
16. Gao L, Sun H, Uttamchandani M et al (2013) Phosphopeptide microarrays for comparative proteomic profiling of cellular lysates. Methods Mol Biol 1002:233–251
17. Gao L, Uttamchandani M, Yao SQ (2012) Comparative proteomic profiling of mammalian cell lysates using phosphopeptide microarrays. Chem Commun 48:2240–2242
18. Asante-Appiah E, Patel S, Desponts C et al (2006) Conformation-assisted inhibition of protein-tyrosine phosphatase-1B elicits inhibitor selectivity over T-cell protein-tyrosine phosphatase. J Biol Chem 281:8010–8015
19. Simoncic PD, Lee-Loy A, Barber DL, Tremblay ML, McGlade CJ (2002) The T cell protein tyrosine phosphatase is a negative regulator of janus family kinases 1 and 3. Curr Biol 12:446–453

Chapter 14

Nanotechnology in the Fabrication of Protein Microarrays

Manuel Fuentes, Paula Díez, and Juan Casado-Vela

Abstract

Protein biochips are the heart of many medical and bioanalytical applications. Increasing interest of protein biochip fabrication has been focused on surface activation and subsequent functionalization strategies for the immobilization of these molecules.

Key words Surface activation, Oriented immobilization, Sample deposition, Pin materials, Microarrays

1 Introduction

Nowadays, the protein microarray is becoming an attractive approach for rapid profiling of the entire proteome. In general, the protein microarray or protein biochip is a versatile platform for the characterization of hundreds to thousands of proteins in a highly parallel and high-throughput manner. An additional driving force for the development of protein biochips emanates from related biotechnological and biomedical fields, such as biosensors, biocatalysis, biomarkers, and drug discovery, among other applications [1, 2].

Typically, protein biochips are prepared by using similar technologies previously developed for the production of DNA microarrays. Thus, thousands of proteins can be spotted onto a chip by using a contact spotter (*see* Fig. 1a) or non-contact spotter such as ink-jet printing (*see* Fig. 1b). The protein microarrays can also be fabricated using photolithographical methods (*see* Fig. 1c). Afterwards, a biological sample can be applied on the chip, and proteins that are bound can be identified by using several detection methods [3].

Protein microarrays are composed of two major classes: analytical and functional, depending on the application [4]. Over the last few years, the number of these applications, especially functional protein microarrays, has dramatically increased for basic and clinical research due mainly to the maturation of the fabrication

Paul C.H. Li et al. (eds.), *Microarray Technology: Methods and Applications*, Methods in Molecular Biology, vol. 1368, DOI 10.1007/978-1-4939-3136-1_14, © Springer Science+Business Media New York 2016

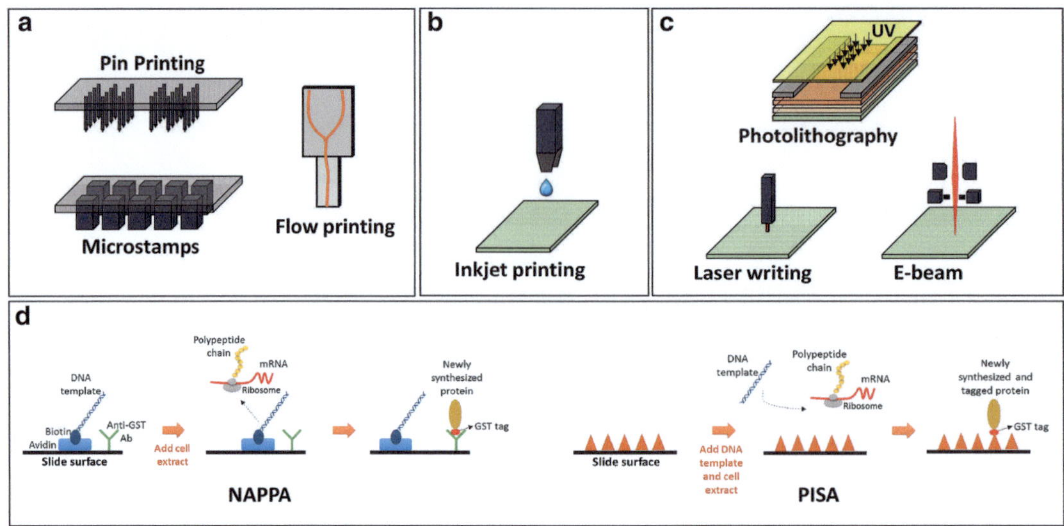

Fig. 1 Various technologies for printing protein microarrays. (**a**) Contact strategies, (**b**) non-contact methods, (**c**) lithography approaches, and (**d**) cell-free methods

technology [5]. In addition, tissue or cell lysates can be fractionated and spotted onto a slide to form reverse-phase protein microarrays. Moreover, it is possible to print DNA molecules on the slide and translate them into proteins in situ by means of NAPPA (Nucleic Acid Programmable Protein Array) or PISA (Protein In Situ Array) technology (*see* Fig. 1d).

Recently, there has been a shift towards generating surfaces that display proteins in highly controlled patterns, preferably with nanometer precision. Common challenges of manufacturing a viable protein chip include the choice of a proper solid surface, and the development of the surface chemistry that is compatible with the immobilization of a diverse set of proteins while maintaining their integrity, native conformation, and biological function. The immobilization can be controlled by using different functional groups or tags of the proteins [6].

The drive to fabricate arrays of protein spots and also to create the arrays on a surface with the feature size in the nanometer regime has spurred the development of various fabrication methods that have been widely employed in nanotechnology and material science. The aim of this chapter is to describe the chemical, biological, and nanotechnological strategies for the generation of protein biochips in two sections. The first section is to describe the current state-of-the-art spotting methods, while the second section is referred to the several surface activation protocols that led to well-functionalized and well-characterized surfaces for microarray fabrication.

1.1 Contact Printing for Microarray Fabrication

The most widely adopted technology for printing microarrays remains contact pin printing (*see* **Note 1**) basically because of the wide adoption in DNA microarrays fabrication, and the direct transfer of lessons learned to later protein printing strategies.

The pin printing hardware includes solid pins, split pins, or quill pins, and also tweezers or other liquid transfer/deposition pin types (*see* **Note 1**). These pins are loaded with protein-containing aqueous solution droplets in printing buffers for transfer onto solid microarray surfaces [3] (*see* **Note 2**). Owing to the nature of contact by the pins during every deposition, there is a chance that the array surface may be damaged.

A robotically controlled print head contains many fluid-dispensing pins (*see* **Note 3**) that are repeatedly dipped into wells of a source microtiter plate containing purified proteins prepared in printing buffers (*see* Fig. 1).

The pin captures a known and controlled volume of protein solutions from each well. Then, the pin lightly touches onto the surface of the functionalized glass or other microarray surface to dispense nanoliter-sized liquid as a droplet [3] (*see* **Note 4**).

The pin printing technology is directly affected by several surface phenomena, such as pin capillarity which depends on the interfacial tensions between solution and pin surface (*see* **Note 5**). Overall, the final spot morphology is dependent of pin-printing solution, substrate chemistry, evaporation process, and pin diameter [3] (*see* **Note 6**).

The liquid to be deposited is drawn into the pin slits on the order of 10–100 μm in diameter. The pin then deposits each liquid droplet of a small volume by a slight contact with the array surface (*see* **Note 7**). Since small volumes of liquid reagents (~ pL) are loaded in the pins, liquid evaporation could become a significant issue in both the deposited final volume and the reagent concentration.

1.2 Lithographical Printing

Lithography, which is a common process in micro and nanodevice fabrications, is adapted for printing biomolecules. This process uses light to transfer a geometric pattern from a photomask to a light-sensitive chemical "photoresist" on the substrate. Biospecific nanopatterns can be fabricated via selective decomposition/activation and a subsequent chemical modification for recruitment of proteins and other biomolecules.

A series of chemical treatments engrave the exposure pattern into or enable deposition of new material in the desired pattern upon the material underneath the photoresist.

Lithography is used because it can create extremely small patterns (down to a few tens-of-nanometers in size). Additionally, this method affords exact control over the shape and size of the objects it creates, and it can create patterns over an entire surface cost-effectively.

The main disadvantages of lithographical printing are (1) that it requires a flat substrate to start with, (2) it is not very effective in creating shapes that are not flat, and (3) it requires extremely clean operating conditions.

Several strategies for lithography have been developed such as photolithography; interference lithography, laser writing, and e-beam (*see* Fig. 1c). All of them present advantages and disadvantages, principally related to the throughput level and dimensions of the patterns [2].

1.3 Printing by Self-Assembled Monolayers (SAMs)

The construction of self-assembled monolayers (SAMs) requires homogeneous surface modification, i.e., control of wetting, control of biocompatibility, lubrication, and corrosion inhibition, as well as metal refining, adhesion, and passivation.

Self-assembled monolayers (SAMs) of bifunctional molecules on metallic and oxide substrates have been extensively studied since the 1980s, and SAMs have become an important form of nanotechnology, including diverse biological applications.

The overall formation of nanoscale assemblies is governed by the intermolecular and interparticle forces [7, 8] that have to be precisely controlled in order to maintain the distinct biological functions and to build a reliable device.

The utility of SAMs is based on their characteristics [9]:

1. SAMs are easy to prepare and they assemble quickly from solutions of the molecules.
2. SAMs are molecularly ordered and robust under many conditions.
3. SAMs are thermodynamically stable.
4. It is possible to control the film thickness of SAMs to within approx. 0.1 nm by varying the length of the constituent molecules.
5. Surface properties of SAMs are controlled through tailoring of exposed surface functional groups.

Since SAMs are nanometer-sized elements in two dimensions (perpendicular to the plane of the surface), patterning of SAMs is the first step toward the realization of devices that involve this class of nanostructures in fabrication, processing, or use. The major advantage of the SAM is Angstrom precision in the positioning of chemical groups at the solid–organic interface. Also, SAMs are relatively stable as compared to other surface-supported molecular platforms because they are strongly (most often covalently) pinned to the substrate via a reactive head group, which are typically thiol, disulfide, or silane [2, 8].

Therefore, SAMs are compatible with the majority of standard and emerging fabrication techniques. The so-called mixed SAMs can be obtained via surface adsorption of a mixture of two or more bifunctional compounds with different terminal chemical groups, an approach that combines chemically and biologically inert molecule modules (such as polyethylene glycol (PEG)) with different terminal functional groups (carboxyl, amino, maleimide, or azide) or biospecific tags, ligands, ssDNAs, or enzymatic cofactors, among others.

Docking of molecules at such biocompatible interface can be precisely controlled in terms of orientation, surface density, conformation, and functional activity [6]. Therefore, the control over the interfacial properties and different interactions between the protein microarrays relies on advanced surface chemistries [6]. Herein, we will briefly present the main chemical groups and the most convenient (easy to apply, ideal features) surface functionalization protocols for the fabrication of protein microarrays.

In general, non-covalent binding chemistries perform similar to covalent chemistries through amine plus cross-linkers or functional groups. Specifically, poor signal intensities are obtained with amine-silane- and poly-lysine-activated surfaces, probably due to the lower density of immobilized antibodies. The ideal surface for antibody arrays should keep antibody function while allowing optimal immobilization efficiency. In such a way, hetero-functional activated chemistries combine both properties and are feasible with different functional groups, from covalent attachment groups (epoxy, aldehyde) to bioaffinity groups (IDA, AMPB) [6].

2 Materials

2.1 Reagents

1. Acetone (HPLC grade), ethanol (HPLC grade), acetonitrile (HPLC grade).
2. Polydimethoxysilane(PDMS)prepolymer,3-(2-amineethylene) propyl-methyldimethoxysilane (AEPDMS), 3-glycidoxypropyl trimethoxysilane (GPTS).
3. Amino-phenyl-boronic (AMPB) acid, iminodiacetic acid (IDA).
4. Sodium periodate ($NaIO_4$), copper sulfate ($CuSO_4$).
5. Tween® 20 (Molecular Grade).
6. Bis[sulfosuccinimidyl]suberate (BS^3), *N*-succinimidyl 3-(2-pyridyldithio)propionate (SPDP).
7. 1 % (v/v) glycidol (glycidol in 100 mM NaOH, pH 10.0).
8. Phosphate buffered saline (PBS).
9. Amino-PEGS-thiol, hydroxy-PEGS-thiol, carboxy-PEGS-thiol (Prochimia).

2.2 Apparatus

1. Glass slides 76×26 mm.
2. Gold-coated slides 25×76×1 mm with coated gold layer of 50 nm (Nanocapture, Gentel Bioscience, USA).
3. LifterSlip coverslips.
4. Plasma chamber (Plasma Etch, USA).

3 Methods

The reproducibility of a good monolayer of materials requires the surface to be cleansed of any contaminants. This cleaning can be achieved by using activated reactive hydroxyl groups in case of silica surface. Unfortunately, there is no universally accepted cleaning agent. Here, the most common methods are described (mainly based on combinations of acids, bases and organic solvents) (*see* **Note 8**).

All the experiments are carried out at room temperature (RT) unless otherwise specified.

3.1 Surface Cleaning

Several protocols are required to clean different surfaces. The common initial protocol is given as the basic protocol.

Basic protocol

1. Carefully wash each slide with MeOH/HCl (dilution 1:1 (v/v)) solution using a cotton wipe.
2. Immerse all slides in MeOH/HCl solution for 30 min.
3. Rinse with deionized water.
4. Dry the slides under nitrogen gas stream immediately prior to the silanization process [10].

Protocol A for cleaning silica surfaces

1. Follow the basic protocol.
2. Heat the slides in concentrated H_2SO_4 for 2 h.
3. Rinse with deionized water.
4. Dry the slides under nitrogen stream immediately prior to silanization process [10].

Protocol B for cleaning silica surfaces

1. Follow basic protocol.
2. Heat the slides in concentrated H_2SO_4 for 2 h.
3. Rinse in deionized water and boil it for 30 min.
4. After rinsing, dry the slides under a stream of nitrogen just before the silanization process [11].

Protocol C for cleaning metallic surfaces

1. Clean each gold metallic surface for 3 min using an oxygen plasma at 40 W in a plasma chamber.
2. Store the slides in a non-oxidative atmosphere such as nitrogen [12].

3.2 Functionalization on Metallic Surface

1. Coat the gold-coated slides with activated-PEG-thiol (HS–$(CH_2)_{11}$–EG_6) to generate a homogeneous self-assembled monolayer (SAM) onto the gold surface. The SAM preparation

should be performed in a tight box filled with ethanol vapour.

2. Clean the gold slides according to protocol C.
3. Functionalize gold surfaces with 1 mM solution (in ethanol) of amino-PEG-thiol (HS–$(CH_2)_{11}$–EG_6–NH_2). After amino-PEG-thiol treatment, place the cover slips and gold slides on a metal rack, and then place it in a close humidified chamber (in ethanol). Incubate for 16 h.
4. Wash the slides with ethanol and dry them using filtered air.
5. Store the slides in a dry place containing silica gel packs until use.
6. Use carboxy-PEG-thiol (HS–$(CH_2)_{11}$–EG_6–COOH) or hydroxy-PEG-thiol (HS–$(CH_2)_{11}$–EG_6–OH) for carboxyl activated surface or hydroxyl activated surface, respectively.

It is recommended to characterize the SAM by using a profilometer, scanning electron microscopy (SEM), or atomic force microscopy (AFM). The AFM image should show that SAMs are uniformly formed on gold coated surfaces. AFM is used to analyze the height of the average monolayer (~2 nm) [1] (*see* Fig. 2).

3.3 Functionalization on the Silica Surface

1. Clean the glass slide surface according to protocol A and B.
2. Cover the glass slides with 300 mL of a solution of polydimethylsiloxane (PDMS) prepolymer dissolved in acetone (concentrations range from 0.2 to 10 %). Once activated with PDMS, the glass slides are considered silanized, and they are functionalized using different chemistries as described below.

A. Amino Silanized Surface

1. Cover the silanized glass surfaces with a solution of 2 % (v/v) 3-(2-amineethylene)-propylmethyldimethoxysilane (AEPDMS) in acetone.
2. Incubate at room temperature (RT) for 30 min with gentle orbital shaking (100 rpm).
3. After the incubation step, wash the slides with acetone and then with distilled water.
4. Dry the slide under a filtered forced air stream.
5. Storage the slides at RT under a dried atmosphere, such as a dried chamber containing silica gel packs.

Further functionalization of the amino-silanized surface with cross-linkers, such as bis[sulfosuccinimidyl]suberate (BS^3) or *N*-succinimidyl-3-(2-pyridylthio)propionate (SPDP), is possible, it is described as follows:

1. Cover the amino-silanized glass slides using 200 μL of 2 nM of BS^3 with coverslips.

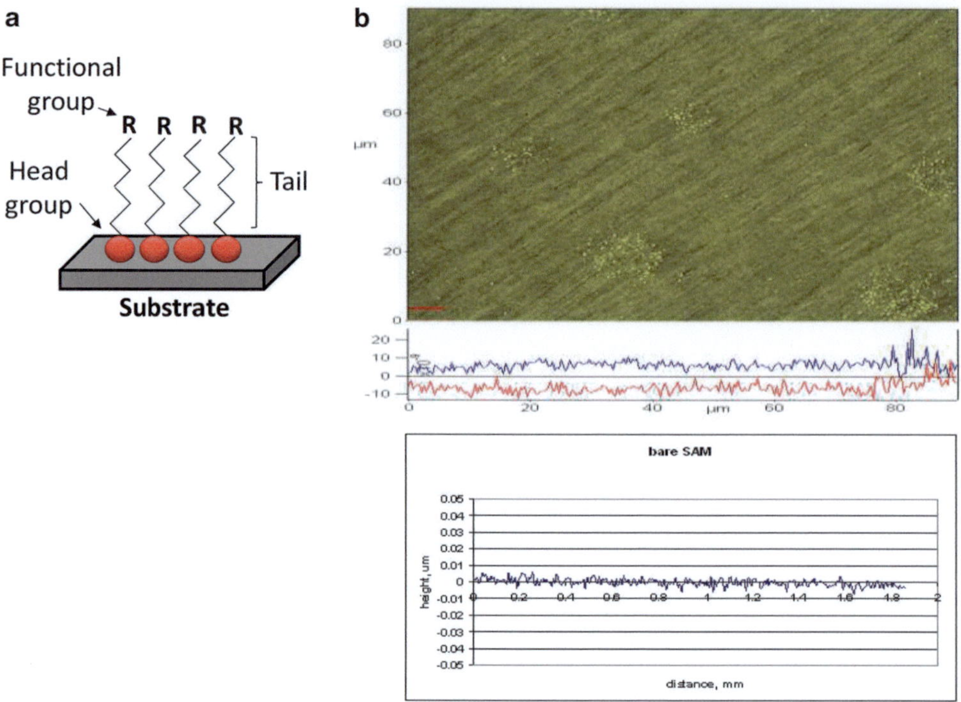

Fig. 2 Self-assembled monolayers (SAMs). (**a**) Representation of the SAM structure formed on a substrate. (**b**) The *upper* SEM image and *middle* profilometer scan represent the uniformity of a gold-coated surface. The *red* and *blue lines* represent the two cross sections indicated on the *upper* diagram. The *lower* diagram is the profilometer scan of a bare SAM of amino-PEG-thiol without protein formed on the gold surface

2. Incubate at RT for 1 h.
3. Wash the slides with water and dry them under a stream of filtered air.

B. Epoxy Silanized Surface

1. Cover the silanized glass surfaces with a solution of 2 % of 3-g lycidoxypropyltrimethoxysilane (GPTS) dissolved in acetone.
2. Incubate at RT for 30 min with gentle orbital shaking (100 rpm).
3. Wash the slides with acetone and distilled water.
4. Dry under a filtered forced air stream.
5. Storage the slides under dry and non-oxidative conditions. The results of using different concentrations of GPTS (0.2, 1, 5, 10 %) are shown in Fig. 3b.

Further functionalization of the epoxy-silanized surface with other chemical groups is possible as described below (*see* **Note 9**).

C. Aldehyde Silanized Surface

1. To create the aldehyde-activated silanized glass surfaces, incubate the epoxy-silanized glass slides for 30 min with HCl (100 mM) acid solution.

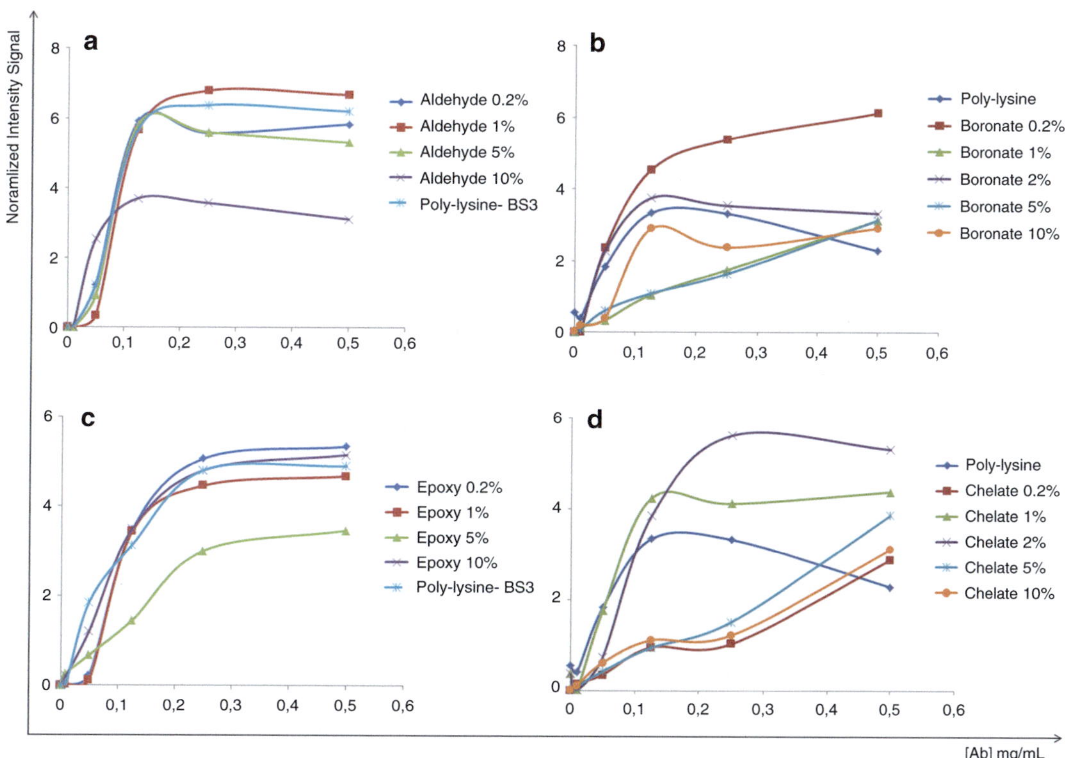

Fig. 3 Different Bioaffinity-activated surfaces. (**a**, **b**) Covalent and bioaffinity-activated surfaces: Saturation curves showing the immobilization capacity of commercial poly-lysine surfaces with BS^3 cross-linker and aldehyde-activated surfaces using different concentrations (0.2, 1, 5, and 10 %) (**a**) and epoxy-activated surfaces at the same concentrations (**b**). (**c**, **d**) Bioaffinity surfaces: Saturation curves showing the immobilization capacity of commercial poly-lysine surfaces and AMPB-activated surfaces (boronate) at different concentrations (0.2, 1, 2, 5, and 10 %) (**c**) and IDA-Cu^{2+}-activated surfaces (metal–ligand chelate) at the same concentrations (**d**). In all of these curves, the *x* axis is the concentration of monoclonal antibody (mAb) printed (ranging from 0 to 0.50 mg/ml), and the *y* axis is the normalized intensity data for 50 antibodies

2. After incubation, wash with distilled water.
3. Treat the slides for 30 min with 100 mL of $NaIO_4$ solution (0.2, 1, 5, or 10 %) at RT and with mild stirring.
4. Wash them dry as described above.
5. Storage the slides at RT under a dried atmosphere until further use (*see* **Note 10**). The results of using different concentrations of $NaIO_4$ (0.2, 1, 5, 10 %) are shown in Fig. 3a.

D. Aminophenylboronic Acid Silanized Surface

1. Incubate freshly activated epoxy-silane-functionalized slides with 2 % (w/v) of aminophenylboronic (AMPB) for 30 min at RT and with mild stirring in a buffer containing 1 % (v/v) acetonitrile in phosphate buffered saline (PBS).

2. Once activated, rinse all slides with distilled water.
3. Dry under a filtered forced air stream.
4. Store in a dried chamber containing silica gel packs until further use. The results of using different concentrations of AMPB (0.2, 1, 2, 5, 10 %) are shown in Fig. 3c.

E. Metal–Ligand Chelate Silanized Surface

1. Incubate freshly activated epoxy-silanized glass slides during 30 min at RT with a solution (0.2, 1, 5, or 10 %) of iminodiacetic acid (IDA) (pH 10.0).
2. Wash with distilled water.
3. Incubate the IDA-activated slides overnight with 10 % (w/v) of $CuSO_4$.
4. After incubation, rinse with distilled water.
5. Dry under a filtered forced air stream.
6. Store the slides in a dried chamber containing silica gel packs until further use. The results of using different concentrations of IDA (0.2, 1, 2, 5, 10 %) are shown in Fig. 3d.

4 Notes

1. The pins for printing must be created with a precise diameter in which the tolerances of the pin heads are in micron and sub-micron. These pins are usually made from stainless steel, titanium, or tungsten.
2. While the pin printing method is suitable for printing many proteins onto the microarrays, some other proteins are not suitable due to their hydrophobic and drying conditions on the array.
3. Large number of pins in the robotic heads is usually accommodated in a pin holder with oversized pin holes that allow the smooth insertion and convenient removal of the pins. However, the problems result from the oversized pin holes are wobbly pins and variation of actual printing positions. The contact of these wobbly pins with the surface also introduces imprecision; since two or more pins are not aligned to contact the surface at the same time as all others in the head, several spots will not print correctly (or not at all) in that particular row or column, leading to array printing errors.
4. The robotic pin head must travel back and forth between the deposition sites, the source plate and the cleaning reservoir. The required time may be reduced by increasing the number of pins and associated fluidics channels interfaced on the print head, which will significantly increase printer complexity and costs.

5. These pin materials, their respective surface finish/treatments, their cleanliness under simple rinse conditions, and the chemical composition of the print solution may affect the solution capillarity, and therefore the final microarray print quality.
6. Careful control of ambient humidity is crucial to spot desiccation, spot morphology, and subsequent image analysis. The deposited microspots may exhibit variable dried morphologies based on environmental conditions. Additionally, a certain risk of reagent evaporation is present even before the pin contacts the surface. The environmental conditions of a typical pin printing setup are usually controlled by enclosing the entire arraying system in a humidity-controlled environmental chamber, also acting to limit ambient dust and atmospheric contamination on the array substrate.

   Compressed air (2–3 psi and filtered) is used to remove excess liquid and dust from the functionalized glass surfaces.
7. Drying of proteins can result in destabilization of secondary and tertiary structures, causing a change or loss of protein functionality.
8. Proper protocols must be used for reagent and plate handling, fluidics, pin loading, printing, and drying to ensure that ambient contamination and air-borne particles are not introduced onto printed substrates. Low signal-to-noise ratios due to factors such as complex surface capture protocols, contamination, and static or no-flow mass transport conditions must be avoided. Poor printing of proteins (i.e., antibody), storage, and shelf-life are issues that should be considered in designing the proper microarray protocol. Nonoptimal protocol conditions such as pH, temperature, or drying, which promote variability in assay kinetics, must be avoided.
9. Epoxy activated glass slides should be prepared freshly for further chemical modifications, especially when epoxy groups are used as precursor of hetero-functional surface activation.
10. Once activated, the slides should be stored for no more than 3 months in order to obtain the expected results.

Acknowledgments

We gratefully acknowledge financial support from the Carlos III Health Institute of Spain (ISCIII, FIS PI11/02114, FIS14/01538, Fondos FEDER-EU) and Junta Castilla-León SA198A12-2. The Proteomics Unit belongs to ProteoRed, PRB2-ISCIII, as supported by grant PT13/0001. P. Díez is supported by a JCYL-EDU/346/2013 PhD scholarship. J. Casado-Vela is a JAE-DOC (CSIC) holder supported by Ministerio de Economía y Competitividad, Spain, co-funded by the European Social Fund (Fondos FEDER).

References

1. Vaisocherová H, Zhang Z, Yang W et al (2009) Functionalizable surface platform with reduced nonspecific protein adsorption from full blood plasma-Material selection and protein immobilization optimization. Biosens Bioelectron 24:1924–1930
2. Valiokas R (2012) Nanobiochips. Cell Mol Life Sci 69:347–356
3. Romanov V, Davidoff SN, Miles AR et al (2014) A critical comparison of protein microarray fabrication technologies. Analyst 139: 1303–1326
4. Casado-Vela J, González-González M, Matarraz S et al (2013) Protein arrays: recent achievements and their application to study the human proteome. Curr Proteomics 10: 83–97
5. Tsarfati-Barad I, Sauer U, Preininger C et al (2011) Miniaturized protein arrays: model and experiment. Biosens Bioelectron 26: 3774–3781
6. Gonzalez-Gonzalez M, Bartolome R, Jara-Acevedo R et al (2014) Evaluation of homo- and hetero-functionally activated glass surfaces for optimized antibody arrays. Anal Biochem 450:37–45
7. Biebricher A, Paul A, Tinnefeld P et al (2004) Controlled three-dimensional immobilization of biomolecules on chemically patterned surfaces. J Biotechnol 112:97–107
8. Jonkheijm P, Weinrich D, Schröder H et al (2008) Chemical strategies for generating protein biochips. Angew Chem Int Ed 47: 9618–9647
9. Tien J, Xia Y, Whitesides GM (1998) Microcontact printing of SAMs. Thin Films 24:227–250
10. Cras JJ, Rowe-Taitt CA, Nivens DA et al (1999) Comparison of chemical cleaning methods of glass in preparation for silanization. Biosens Bioelectron 14:683–688
11. Krasnoslobodtsev AV, Smirnov SN (2002) Effect of water on silanization of silica by trimethoxysilanes. Langmuir 18:3181–3184
12. Kyprianou D, Guerreiro AR, Chianella I et al (2009) New reactive polymer for protein immobilisation on sensor surfaces. Biosens Bioelectron 24:1365–1371

Chapter 15

Epitope Mapping Using Peptide Microarray in Autoantibody Profiling

Sebastian Henkel, Robert Wellhausen, Dirk Woitalla, Katrin Marcus, and Caroline May

Abstract

The use of peptide microarrays for epitope mapping of autoantibodies greatly facilitates the early diagnosis of allergic, cytotoxin-associated diseases and especially inflammatory diseases. A common approach to create the microarrays utilizes nitrocellulose-coated glass slides for peptide probe binding, which is based on surface adsorption. Advantages of this method include excellent peptide binding capacity and long-term stability. To ensure equal accessibility to all antibodies on the peptide microarray during epitope mapping, all probes are immobilized in a random manner, thus avoiding concentration-dependent effects on signal intensity.

In this chapter, we provide a step-by-step protocol on how to construct the peptide microarrays and perform epitope mapping of autoantibodies using them. Finally we present a comparative approach for the evaluation of the data.

Key words Peptide microarray, Autoantibody profiling, Epitope mapping, Nitrocellulose-coated glass slides

1 Introduction

Peptide and protein microarrays are effective time-saving methods for the simultaneous analysis of greater than 20,000 different biomolecules in a nearly automated manner. This high-throughput method has been used for the identification of new drugs and biomarkers [1]. Numerous applications are available and they can be used to address many scientific questions. Typical applications include enzymatic activity determination and pathway mapping. Furthermore, one of the more prominent applications is immune response profiling, particularly in the field of autoimmune disease diagnostics, to evaluate biomarker efficacy as well as in vaccine design [2, 3]. A particular application is epitope mapping which is able to identify the polyclonal antibody repertoire elicited by

Paul C.H. Li et al. (eds.), *Microarray Technology: Methods and Applications*, Methods in Molecular Biology, vol. 1368, DOI 10.1007/978-1-4939-3136-1_15, © Springer Science+Business Media New York 2016

vaccination [4], produced due to the immunogenicity of cytotoxins [5] and of allergens [6], and/or produced in neuronal diseases [3] and in inflammation [7, 8].

A key mechanism of the demonstrated microarray approach is the antibody–antigen reaction that is similar to the one used in Western blot analysis and that is based on the unique recognition of an antigen (i.e., a peptide) by its specific antibody. We used a non-contact micro-dispensing instrument to spot the peptide solutions onto a nitrocellulose-coated slide. A major advantage of peptide microarrays in comparison to the commonly used Western blot analysis is the direct immobilization of the peptides to the nitrocellulose surface [9]. In contrast to interaction studies wherein a controlled orientation of the peptides is preferred [10], in immune response profiling, a randomly oriented environment of peptides is preferred to facilitate equal access to functional domains for the antibodies [9]. All possible immunogenic epitopes and various structures are represented by a uniform distribution of peptides on the nitrocellulose surface [11]. We describe the application of antibody-containing human serum to a nitrocellulose-coated peptide microarray for the identification of immunogenic peptides. The major advantages of this approach include the relatively high load of peptide probes in each spot and the long-term stability of the probes [12].

We describe the major steps for the successful epitope mapping of a polyclonal antibody using the amino acid sequence of a corresponding immunogenic protein that has been divided into peptides with a length of 21 residues and an overlap of 16 residues. The peptides were synthesized and subsequently lyophilized for storage. The subsequent steps consist of: (a) sample preparation with buffer handling, in which the peptides are dissolved; (b) the spotting process, which includes the transfer of peptides onto the nitrocellulose membrane; (c) the incubation of the microarrays with human serum, and (d) data acquisition (Fig. 1).

The epitope mapping method described in this chapter includes general recommendations from our experience [3, 13] and notes for quality assurance. In addition, we show the dynamic range of detecting the test epitope (Fig. 4). It should be noted that this protocol varies between proteins and peptides and must be optimized for each application and biomolecule.

2 Materials

2.1 Blood Sampling

1. Serum separator tubes (9 mL).
2. Safety syringe.
3. Safety holder to connect the syringe to the tube.
4. Needle and blade waste container.

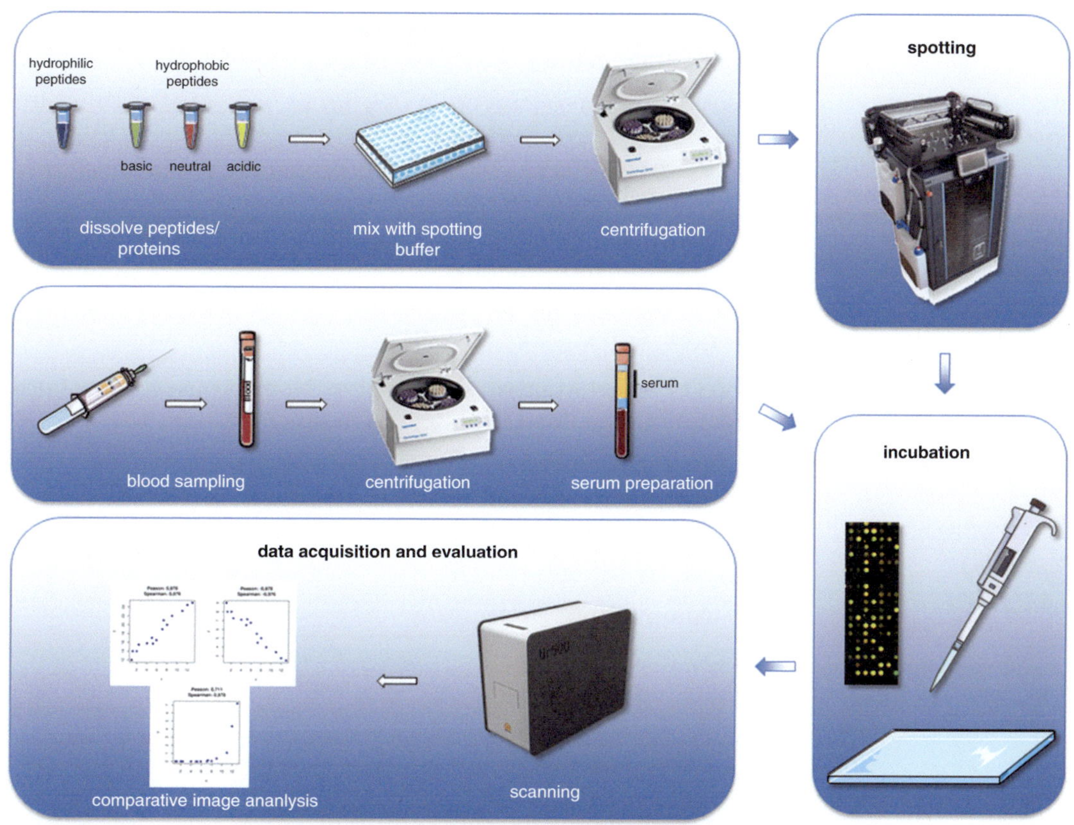

Fig. 1 Overview of the epitope mapping workflow using nitrocellulose-coated microarray slides and a non-contact micro-dispensing instrument. (1) The peptides and proteins are dissolved based on their physico-chemical characteristics. (2) Simultaneously, the collection of the desired sera for a comparative study should be completed. (3) After spotting, the microarrays are ready for incubation with the sera and subsequent incubation with a fluorescently labeled secondary antibody. (4) The final step consists of data acquisition to detect and measure the fluorescence signals resulting from antibodies produced during the immune response and data evaluation to compare the different human subject groups (i.e., diseased vs. control)

2.2 Serum Preparation

1. 1 mL cryogenic tubes.

2.3 Peptide and Protein Preparation

1. 0.1 M (w/v) ammonium bicarbonate.
2. >99.0 % (v/v) trifluoroacetic acid.
3. PBS (phosphate-buffered saline, 10×), pH = 7.4; without calcium chloride and without magnesium chloride.
4. >99.0 % (v/v) *N*,*N*-dimethylformamide.
5. 100 % (w/v) glycerol.
6. 100 % (v/v) Triton X-100.

2.4 Control Samples

1. Control spot antibody (Alexa Fluor 555 Goat Anti Rabbit IgG, Life Technologies, Darmstadt).
2. Positive control antigen (HA-epitope extended with neutral amino acids and present as the following amino acid frequency SNCYPYDVPDYASLRSLVASS, Peps4Life, Heidelberg).

2.5 Peptide Array Preparation

1. Non-contact micro-dispensing instrument (instrumentONE; M2-Automation, Berlin).
2. Piezo-element-driven micro-dispensing nozzle (instrumentONE; M2-Automation, Berlin).
3. PATH® plus Microarray Ultrathin Nonporous Nitrocellulose Slides (Grace Bio-Labs, Bend, Oregon).
4. Glass slides coated with water-sensitive paper.

2.6 Peptide Array Processing

1. Circular Microtiter Plate (MTP) shaker.
2. Table centrifuge.
3. Vacuum pump.
4. Fluorescence scanner.
5. Four-cell tray with lid.
6. Forceps.
7. 500 mL vacuum filtration unit containing a filter disk with a pore size of 0.45 μm (Filtropur V50 500 ml, 0.45 μm, Sarstedt, Nümbrecht).
8. Polyacetal rack.
9. Polyacetal trough.
10. Secondary antibody (Alexa Fluor® 647 Goat Anti-Human IgG, Life Technologies, Darmstadt).
11. Synthetic block 10× (from Invitrogen™).
12. Blocking buffer (50 mM HEPES, pH = 7.5; 25 % (w/v) glycerol; 0.08 % (w/v) Triton X-100; 200 mM NaCl; and 20 mM reduced glutathione).
13. 1 M (w/v) dithiothreitol.
14. Tween 20 (50 % (w/v) solution).

2.7 General

1. Laminar flow hood, HEPA filtered and UV sterilized.
2. Safety workbench.
3. Centrifuge, swinging bucket rotor, adjustable acceleration and braking, cooling option.
4. Containment for biowaste.
5. pH meter.
6. Water squirt bottle.

7. 0.1–2.5, 0.5–10, 2–20, 20–200, 100–1000, and 500–5000 μL micropipettors.
8. 2.5 μL, 10 μL, 20 μL, 200 μL 1000 μL, and 5 mL sterile micropipette tips.
9. Sterile 15 and 50 mL conical centrifuge tubes.
10. Stand for centrifuge tubes.
11. 0.5 mL microtubes.
12. 384-well microtiter plates.
13. Ultrapure deionized water.
14. ≥96 % isopropyl alcohol (technical grade).
15. Disinfectant.
16. Ice box.
17. 4 °C refrigerator, −20 °C freezer, −80 °C freezer.

3 Methods

3.1 Blood Sampling

1. Perform blood withdrawal according to local legislative regulations by authorized medical personal only.
2. Assemble 4×9 mL serum separator tubes with safety syringes using appropriate adapters (*see* **Note 1**).
3. Fill each tube with blood obtained by venipuncture of a peripheral vein (*see* **Note 2**).
4. Slowly invert tubes three times to ensure homogenous mixing of blood and uniform coagulation (*see* **Note 3**).

3.2 Serum Preparation

1. Centrifuge the serum tubes at 3000×*g* for 10 min at room temperature.
2. Label the cryo tubes with a unique serial identification number.
3. Prepare 500-μL serum aliquots in cryo tubes using micropipettes under a safety workbench and freeze them immediately.
4. Aliquots are stored at −80 °C. Avoid freeze–thaw cycles (*see* **Note 4**).

3.3 Peptide and Protein Preparation

Samples are prepared from lyophilized synthetic peptides using a total quantity of 300 μg. The peptide sequences should partially overlap to cover all possible sequences and ensure that potential epitopes are not separated. We used peptides consisting of 21 residues with an overlap of 16 residues.

1. Subdivide all lyophilized peptides into four groups according to their solubility and isoelectric characteristics (*see* **Note 5**).
2. The first group includes all water-soluble peptides, regardless of isoelectric properties (WS-group). Pipette 360 μL of water

into a microtube with 300 μg of peptide, and pipette an additional 40 μL of PBS (10×) into the tube (*see* **Note 6**).

3. The second group includes all insoluble peptides with an acidic pH value. Pipette 40 μL of PBS (10×) into a microtube with 300 μg of peptide. If the peptide does not dissolve, add 5 μL of 0.1 M ammonium bicarbonate followed by water giving a total volume of 400 μL (*see* **Note 6**).
4. The third group includes all insoluble peptides with basic pH values. Use 5 μL of 100 % TFA to dissolve these peptides, and add 355 μL of water and 40 μL of PBS (10×) to each microtube (*see* **Note 6**).
5. The fourth group includes all insoluble peptides with neutral pH values. Dissolve these hydrophobic peptides with 5 μL of 100 % DMF (*see* **Note 7**). Pipette an additional 355 μL of water and 40 μL of PBS (10×) into each microtube (*see* **Note 6**).
6. Prepare 5 mL of peptide spotting buffer, which is a mixture of 3.75 mL of water, 1.238 mL of glycerol and 12 μL of Triton X-100 (*see* **Note 8**).
7. Equimolar concentrations of 0.75 μg/μL should be used for all peptides. Subsequently, add 50 μL of spotting buffer into each microtube.
8. Pipette 25 μL of each peptide into each well in a 384-well microtiter plate. Prepare a table that includes the information for the peptide present in each well (*see* **Note 9**). This information can be subsequently uploaded for analysis.
9. Centrifuge the microtiter plate at 1000 × *g* for 10 s, and seal the plates with an airtight seal until subsequent analysis.
10. Store the microtiter plates at 4 °C if they will be analyzed the following day. Otherwise, the plates can be stored at −20 °C for approximately 3 months.

3.4 Control Samples

For quality assurance, control samples should be included within the peptide array.

1. Negative control samples should consist of all of the different solutions representing the different buffers used for peptide dissolution. Perform the various preparation steps (*see* Subheading 3.3) without the peptides, and mix the solutions with the spotting buffer. If there is no contamination, the spots should not exhibit a fluorescence signal on the final peptide array after all incubation steps.
2. Use the HA-epitope as a positive control sample. Dissolve this antigen and immobilize it on the slide in a similar fashion as the peptides (*see* **Note 10**).

3. Immobilize the control spot antibody, which is labeled with a fluorescent probe that is monitored at a different wavelength than the secondary antibody for incubation (*see* Subheading 3.6, **step 12**), in a similar fashion as the peptides (*see* Subheading 3.3). The spots obtained from this sample serve as guide dots to ensure correct array alignment during the entire assay process.

3.5 Peptide Array Preparation

The peptide arrays are prepared using a piezoelectric non-contact micro-dispensing instrument that is able to dispense liquid droplets with a volume of 20–100 pL (*see* **Note 11**).

1. Clean the entire system and rinse the hoses with isopropyl alcohol. The entire process should be performed under low-dust and high-purity conditions (*see* **Note 12**).
2. Place the microtiter plate with the samples (peptides) in the MTP holder of the circular MTP shaker. Shake at 150 rpm for 5 min.
3. Adjust the micro-dispensing parameters according to the physicochemical characteristics of the peptide solutions in contrast to water with the software provided by the manufacturer. A longer pulse duration (μs) and higher pulse amplitude (voltage) may be necessary for solvents, which are more viscous than water (*see* **Note 13**).
4. A washing step, which is performed after each new biomolecule has been spotted, is established via the software. Dispense 100 μL of solvent and 200 droplets, and wash the nozzle for 3 s at the wash station. The external wash consists of a 6-s wash at the wash station and a 4-s pause interval. Individualized parameters for washing are also possible (*see* **Note 14**).
5. Pre-dispense 150 droplets to discard any liquid that gets into the nozzles due to the washing fountain.
6. Use a nonporous nitrocellulose film-coated glass slide with low fluorescence for improved detection at a high signal-to-noise ratio (*see* **Note 15**). Place the slides in the slide holder. The first and last slides are coated with water-sensitive paper. These slides are important for a quick quality assurance. After the spotting step, an indicator for the complete spotting process is the visualization of these slides. However, spotting of a loading control of all spotted peptides labeled with an antibody or fluorescent dye is strongly recommended.
7. On the software, load the content table and create a grid by enabling various fields with blocks, which are indicated by guide dots, to structure the microarray. This step facilitates dot orientation after scanning. Furthermore, choose a randomized setting of triplicates for each peptide (*see* **Note 16**).

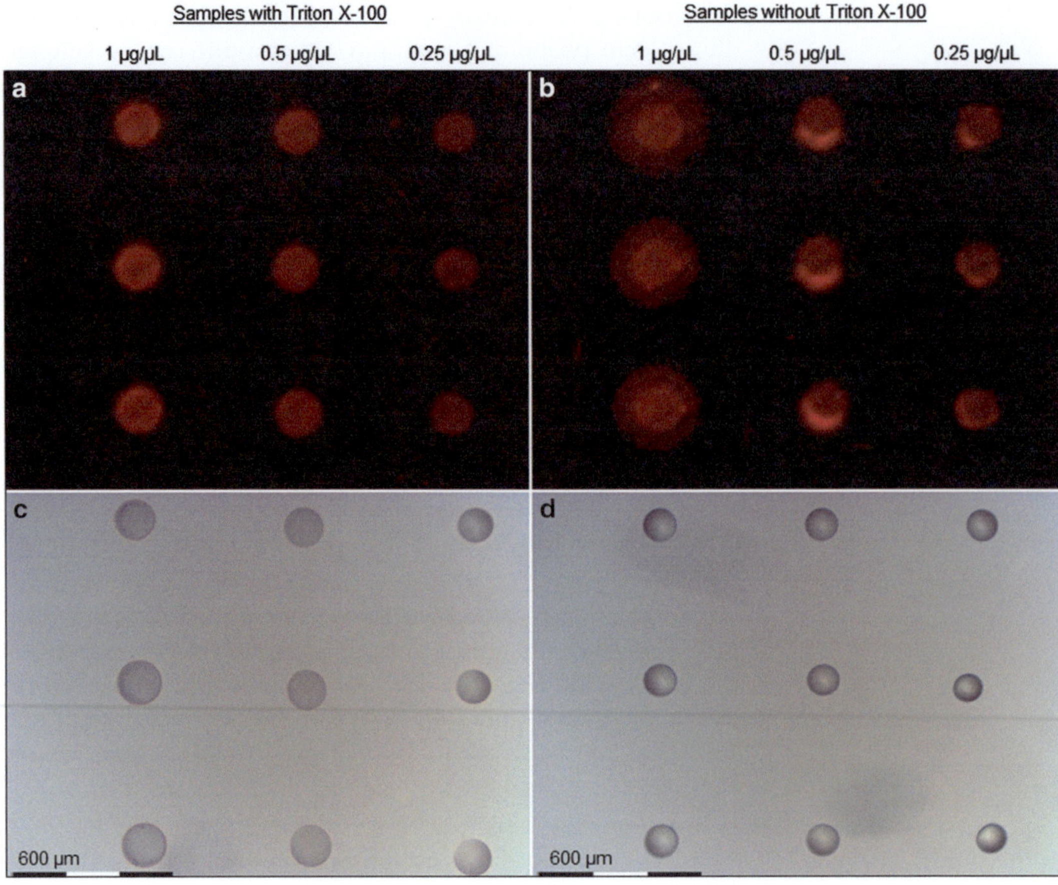

Fig. 2 Comparison of a water-soluble peptide with 0.08 % Triton X-100 (**a** and **c**) and a hydrophobic peptide in buffer without Triton X-100 (**b** and **d**). The peptides are spotted in triplicate at concentrations of 1, 0.5, and 0.25 µg/µL. Part(**a** and **b)** show the fluorescence signals detected with a high sensitivity interline charge-coupled scanner at a resolution of 10 μm^2/pixel with a dynamic range larger than 12-bit and an exposure time of 10,000 ms. Part(**c** and **d**) show the spots prior to incubation using a microscope with fivefold magnification. In (**a**), homogenous signal intensity from the spots due to the presence of Triton X-100 in the spotting buffer is shown. Enhanced adsorption of the spots onto the nitrocellulose membrane is also apparent in (**c**) in comparison with (**d**). The light microscope image without Triton X-100 (**d**) shows raised spots because of higher surface tension compared with the samples containing Triton X-100 (**c**). The *spots* in (**b**) exhibit a halo-shaped structure and an inhomogeneous fluorescence signal; this is often observed in samples without detergent and derives from high surface tension with subsequent irregular drying from the inside to the outside [10]. Within the spot, immobilization and accumulation of the biomolecules is favored at the outer rim [2]

8. Adjust the spacing distance between the spots to 600 µm (*see* **Note 17**). A typical image of these spots dispensed on a slide is shown in Fig. 2.
9. When the settings are complete, begin a test run by using water or a cheap test peptide prior to using the samples to produce the arrays. If there are any errors in the automated workflow, the wastage of the samples is prevented (*see* **Note 18**).

10. If necessary, optimize the parameters and produce the peptide arrays under standardized conditions (low-dust environment, room temperature and 40–50 % humidity).
11. If some samples could not be spotted, they can be re-spotted as long as the array slide has not been moved.
12. Dry the spotted microarray slides for 48 h at room temperature under the dust-free workbench and do not move them. Then, the arrays can be stored at –20 °C for 3 months.

3.6 Peptide Array Processing

The following peptide array processing procedures have been modified from the instructions from Invitrogen™ for ProtoArray® Human Protein Arrays and the modified procedures have been previously tested in various applications for immune profiling [3, 14–16]. First, prepare the blocking buffer and washing buffer solutions. The step-by-step instructions provide an example of how to incubate four peptide arrays.

1. All reagents and vessels should be cooled to 4 °C prior to use.
2. Prepare 50 mL of blocking buffer working solution by mixing 5 mL of synthetic block solution (10×) and 45 mL of blocking buffer (*see* **Note 19**) and 0.1 mL of 1 M DTT.
3. Prepare washing buffer as a mixture of 60 mL of PBS (10×), 3 mL of Tween 20 (10×), 60 mL of synthetic block (10×), and deionized water to a total volume of 600 mL.
4. Filter the washing buffer using the vacuum filtration unit together with a pump to remove lint by vacuum suction.
5. Thaw the self-made peptide arrays at 4 °C for 15 min, then at room temperature for an additional 15 min. During this time, cool the required volume of deionized water at 4 °C (*see* **Note 20**).
6. Scan the arrays at the wavelength at which the control sample spots fluoresce prior to incubation with antibodies (*see* **Note 21**).
7. Place the arrays in the tray (*see* **Note 22**), and pipette 5 mL of the prepared blocking buffer into each tray chamber.
8. Incubate the arrays with the blocking buffer at 4 °C for 60 min with horizontal rotation using the circular MTP shaker (150 rpm). During this time, dilute the human serum with washing buffer at a ratio of 1:250 (*see* **Note 23**).
9. Remove the blocking buffer carefully and rapidly using a vacuum pump. Pipette 5 mL of washing buffer into the chamber, and incubate the array for 5 min at 4 °C with horizontal rotation (*see* **Note 24**).
10. Remove the washing buffer and add the diluted serum to each chamber. Incubate the arrays for 90 min at 4 °C with horizontal rotation.

11. Remove the sera and pipette 5 mL of washing buffer into each chamber. Incubate the arrays for 5 min at room temperature, and repeat this washing step for a total of five times.
12. Dilute the secondary antibody at a ratio of 1:2000 with washing buffer, and pipette 5 mL into each chamber. Incubate the arrays with these antibody for 90 min at 4 °C with horizontal rotation.
13. Remove the antibody and repeat the five washing steps (*see* Subheading 3.6, **step** 7).
14. Perform the last washing step with water. The microarrays will be sequentially sorted into a prepared rack that is located in a watertight box (*see* **Note 25**).
15. Place the racks into dry boxes without lids and centrifuge at 3000 × *g* for 2 min.
16. After centrifugation, take images of the microarrays, and digitalize the different images directly for optimal comparability of the results using a fluorescence scanner (*see* **Note 26**).
17. After scanning, the arrays may be stored for a longer time under a modified atmosphere (argon) in sealed opaque pouches at 4 °C.
18. Import the data into R (a free software environment for statistical computing) *via* Bioconductor using the Limma package, and choose cyclic local regression (cyclic loess) as the normalization method to preprocess the arrays (*see* **Note 27**). Subsequent selection and sorting of peptide candidates can be performed based on the "minimum M-Statistic" *p*-value ("*M*-score" [13, 17]).

4 Notes

1. The appropriate blood sampling system/adapters should be chosen by authorized and experienced medical personal.
2. Coagulation times may differ according to the instruction manual for the serum tubes.
3. Ensure that the blood is coagulated and non-hemolytic.
4. Although human serum can be frozen for years without a significant loss in antibody activity [18], for a comprehensive analysis, the subsequent incubation of serum on the peptide microarray should be performed over a short period of time under standardized conditions (particularly, temperature and humidity).
5. If the information is not available for the peptides, use a free peptide property calculator (e.g., Innovagen) to obtain biophysical properties and an estimation of the water solubility.

If possible, use water to dissolve the peptides. In most cases, peptides are readily soluble in water; however, experimental verification may be required.

6. We recommend the use of low bind tubes (e.g., Eppendorf LoBind Tubes), to prevent loss of peptide by sample-to-surface binding. If the quantity of the peptides is not known, perform an analysis using UPLC. Adjust all concentrations to 0.75 μg/μL.
7. Use glass as the reaction container, and do not store DMF-containing liquids in polymeric containers. DMF is a polar organic solvent that dissolves polyacrylic fibers, vinyl resins and polyacetylene. DMF also causes first degree burns upon short exposure and it is toxic to the liver [19].
8. Triton X-100 is required to reduce the surface tension of water to prevent annular (or donut-shaped) inhomogeneous fluorescent spots (*see* Fig. 2). Glycerol is primarily used to increase physical adsorption, and hence, to increase the binding capacity [20].
9. The template of the table with the peptide information depends on the software used.
10. There are various options in the selection of positive controls. The specified option is only a technical control to verify that the incubation was properly performed and that all solutions and antibodies function well. Different peptides with physicochemical properties that are similar to those of the sample peptides (particularly in their affinity for the antibody) may be used as positive controls. However, this control approach is difficult to implement for an immune response profile because of the magnitude of interindividual differences. Furthermore, there are also intra-individual fluctuations in antibody titer [21]. Therefore, this positive control approach is more suitable for peptide–protein and protein–protein interactions.
11. The dispensed volume usually varies between different nozzles obtained from the manufacturer. Therefore, always make an individualized adaption of dispensing parameters according to the data sheet, which the nozzle included.
12. Dust or other particles that are visible after scanning may lead to false-positive results.
13. The micro-dispensing parameters must be adjusted according to the viscosity of the spotting buffer and samples. All samples should be handled using identical global parameters once the settings have been tuned. If the characteristics of the peptides substantially differ, individual parameter settings are also possible. Figure 3 shows the perfect trajectory of a drop (stable size, linear and vertical trajectory without satellite droplets). Note that these parameters must always be evaluated in two mutually orthogonal planes, to ensure a three-dimensional assessment of the mentioned drop morphology and trajectory.

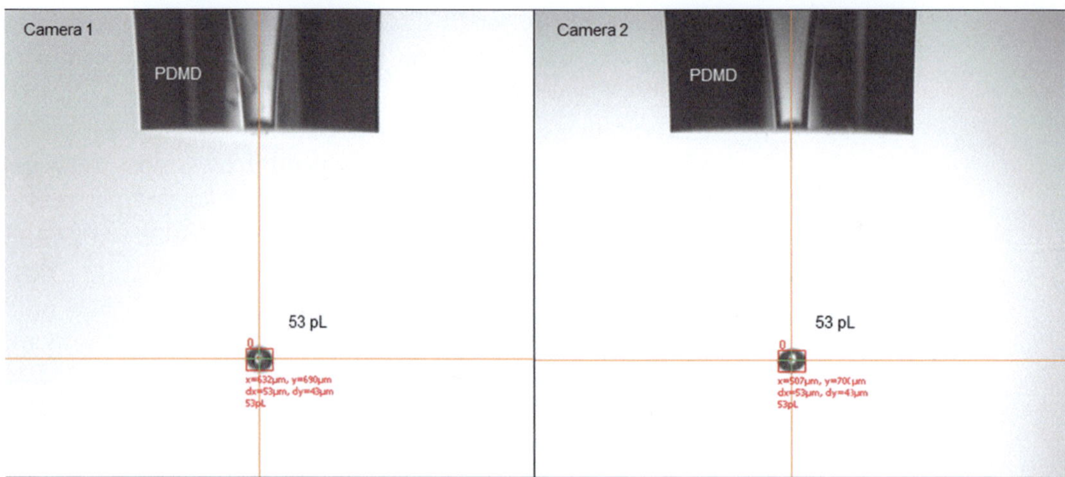

Fig. 3 Sample trajectory of a 53-pL droplet during micro-dispensing recorded with two orthogonally oriented cameras. The droplet was formed using the following settings of a piezo-element-driven micro-dispensing (PDMD) nozzle: pulse duration = 10 μs, voltage = 87 V, frequency = 200 Hz. The droplet volume may vary and depends on both the nozzle and the physicochemical properties of the solutions (peptides and buffer)

14. The wash should consist of a complete wash (i.e., a wash after the sample is spotted and an external wash conducted to prevent cross-contamination). Individualized parameters are necessary if you use peptides with specific physicochemical properties and particularly viscous buffers so that the nozzle does not get clogged.
15. Background noise is reduced when the slide surface is coated with ultrathin, nonporous dark nitrocellulose (Fig. 4) [22]. We recommend the use of PATH plus Microarray slides from Grace Bio-Labs.
16. The randomized settings are necessary to avoid cross-reaction and mixing of the various peptide samples.
17. When more droplets are dispensed, the distance should be carefully adapted to avoid bleeding and cross-reactivity. For instance, for the deposition of five droplets with a volume of approximately 80 pL, the center-to-center spacing is 600 μm.
18. A test run should be performed to verify that the micro-dispensing parameters apply to all of the samples, so that the wash is effective, so that the dispensing nozzle does not get clogged, and so that possible exceptions are spotted. In this occasion, possible exceptions are peptides with special or different physicochemical properties in contrast to the remaining peptides. Adjust the micro-dispensing parameters and save the settings for these peptides in your software for an automated production of microarrays.

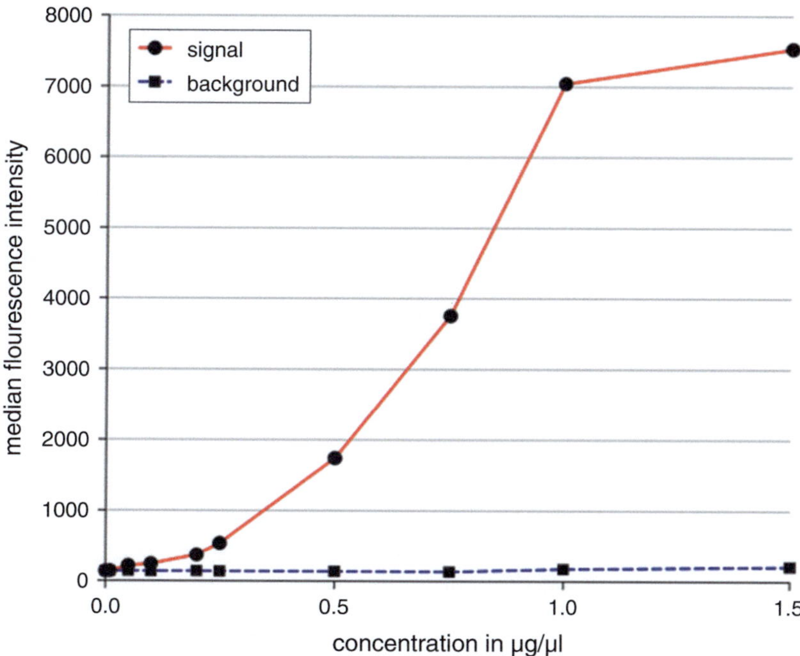

Fig. 4 Dynamic range of the test sample (modified HA-epitope). We probed five separate arrays with the modified HA-epitope. The fluorescence signal is obtained from the secondary antibody (Alexa Fluor® 647 Goat Anti-Human IgG) measured by a fluorescence scanner with a 635 nm wavelength laser diode. Graphs displaying median fluorescence signal of a dilution series of the modified HA-epitope ranging from 0.01 to 1.5 μg/μl and the corresponding background signal, in triplicates

19. The specified buffer has a low fluorescent background and it also protects the biomolecules on the surface of the array [20].
20. During the entire experiment, all materials and solutions should be cooled to 4 °C to ensure a standardized environment and to avoid rearrangements of the molecules (peptides) at high temperatures. Additionally, tighter binding of human IgG antibodies and peptides is observed at 4 °C [23, 24]. After the solutions have been prepared, the following guidelines for handling microarrays should be observed: (a) Use gloves in handling the microarray slide, and it should only be touched on the edge or on unspotted locations (never touch the spotted area!). (b) The microarray surface should only be pointed upwards. The only exception is during the scanning process, wherein the microarray orientation depends on the manufacturer.
21. The quality control step detects the guide dots and checks for impurities on the array. Prior to scanning, the scanner should be started 30 min to warm up.
22. After this step, the entire experiment will be performed under a clean beach. Solutions should not be directly pipetted onto the slide. Add solutions next to the array inside the chamber using only a 5 mL micropipette.

23. A total of 5 mL of diluted serum is required for each chamber of the tray. Therefore, mix 20 μL of the four different sera with 5 mL of washing buffer in four different 15 mL tubes.
24. The surface of the array must be wet throughout the procedure. The microarray should only be dried at the end of the incubation prior to scanning as described in the protocol (a short immersion in deionized water and subsequent centrifugation). This drying step ensures low background from solvents and artificial residues, which can increase the background signal.
25. Prevent drying of the spotted surface before and during microarray sorting by working quickly and maintaining the humidity at approximately 50 %.
26. Choose excitation wavelengths and emission filters according to the control sample spots and the secondary antibodies for detection. Scanner settings are as follows: detection: high sensitivity interline charge-coupled device; resolution: 10 μm^2/pixel; dynamic range: larger than 12-bit; and exposure time: 10,000 ms.
27. Limma [25] is one of the few R [26] packages that provides full non-Affymetrix single-channel microarray support including generic data objects, background correction, log transformation and three different normalization approaches (i.e., variance stabilization, quantile and cyclic loess normalization). This method is adopted by Jiang et al. [27] and described by May et al. [3] for highly immune-reactive IgG antibody selection of biomarkers in amyotrophic lateral sclerosis patients.

Acknowledgments

This work was supported by P.U.R.E. (Protein Unit for Research in Europe), which is a project of the federal state of Nordrhein-Westfalen, Germany, and by the Ministerium für Innovation, Wissenschaft und Forschung des Landes Nordrhein-Westfalen and WTZ Brasilien, which is a project of the German Ministry of Education and Research (BMBF), Germany (01DN14023).

References

1. Abel L, Kutschki S, Turewicz M, Eisenacher M, Stoutjesdijk J, Meyer HE, Woitalla D, May C (2014) Autoimmune profiling with protein microarrays in clinical applications. Biochim Biophys Acta 1844(5):977–987
2. Wellhausen R, Seitz H (2012) Facing current quantification challenges in protein microarrays. J Biomed Biotechnol 2012:831347
3. May C, Nordhoff E, Casjens S, Turewicz M, Eisenacher M, Gold R, Brüning T, Pesch B, Stephan C, Woitalla D, Penke B, Janáky T, Virók D, Siklós L, Engelhardt JI, Meyer HE (2014) Highly immunoreactive IgG antibodies directed against a set of twenty human proteins in the sera of patients with amyotrophic lateral sclerosis identified by protein array. PLoS One 9(2), e89596

4. Cohen O, Mechaly A, Sabo T, Alcalay R, Aloni-Grinstein R, Seliger N, Kronman C, Mazor O (2014) Characterization and epitope mapping of the polyclonal antibody repertoire elicited by ricin-holotoxin based vaccination. Clin Vaccine Immunol 21(11):1534–1540
5. Akada J, Okuda M, Hiramoto N, Kitagawa T, Zhang X, Kamei S, Ito A, Nakamura M, Uchida T, Hiwatani T, Fukuda Y, Nakazawa T, Kuramitsu Y, Nakamura K (2014) Proteomic characterization of Helicobacter pylori CagA antigen recognized by child serum antibodies and its epitope mapping by peptide array. PLoS One 9(8):e104611
6. Lin J, Bardina L, Shreffler WG, Andreae DA, Ge Y, Wang J, Bruni FM, Fu Z, Han Y, Sampson HA (2009) Development of a novel peptide microarray for large-scale epitope mapping of food allergens. J Allergy Clin Immunol 124(2):315–322, 322.e311–313
7. Ng JK, Malotka J, Kawakami N, Derfuss T, Khademi M, Olsson T, Linington C, Odaka M, Tackenberg B, Prüss H, Schwab JM, Harms L, Harms H, Sommer C, Rasband MN, Eshed-Eisenbach Y, Peles E, Hohlfeld R, Yuki N, Dornmair K, Meinl E (2012) Neurofascin as a target for autoantibodies in peripheral neuropathies. Neurology 79(23):2241–2248
8. Prüss H, Höltje M, Maier N, Gomez A, Buchert R, Harms L, Ahnert-Hilger G, Schmitz D, Terborg C, Kopp U, Klingbeil C, Probst C, Kohler S, Schwab JM, Stoecker W, Dalmau J, Wandinger KP (2012) IgA NMDA receptor antibodies are markers of synaptic immunity in slow cognitive impairment. Neurology 78(22):1743–1753
9. Price JV, Haddon DJ, Kemmer D, Delepine G, Mandelbaum G, Jarrell JA, Gupta R, Balboni I, Chakravarty EF, Sokolove J, Shum AK, Anderson MS, Cheng MH, Robinson WH, Browne SK, Holland SM, Baechler EC, Utz PJ (2013) Protein microarray analysis reveals BAFF-binding autoantibodies in systemic lupus erythematosus. J Clin Invest 123(12):5135–5145
10. Guo A, Zhu X-Y (2007) The critical role of surface chemistry in protein microarrays. In: Predki PF (ed) Functional protein microarrays in drug discovery. Drug discovery series, vol 8. CRC, Boca Raton, pp 53–72
11. Cretich M, Damin F, Pirri G, Chiari M (2006) Protein and peptide arrays: recent trends and new directions. Biomol Eng 23(2–3):77–88
12. Tonkinson JL, Stillman BA (2002) Nitrocellulose: a tried and true polymer finds utility as a post-genomic substrate. Front Biosci 7:c1–12
13. Turewicz M, May C, Ahrens M, Woitalla D, Gold R, Casjens S, Pesch B, Brüning T, Meyer HE, Nordhoff E, Böckmann M, Stephan C, Eisenacher M (2013) Improving the default data analysis workflow for large autoimmune biomarker discovery studies with ProtoArrays. Proteomics 13(14):2083–2087
14. Schweitzer B, Meng L, Mattoon D, Rai AJ (2010) Immune response biomarker profiling application on ProtoArray protein microarrays. Methods Mol Biol 641:243–252
15. Le Roux S, Devys A, Girard C, Harb J, Hourmant M (2010) Biomarkers for the diagnosis of the stable kidney transplant and chronic transplant injury using the ProtoArray® technology. Transplant Proc 42(9):3475–3481
16. Nguyen MC, Tu GH, Koprivnikar KE, Gonzalez-Edick M, Jooss KU, Harding TC (2010) Antibody responses to galectin-8, TARP and TRAP1 in prostate cancer patients treated with a GM-CSF-secreting cellular immunotherapy. Cancer Immunol Immunother 59(9):1313–1323
17. Sboner A, Karpikov A, Chen G, Smith M, Mattoon D, Dawn M, Freeman-Cook L, Schweitzer B, Gerstein MB (2009) Robust-linear-model normalization to reduce technical variability in functional protein microarrays. J Proteome Res 8(12):5451–5464
18. Linke RP (2012) Immunologische Techniken. In: Lottspeich F, Engels JW (eds) Bioanalytik, 3rd edn. Spektrum Akademischer Verlag, München, pp 77–124
19. Redlich CA, Beckett WS, Sparer J, Barwick KW, Riely CA, Miller H, Sigal SL, Shalat SL, Cullen MR (1988) Liver disease associated with occupational exposure to the solvent dimethylformamide. Ann Intern Med 108(5):680–686
20. Luttmann W, Bratke K, Küpper M, Myrtek D (2014) Der experimentator: immunologie. Spektrum Akademischer Verlag, Heidelberg
21. Isacson J, Trollfors B, Hedvall G, Taranger J, Zackrisson G (1995) Response and decline of serum IgG antibodies to pertussis toxin, filamentous hemagglutinin and pertactin in children with pertussis. Scand J Infect Dis 27(3):273–277
22. Balboni I, Limb C, Tenenbaum JD, Utz PJ (2008) Evaluation of microarray surfaces and arraying parameters for autoantibody profiling. Proteomics 8(17):3443–3449
23. Shopes B (1995) Temperature-dependent binding of IgG1 to a human high affinity Fc receptor. Mol Immunol 32(5):375–378

24. Dorrington KJ (1976) Properties of the Fc receptor on macrophages. Immunol Commun 5(4):263–280
25. Smyth GK (2005) Limma: linear models for microarray data. In: Gentleman R, Carey V, Dudoit S, Irizarry R, Huber W (eds) Bioinformatics and computational biology solutions using R and bioconductor. Springer, New York, pp 397–420
26. Rdc T (2011) R: a language and environment for statistical computing. R Foundation for Statistical Computing, Vienna, Austria
27. Jiang H, Deng Y, Chen HS, Tao L, Sha Q, Chen J, Tsai CJ, Zhang S (2004) Joint analysis of two microarray gene-expression data sets to select lung adenocarcinoma marker genes. BMC Bioinformatics 5:81

Chapter 16

Preparation of Glycan Arrays Using Pyridylaminated Glycans

Shin-ichi Nakakita and Jun Hirabayashi

Abstract

We describe the method to prepare neoglycoproteins from the conjugation of bovine serum albumin and pyridylaminated glycans. Large quantities of glycans (>1 mg) can be pyridylaminated and then converted to their 1-amino-1-deoxy derivatives by reaction with hydrogen followed by hydrazine. These pyridylaminated glycans can then be conjugated to bovine serum albumin via esterification with *N*-(*m*-maleimidobenzoyloxy)succinimide to form a neoglycoprotein, e.g., glycosylated bovine serum albumin. As a demonstration, we prepared High-mannose bovine serum albumin, which was immobilized on an activated glass slide. Then, we showed that the neoglycoprotein bind to Cy3-labeled *Lens culinaris* agglutinin, a mannose-specific plant lectin, as detected using an evanescent-field-activated fluorescence scanner system.

Key words Pyridylamination, 1-Amino-1-deoxy derivative, Neoglycoprotein, Glycan array, Evanescent-field-activated fluorescence scanner

1 Introduction

DNA and protein arrays have been widely used in genomic and proteomic investigations. For instance, DNA arrays were used for expression profiling, whereas protein arrays were employed for identification of drug targets [1, 2]. More recently, glycan arrays have been used in glycomic studies [3–7]. Glycans, which are sugars found on the surface of all cells, participate in various biological phenomena, e.g., microbial infection, immune responses, signaling for cell differentiation, and tumor-cell metastasis [8–10]. Such glycomic studies require large quantities (mg amounts) of a structurally defined glycan array, and their preparation can be a costly and labor-intensive undertaking. To date, most glycans that have been used in glycan arrays were prepared by complicated organic synthetic methods [11–13]. Herein, we report a simple procedure to prepare a synthetically modified glycoprotein called

Paul C.H. Li et al. (eds.), *Microarray Technology: Methods and Applications*, Methods in Molecular Biology, vol. 1368,
DOI 10.1007/978-1-4939-3136-1_16,

Biomaterial
⬇ Hydrazinolysis (see 3.1 1-5)
⬇ Pyridylamination (see 3.1 12-20)
⬇ Anion exchange (see 3.1 21-23)
⬇ Reversed-phase (see 3.1 24-25)
⬇ Reversed-phase (see 3.1 26-27)
PA-glycan

Fig. 1 Flow diagram for purification of a PA-glycan from biomaterials

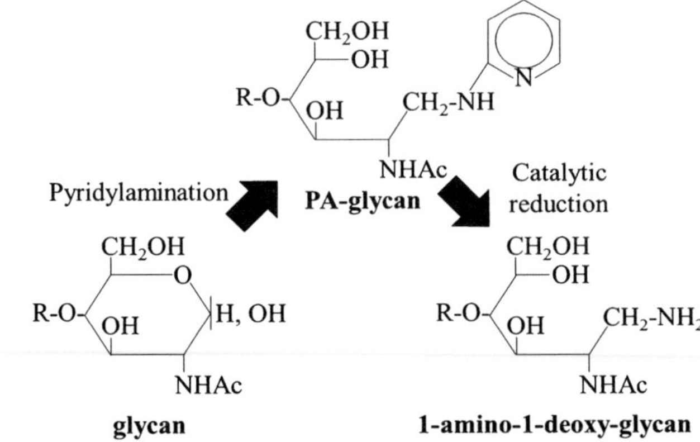

Fig. 2 Pyridylamination of a glycan and its conversion to 1-amino-l-deoxy-glycan

a neoglycoprotein (*see* Fig. 1). Large amounts of a specific glycan or mixture of glycans can be obtained by hydrazinolysis of an appropriate biomaterial, such as hen egg yolk [14–17]. After the glycan is converted to its pyridylaminated (PA) derivative (*see* Fig. 2), the purification of PA-glycan is carried out by conventional HPLC. After purification, the PA-glycan is converted to its 1-amino-1-deoxy derivative [18], which is then directly coupled to a protein carrier (*see* Fig. 3). As an example, bovine serum albumin (BSA) is used and it is conjugate to the PA-glycan via an appropriate bifunctional cross-linker, e.g., *N*-(*m*-maleimidobenzoyloxy) succinimide (MBS) to generate the neoglycoprotein, e.g., M8A-BSA. As a demonstration of this method, the glycan is bound to a lectin which is mannose-specific. We show that M8A-BSA can bind the (*see* Fig. 4) Cy3-labeled plant lectin, *Lens culinaris* agglutinin (LCA) as documented using an evanescent field-activated fluorescence scanner [19]. We have also studied influenza virus binding by using the appropriate neoglycoprotein (*see* Fig. 5).

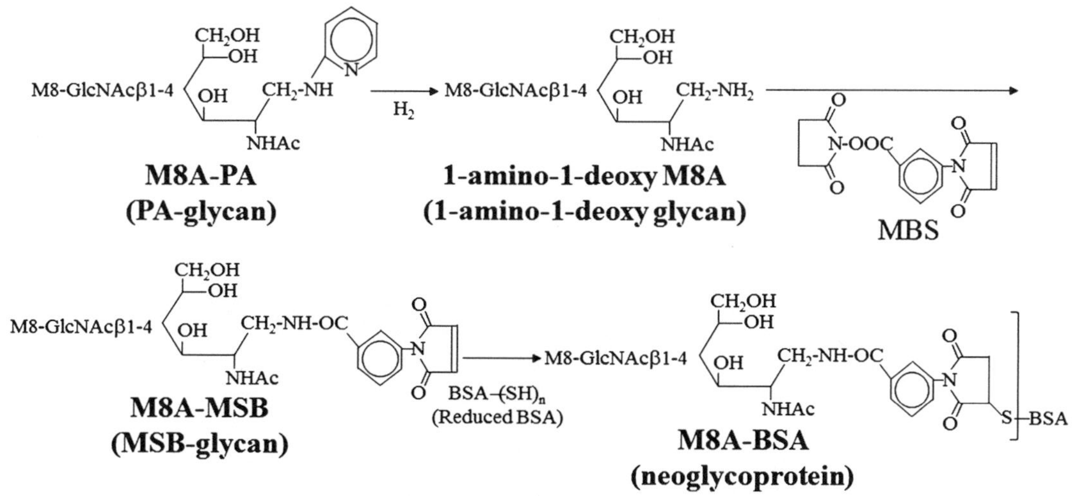

M8A-PA (PA-glycan) → (H_2) 1-amino-1-deoxy M8A (1-amino-1-deoxy glycan) → (MBS) M8A-MSB (MSB-glycan) → (BSA-(SH)$_n$ (Reduced BSA)) M8A-BSA (neoglycoprotein)

Fig. 3 Reduction of PA-glycan to 1-amino-l-deoxy-glycan, preparation of MSB-glycan, and its coupling to reduced BSA

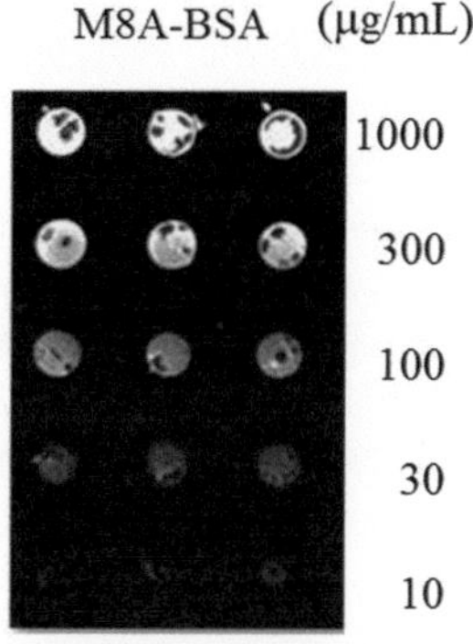

Fig. 4 Assay for the binding of biotinylated LCA on a glycan array immobilized with M8-BSA. Varying concentrations of M8-BSA (10–1000 μg/mL) were spotted on an epoxy-coated glass slide. Ligand coupling was measured using 10 ng/mL of biotinylated LCA and 1 ng/mL Cy3-labeled streptavidin. Fluorescent images were immediately acquired using the evanescent field-activated scanner

2 Materials

2.1 Preparation of PA-Glycans from Glycoproteins

1. Lyophilized biomaterials (e.g., glycoprotein or tissue).
2. Anhydrous hydrazine.
3. Screw-cap test tubes with a Teflon seal (10 × 100 mm: 7.85 mL, 20 × 250 mm: 78.5 mL).
4. Saturated sodium bicarbonate solution.
5. Acetic anhydride.

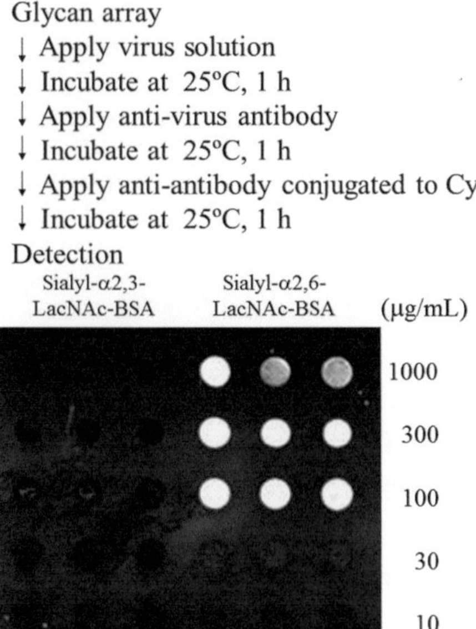

Fig. 5 Binding assay of influenza virus (H1N1 Narita 2009 strain) against sialyl-α-2,3LacNAc BSA and sialyl-α-2,6LacNAc BSA. Various concentrations of sialyl-α-2,3 and 2,6LacNAc-BSA (10–1000 μg/ml) were spotted into the wells of an epoxy-coated glass slide (6 nL/1 spot). H1N1 influenza virus (Narita 2009 strain) in binding buffer (2^2 hemagglutinin units) was applied to each of the arrays and incubated at 25 °C for 1 h. Then the arrays were incubated with anti-influenza virus rabbit antibody at 25 °C for 1 h, and then with anti-rabbit antibody conjugated to Cy3 at 25 °C for 1 h. Fluorescent images were immediately acquired using the evanescent-field-activated scanner. The H1N1 influenza virus bound only to sialyl-α-2,6LacNAc

6. Dowex cation exchanger 50 W-X2, (H^+ form; 200–400) (Bio-Rad Laboratories, Hercules, CA).
7. 2-Aminopyridine as the coupling reagent: 5.52 g of 2-aminopyridine dissolved in 2 mL acetic acid Borane-dimethylamine complex as the reducing reagent: 2 g of borane-dimethylamine complex dissolved in 0.8 mL of acetic acid and 0.5 mL of water.
8. Anionic exchange column Mono-Q HR 5/5 (0.5 × 5.0 cm, GE Healthcare, Pittsburgh, PA).

   Mono-Q elution system *A*: $ammonia_{(aq)}$, pH 9.0.
   Mono-Q elution system *B*: 500 mM ammonium acetate (adjusted to pH 9.0 with $ammonia_{(aq)}$).
9. Reversed-phase HPLC column Cosmosil 5C18P (1.0 × 25 cm; Nacalai Tesque, Kyoto, Japan).

   50 mM ammonium acetate (adjusted to pH 4.0 with acetic acid), 0.5 % (v/v) 1-butanol, 0.1 % (v/v) 1-butanol.

2.2 Conversion of a PA-Glycan to Its 1-Amino-l-deoxy Derivative

1. $Man_8GlcNAc_2$ (M8A), where Man represents mannose.
2. Sialyl-α-2,3-LacNAc and Sialyl-α-2,6-LacNAc.
3. Palladium black.
4. Hydrogen gas.
5. Thin layer chromatographic (TLC) plates.
6. Chromatographic column TSK-gel HW-40F (1.6×90 cm; Tosoh Co., Tokyo, Japan).
7. Ninhydrin.

2.3 Conjugation of the 1-Amino-l-deoxy PA-Glycan Derivative and BSA to Form a Neoglycoprotein

1. Bovine serum albumin (BSA).
2. *N*-(*m*-maleimidobenzoyloxy)succinimide (MBS).
3. $NaBH_4$.

2.4 Binding Assay Using the Neoglycoprotein

1. Silicone rubber sheet (7-chamber type) attached to an epoxy-coated glass slide (Rexxam Co., Ltd., Osaka, Japan).
2. Binding buffer: 25 mM Tris–HCl, pH 7.4, 0.8 % NaCl, 1 % (v/v) Triton X-100, 1 mM $MnCl_2$, $CaCl_2$.
3. Biotinylated *Lens culinaris* agglutinin (LCA).
4. Streptavidin-conjugated Cy3.
5. H1N1 influenza virus (Narita 2009 strain).
6. Anti-influenza virus rabbit antibody, anti-rabbit antibody conjugated to Cy3 (Cosmobio Co., Ltd., Tokyo, Japan).
7. Evanescent-field-activated scanner (Bio-Rex Scan 200; Rexxam Co., Ltd.).

3 Methods

3.1 Preparation of PA-Glycans from Glycoproteins

The following procedure has been reported in our previous work [15].

1. Place a lyophilized sample of a biomaterial, such as glycoprotein or tissue (<150 mg),at the bottom of a screw-cap test tube.
2. Add 5 mL of anhydrous hydrazine into the tube (*see* **Notes 1** and **2**).
3. Close the screw-cap.
4. Incubate the tube at 100 °C, 10 h.
5. Remove hydrazine in vacuo using a rotary pump connected to a glass cold trap.

6. Add 10 mL of saturated sodium bicarbonate solution and 0.4 mL of acetic anhydride.
7. Leave the tube at 0 °C for 30 min.
8. Add 25 g of cationic exchanger: Dowex 50W-X2 (H^+ form) to a beaker.
9. Mix the content of the tube to the cationic exchanger, and then pour the mixture into a glass column.
10. Wash the column with five volumes of water.
11. Collect the outflow that contains the free glycan. Then, concentrate and lyophilize it.
12. Perform pyridylamination by adding 0.5 mL of the coupling reagent: 2-aminopyridine (*see* **Note 3**) to the lyophilized glycan. After complete mixing of the reagents, heat the mixture at 90 °C for 1 h.
13. Perform reduction by adding 1.8 mL of the reducing reagent: borane-dimethylamine complex.
14. Heat at 80 °C, 35 min.
15. Add 0.9 mL of water.
16. Add 0.9 mL of phenol–chloroform (1:1 v/v) and mix well.
17. Centrifuge the solution at 200 × *g* for 1 min.
18. Carefully collect the upper (aqueous) phase and transfer it to a new screw-cap tube.
19. Repeat **steps 15–17** for two more times.
20. Recover the water phase and lyophilize the pyridylaminated glycan.
21. Dissolve the lyophilized sample in water and perform anionic exchange chromatography using a Mono-Q column.
22. Equilibrate the Mono-Q column with Solvent A at a flow rate of 1.0 mL/min, and column temperature at 25 °C. Inject the PA-glycan sample and elute using 100 % Solvent A (0 % Solvent B). Fluorescent detection was conducted using an excitation wavelength of 310 nm and an emission wavelength of 380 nm.
23. After 5 min., increase the proportion of Solvent B with a linear gradient from 0 to 10 % B at 10 min., to 30 % B at 22 min., and then to 100 % B at 27 min.
24. Collect fractions that contain PA-glycans based on the fluorescent detector signal.

    Perform reversed-phase HPLC using the Cosmosil 5C18P column use the elution solvent: 50 mM ammonium acetate (adjusted to pH 4.0 with acetic acid), 0.5 % (v/v) 1-butanol (*see* **Note 4**). Use a flow rate of 3.0 mL/min. and column temperature at 25 °C. Detection of PA-glycans was conducted using an excitation wavelength of 320 nm and an emission wavelength of 400 nm.

25. Collect the fractions that contain the PA-glycans.
26. Purify the PA-glycan of interest by performing a second-time of reversed-phase HPLC (*see* **Note 4**). The elution solvent is 50 mM ammonium acetate (adjusted to pH 4.0 with acetic acid) and 0.1 % (v/v) 1-butanol. Use a flow rate of 3.0 mL/min. and column temperature at 25 °C. Detection of PA-glycans was performed using an excitation wavelength of 320 nm and an emission wavelength of 400 nm.
27. Collect the fractions that contain the PA-glycans.
28. Lyophilize and store the PA-glycans.

3.2 Conversion of a PA-Glycan to 1-Amino-l-deoxy Glycan (See Fig. 2)

The following procedure has been briefly described in a previous publication [18].

1. Lyophilize >100 nmol of the PA-glycan in a test tube.
2. Add 5 mL of 0.1 % (v/v) acetic acid to the tube and mix well.
3. Add 5 mg of palladium black powder.
4. Reduce the PA-glycan by bubbling into the tube with hydrogen gas at atmospheric pressure and room temperature for 3 h (*see* **Note 5**).
5. Reduction of a PA-glycan can be monitored using thin layer chromatography (TLC) (*see* **Note 6**).
6. Filter the reaction mixture through a 0.22-μm Milex filter into a new test tube to remove palladium black (*see* **Note** 7).
7. Evaporate the filtrate in the tube to dryness to obtain a solid residue.
8. Add 1 ml of anhydrous hydrazine to the tube.
9. Seal the tube and heat the solution at 70 °C for 2 min.
10. Remove hydrazine by vacuo and dry the residue.
11. Add 0.5 mL of 10 mM acetic acid.
12. Purify the 1-amino-1-deoxyglycan using a HW-40F chromatography column. Elute with 10 mM acetic acid using gravitational flow and column temperature at 25 °C.
13. Test the fractions with ninhydrin and pool the ninhydrin-positive fractions and lyophilize.

3.3 Preparation of the Neoglycoprotein (See Fig. 3)

The following procedure is based on that found in a previous publication [20, 21].

1. Place >10 μmol of a lyophilized 1-amino-l-deoxy-glycan into a test tube (*see* **Note 8**).

2. Add 0.2 mL of 0.5 M sodium phosphate (pH 7.0) to the tube and mix well.
3. Add 50 mg of *N*-(*m*-maleimidobenzoyloxy)succinimide (MSB) which has been dissolved in 0.2 mL DMF to convert 1-amino-l-deoxy-glycan to MSB-glycan.
4. Heat at 30 °C for 30 min.
5. To remove unreacted MSB, extract the mixture in the tube with 0.2 mL of CH_2Cl_2 for three times.
6. Carefully collect the aqueous phase that contains the MSB-glycan and place it into a new test tube.
7. The MSB-glycan will be coupled to BSA to produce the neo-glycoprotein. BSA has been reduced to remove disulfide linkage(*see* **Note 9**).
8. Add the MSB-glycan solution (1 μmol/0.4 mL) dropwise into the solution containing reduced BSA (100 nmol BSA/1.24 mL).
9. Mix and incubate at room temperature for 2 h.
10. Dialyze the mixture extensively against 50 mM Tris–HCl, pH 7.4, 0.15 M NaCl (*see* **Note 10**).
11. Concentrate the sample to produce the neoglycoprotein at an equivalent of 2 mg BSA/mL.

3.4 Binding Assay Using the Neoglycoprotein

The following procedure is based on a previous report [12, 20, 21].

1. Attach the 7-chamber silicone rubber sheet on an epoxy-coated glass slide.
2. Individually, place a drop (6 nL) of the neoglycoprotein into the chambers on top of an epoxy-coated glass slide.
3. Place the slide into a humidity-controlled box at 25 °C for 15 h.
4. Wash the slide with the binding buffer three times.
5. Apply 80 μL of lectin solution into each chamber. For lectin assay, use 10 ng/mL of biotinylated LCA. For influenza virus assay, use 2^2 hemagglutinin units of H1N1 influenza virus.
6. Incubate at 25 °C, 1 h.
7. Remove the lectin solution from the chamber.
8. For LCA, apply 1 ng/mL of Cy3-conjugated streptavidin (80 μL) in the chamber on the slide, and incubate at 25 °C for 1 h.
9. For H1N1 influenza virus, incubate with anti-influenza virus rabbit antibody at 25 °C for 1 h, and then with anti-rabbit antibody conjugated to Cy3 at 25 °C for 1 h.

10. Remove the solution from the chambers. Fluorescent images were immediately acquired using the evanescent field-activated scanner (*see* **Note 11**).
11. The results for the lectin assay are shown in Fig. 4.
12. The data for the influenza virus assay is depicted in Fig. 5. The H1N1 influenza virus bound only to sialyl-α-2,6LacNAc (*see* **Note 12**).

4 Notes

1. Anhydrous hydrazine is a strong base/reducing reagent. Anhydrous hydrazine is highly explosive too (*see* **Note 2**). Care should be taken when handling the liquid. Safety equipment, e.g., goggles and gloves, should be worn, and the operation should be carried out in a fume hood.
2. Hydrazine monohydrate, which is not explosive, can be used in place of explosive anhydrous hydrazine to prepare mg amounts of *N*-glycans. For example, 0.58 mg of disialo-biantennary *N*-glycan (240 nmol) was prepared from 600 mg of hen egg yolk powder using 60 mL of hydrazine monohydrate [17].
3. The coupling reagent 2-aminopyridine is a very viscous liquid and it is not easy to mix it with glycans. To completely mix and dissolve a glycan in acetic anhydride, the full length of the reaction tube is warmed with a hot air blower with setting on High and the tube is also vortexed until the glycans are completely dissolved.
4. The first HPLC is performed to bulk separate a glycan mixture, whereas the second HPLC is conducted to separate individual glycans with good purity. In our experience, almost all *N*-glycans obtained from vertebrates can be separated in this manner.
5. Hydrogen gas is highly flammable in air under a wide range of hydrogen-air compositions. The work should be done in a fume hood.
6. Reduction of a PA-glycan can be monitored as the disappearance of the fluorescence due to the loss of pyridine. To achieve this, periodically take a drop of the reaction mixture and spot it onto a TLC plate and observe the fluorescence induced by UV irradiation.
7. Palladium black is a fine metal powder and it may easily ignite, particularly when dry. The powder should be kept under water when not in use.

8. The amount of glycan is adjusted so that the molar ratio of BSA to glycan is 1:10.
9. Before coupled with MSB-glycan, BSA is first reduced. Into a screw cap tube (10 × 100 mm), mix 6.9 mg of BSA and 0.3 mL of 6 M urea in 0.01 M EDTA to completely dissolve the BSA. Add 12 mg of $NaBH_4$ and incubate at room temperature for 10 min. Add 0.12 mL of 1-butanol, mix, and incubate at room temperature for 20 min. Finally, add 0.6 ml of 0.1 M NaH_2PO_4 and 0.24 mL of acetone. This reduced BSA sample has to be used immediately after preparation.
10. Dialysis should be performed at a low temperature (i.e., 10–20 °C) to prevent precipitation of the neoglycoprotein.
11. For the glycoprotein–lectin interaction assay, we usually use an evanescent field-activated fluorescence scanner [12], which allows for the highly sensitive and reproducible detection of the fluorescence.
12. Detection limit of influenza virus using the neoglycoprotein array is 2^2 hemagglutinin units, whereas that for a conventional assay is 2^8–2^9 units (or ~10 μg/mL) [22].

Acknowledgement

We are grateful to Professor Yasuo Suzuki (Chubu University, Japan) for providing the H1N1 influenza virus sample

References

1. Trevino V, Falciani F, Barrera-Saldana HA (2007) DNA microarrays. A powerful genomic tool for biomedical and clinical research. Mol Med 13:527–541
2. Kricka LJ, Master SR, Joos TO, Fortina P (2006) Current perspectives in protein array technology. Ann Clin Biochem 43:457–467
3. Smith DF, Song X, Cummings RD (2010) Use of glycan microarrays to explore specificity of glycan-binding proteins. Methods Enzymol 480:417–444
4. Rillahan CD, Paulson JC (2011) Glycan microarrays for decoding the glycome. Annu Rev Biochem 80:797–823
5. Oyeleran O, Gildesleeve JC (2009) Glycan array. Recent advances and future challenges. Curr Opin Chem Biol 13:406–413
6. Fukui S, Feizi T, Galustian C, Lawson AM, Chai W (2002) Oligoglycan microarrays for high-throughput detection and specificity assignments of carbohydrate-protein interaction. Nat Biotechnol 20:1011–1017
7. Wang D, Liu S, Trummer BJ, Deng C, Wang A (2002) Carbohydrate microarrays for the recognition of cross-reactive molecular markers of microtubes and host cells. Nat Biotechnol 20:275–281
8. Robinson MJ, Sancho D, Slack EC, LeibundGut-Landmann S, Reis e Sousa C (2006) Myeloid C-type lectins in innate immunity. Nat Immunol 7:1258–1265
9. Crocker PR, Paulson JC, Valki A (2007) Siglecs and their roles in the immune system. Nat Rev Immunol 7:255–266
10. van Vliet SJ, den Dunnen J, Gringhuis SI, Geijtenbeek TB, van Kooyk Y (2007) Innate signaling and regulation of dendric cell immunity. Curr Opin Immunol 19:189–194
11. Blixt O, Head S, Mondala T, Scanlan C, Huflejt ME, Alvarez R, Bryan MC, Fazio F, Calarese D, Stevens J, Razi N, Stevens DJ, Skehel JJ, van Die I, Burton DR, Wilson IA, Cummings R, Bovin N, Wong CH, Paulson JC (2004) Printed covalent glycan array for ligand profil-

ing of diverse glycan binding proteins. Proc Natl Acad Sci U S A 101:17033–17038

12. Tateno H, Mori A, Uchiyama N, Yabe R, Iwaki J, Shikanai T, Angata T, Narimatsu H, Hirabayashi J (2008) Glycoconjugate microarray based on an evanescent-field fluorescence-assisted detection principle for investigation of glycan-binding proteins. Glycobiology 18:789–798
13. Padler-Karavani V, Song X, Yu H, Hurtado-Ziola N, Huang S, Muthana S, Chokhawala HA, Cheng J, Verhagen A, Langereis MA, Kleene R, Schachner M, de Groot RJ, Lasanajak Y, Matsuda H, Schwab R, Chen X, Smith DF, Cummings RD, Varki A (2012) Cross-comparison of protein recognition of sialic acid diversity on two novel sialoglycan microarrays. J Biol Chem 287:22593–22608
14. Hase S, Ikenaka T, Matsushima Y (1978) Structure analyses of oligoglycans by tagging of the reducing end sugars with a fluorescent compound. Biochem Biophys Res Commun 85:257–263
15. Sumiyoshi W, Nakakita S, Miyanishi N, Hirabayashi J (2009) Strategic glycan elution map for the production of human-type N-linked oligoglycans: the case of hen egg yolk and white. Biosci Biotechnol Biochem 73:543–551
16. Yashizawa Z, Sato T, Schmid K (1966) Hydrazinolysis of alpha-1-acid glycoprotein. Biochim Biophys Acta 121:417–420
17. Nakakita S, Sumiyoshi W, Miyanishi N, Hirabayashi J (2007) A practical approach to N-glycan production by hydrazinolysis using hydrazine monohydrate. Biochem Biophys Res Commun 362:639–645
18. Hase S (1992) Conversion of pyridylamino sugar chains to 1-amino-1-deoxy derivatives, intermediates for tagging with fluorescein and biotin. J Biochem 112:266–268
19. Kuno A, Uchiyama N, Koseki-Kuno S, Ebe Y, Takashima S, Yamada M, Hirabayashi J (2005) Evanescent-field fluorescence-assisted lectin microarray: a new strategy for glycan profiling. Nat Methods 2:851–856
20. Lobl TJ, Mitchell MA, Maggiora LL (1990) SV40 large T-antigen nuclear signal analogues: successful nuclear targeting with bovine serum albumin but not low molecular weight fluorescent conjugates. Biopolymers 29:197–203
21. Takahashi T, Kawakami T, Mizuno T, Minami A, Uchida Y, Saito T, Matsui S, Ogata M, Usui T, Sriwilaijaroen N, Hiramatsu H, Suzuki Y, Suzuki T (2013) Sensitive and direct detection of receptor binding specificity of highly pathogenic avian influenza A virus in clinical samples. PLoS One 8:e78125
22. Xu R, de Vries RP, Zhu X, Nycholat CM, McBride R, Yu W, Paulson JC, Wilson IA (2013) Preferential recognition of avian-like receptors in human influenza A H7N9 viruses. Science 342:1230–1235

Chapter 17

Competitive Immunoassays Using Antigen Microarrays

Zhaowei Zhang, Weihua Hu, Qi Zhang, Peiwu Li, and Changming Li

Abstract

In this work, a non-fouling antigen competitive immunoassay microarray based on the polymer brush is reported to detect multiple mycotoxins. The detection is achieved by utilizing highly specific monoclonal antibodies produced in our laboratory. The polymer brush, poly[oligo(ethylene glycol) methacrylate-co-glycidyl methacrylate] (POEGMA-*co*-GMA), is synthesized via surface-initiated atom transfer radical polymerization (SI-ATRP) on standard glass slides. In the polymer brush, the epoxy groups of glycidyl methacrylate (GMA) residues provide covalent binding sites for spotted antigens. Moreover, the abundant poly(ethylene glycol) (PEG) side chains in the brush are able to ultimately suppress the nonspecific protein adsorption in solution (non-fouling). The polymer brush shows a high and uniform protein loading, along with a high resistance to nonspecific protein absorption that are both important to achieve a highly sensitive immunoassay. As a demonstration of a multiplex assay, aflatoxin B1 (AFB1), ochratoxin A (OTA), and zearalenone (ZEN) are selected as antigen targets for simultaneous detections using the microarray.

Key words Competitive immunoassay, Antigen microarray, Mycotoxins, Polymer brush

1 Introduction

The antigen microarray provides a novel strategy to realize high-sensitivity, high-specificity, and high-throughput immunoassay [1, 2]. Generally, the targets of interest detected using antigen microarrays are pathogens (i.e., viruses or bacteria) found in food, environmental, and medical samples. Immunoassays by the antigen microarray have emerged for detection of multiplexed targets such as chemical contaminants because this method can provide quantitative data of targets on larger numbers of samples. It is often desirable to analyze a number of proteins in samples, but usually, the information about a focused set of target molecules is sufficient. In recent years, antigen microarrays have evolved as powerful tools to address the high-throughput requirements by transferring traditional immunoassay, e.g., ELISA, into a miniaturized format. On our antigen microarrays, purified capture antigen/antibody are immobilized at spatially defined locations. Typically, these reagents

Paul C.H. Li et al. (eds.), *Microarray Technology: Methods and Applications*, Methods in Molecular Biology, vol. 1368, DOI 10.1007/978-1-4939-3136-1_17,

are arranged as arrays of microspots on flat substrates where the biochemical reaction binding is performed. This chapter focuses on a recent research in our group to demonstrate the immunoassay used for multiplex detection of mycotoxins in spiked peanut samples using the antigen microarray [3].

The antigen microarray, which is a renowned tool for multiplexed immunoassay, has attracted great interest in recent years. Since mycotoxins are small molecules with molecular weights less than 1000, direct binding of mycotoxins in a sandwich immunoassay format is not feasible. Therefore, their detections mainly rely on a competitive format rather than a direct one. In the competitive immunoassay, the target antigen competes with the analog antigen attached on a substrate for binding with the antibody, and the target is quantified by a signal that is related to the amount of antibody remained bound on the substrate-attached analog antigen. This leads to the detected antibody signal being inversely proportional to the target concentration. Unfortunately, the requirement of competitive assay intrinsically causes great difficulties to fabricate an antigen microarray to achieve highly sensitive immunoassays. The performance of the microarray will demand the significant improvements (e.g., more effective probe immobilization) to enhance sensitivity. An antigen microarray based on the commercial agarose-modified substrate has been developed to simultaneously detect multiple mycotoxins, but only a detection limit of tens pg/mL to several ng/mL mycotoxins is achieved [4]. A displacement-based microarray has also been used to detect a mycotoxin called patulin, but only a detection limit of 10 ng/mL is realized [5].

In order to achieve high sensitivity and broad detection range to analyze small target toxins in high throughput, an antigen microarray should possess high probe loading capacity, as well as low spot-to-spot signal variation and low nonspecific protein absorption in order to achieve a high signal-to-noise ratio. At the same time, monoclonal antibodies with high affinity and excellent specificity for mycotoxins, which are also very critical in their multiplexed microarray detection, were developed with the hybridoma technology. It has been discovered by our research that the polymer brush, POEGMA-*co*-GMA, possesses a porous structure with many bundles separated by gaps, which are formed by aggregation of the polymer chains during drying [6]. The spotted protein solution can penetrate into the gaps of the brush in a dry environment by capillary action, which will result in high protein loading for high sensitivity. In addition, the dense PEG side chains are able to block nonspecific protein adsorption. This will create a non-fouling surface to reduce the background in the immunoassay detection to achieve a detection limit of 3–4 pg/mL.

2 Materials

All chemicals were of analytical reagent grade, unless noted otherwise.

2.1 Reagents

1. Deionized (DI) water from a Millipore Milli-Q system.
2. Mycotoxins, including aflatoxin B1 (AFB1), ochratoxin A (OTA), and zearalenone (ZEN), from Sigma-Aldrich.
3. Mycotoxin conjugated with (bovine serum albumin) BSA to increase antigenicity: AFB1-BSA conjugate and OTA-BSA conjugate were obtained from Sigma-Aldrich; ZEN-BSA conjugate came from Aokin (Germany).
4. Cy3-labeled goat anti-mouse IgG from Invitrogen.

2.2 Immunization of Mice

1. Balb/c female mice (8–10 weeks old).
2. Freund's complete adjuvant (FCA), and Freund's incomplete adjuvant (FIA) which lacks the mycobacterial components.
3. SP 2/0 murine myeloma cells (from China Center for Type Culture Collection).
4. RPMI-1640 media.
5. Fetal bovine serum (FBS).
6. Polyethylene glycol (PEG) 1500.
7. Feeding cells from peritoneal macrophages and spleen cells using untreated Balb/c mouse.
8. Hypoxanthine/aminopterin/thymidine (HAT).

2.3 Slide Preparation

1. 0.1 M potassium hydroxide (KOH) solution.
2. Ethanol, methanol.
3. (3-aminopropyl)triethoxysilane (APTES).
4. Tetrahydrofuran (THF).
5. α-bromoisobutyryl bromide (BIB, 64 μL per 10 mL).
6. Triethylamine (TEA, 77 μL per 10 mL).
7. Oligo(ethylene glycol) methacrylate (OEGMA, MW = 360).
8. Glycidyl methacrylate (GMA).
9. Copper (I) bromide (CuBr).
10. 2,2′-bipyridine (BiPY).

2.4 Equipment

1. Centrifuge.
2. Standard glass slides.
3. Ultrasonic bath.
4. VersArray Chipwriter™ Compact System (Bio-Rad).

5. Stealth microspotting pins (SMP3).
6. Proscanarray microarray scanner (PerkinElmer, USA).
7. ProScanArrays Analysis software.
8. HPLC. Conditions: The HPLC separation was conducted with a Thermo Scientific C_{18} column (Hypersil Gold, 100 mm × 2.1 mm, 3.0 μm) at 35 °C. Elution A consisted of water with 0.05 % formic acid while elution B was acetonitrile with 0.05 % formic acid. A linear gradient elution program was as follows: 0–10 min, 15–100 % B; 11–13 min, 100 % B; 14–15 min, 100–15 % B; 16–18 min, 15 % B. The elution flow rate was set at 250 μL/min and 10 μL sample solution was injected into HPLC.

3 Methods

3.1 Production of Monoclonal Antibodies

Antigens (AFB1-BSA, OTA-BSA and ZEN-BSA) were multiple-site subcutaneously injected into Balb/c mouse separately and four immunizations were carried out to select the hybridomas.

1. Immunize two BALB/c female mice with AFB1-BSA conjugate Intraperitoneally inject the first dose of 50 μg of conjugate as an emulsion of PBS and Freund's complete adjuvant (FCA). Repeat this procedure with OTA-BSA on two mice, and ZEN-BSA on another two mice.
2. Perform two subsequent injections on the six mice at 3-week intervals using the conjugate emulsion in Freund's incomplete adjuvant (FIA) which lacks the mycobacterial components. One week after the last injection, tail-bleed the mice. Collect the sera and test them for anti-hapten antibody titer by indirect ELISA [7]. Test the sera for analyte recognition properties by competitive indirect ELISA [7].
3. After the mice undergo the resting period of at least 3 weeks from the last injection in adjuvants, select the mice to be spleen donors to receive a final soluble intraperitoneal injection of 80 μg of conjugate in PBS, 3 days prior to cell fusion for hybridoma production.
4. Culture SP 2/0 murine myeloma cells in RPMI-1640 media supplemented with 20 % FBS. Cell fusion was carried out in the following **steps 5–7** as described in ref. [7].
5. Mix mice splenocytes (contains antibody-secreting plasma cells) with the myeloma cells in a centrifuge tube (50 mL) at the ratio of 5–10 to 1. Centrifuge at 1000 × *g* for 5 min.
6. Add 1 mL of PEG 1500 at 37 °C to the cell pellet over 1 min. to initiate cell fusion.

7. Add 30 mL of RPMI-1640 (with no FBS) over 5 min. Leave the cells aside for 5–10 min at 37 °C.
8. Spin down the fused cells and remove the RPMI-1640 media.
9. After that, resuspend the cells with 20 ml of RPMI-1640 supplemented with 20 % FBS, and culture the cells overnight.
10. Count the cells and put an average amount of 10^7 into a selective hemi-solid media containing HAT and feeding cells, and culture for over 2 weeks.
11. After seeing by naked eyes the many white cell colonies which contain monoclonal hybridomas, remove the hybridomas and put them onto the wells of a 96-well culture plate.
12. In the hybridoma selection process, use a two-step ELISA screening procedure [7] to select hybridomas with high sensitivity (IC_{50} value of AFB, OTA, ZEN, respectively) and good specificity.
13. Intraperitoneally inject the hybridoma cells into the Balb/c mice treated Freund's incomplete adjuvants (FIA) to prepare ascites, which were used to produce monoclonal antibodies.

3.2 Preparation of Slides Modified with the POEGMA-co-GMA Polymer Brush

1. Clean standard glass slides ultrasonically in KOH solution (0.1 M), DI water, and ethanol for 10 min, successively.
2. Dry the slides with compressed air. Incubate them in 5 % (v/v) (3-aminopropyl)triethoxysilane (APTES) solution in ethanol for 2 h at room temperature. Then, rinse the slides with ethanol.
3. Subject the APTES-modified slides to thermal curing at 110 °C in a vacuum oven for 2 h. Then, incubate the slides with an ice-cold tetrahydrofuran (THF) solution containing the initiator α-bromoisobutyryl bromide (BIB) and triethylamine (TEA), for 2 h to attach the initiator. Wash the slides thoroughly with THF.
4. To grow the POEGMA-*co*-GMA polymer brush, immerse the initiator-attached slides in the deoxygenized methanol/H_2O (1:1) growth solution, which contains 20 % v/v oligo(ethylene glycol) methacrylate (OEGMA), 0.5 % v/v glycidyl methacrylate (GMA), 2.9 mg/mL of CuBr and 6.25 mg/mL of 2,2′-bipyridine (BiPY).
5. Allow the polymerization to continue for 6 h in an inert atmosphere at room temperature. Then, remove the slides from the growth solution.
6. Wash the slides intensively with ethanol and DI water. Then, dry them by gentle nitrogen flow and store them at 4 °C until use.

3.3 Printing of Antigen Microarrays

1. Spot three mycotoxin antigens (AFB1-BSA, OTA-BSA and ZEN-BSA) on the slides modified with the POEGMA-*co*-GMA polymer brush to form three 4×4 subarrays using a VersArray Chipwriter™ Compact System equipped with stealth microspotting pins (SMP3).
2. Print a 4×4 BSA subarray as a negative control on the slides.
3. Store the spotted slides in a dry cabinet at room temperature for overnight incubation. Then, wash the slides with 0.01 M of PBS to remove any unattached antigens.
4. Dry the slides under a gentle nitrogen gas flow and the slides were ready for use.
5. A batch of such slides was prepared with the same manner (please *see* Subheading 3.2 and 3.3) for evaluation of reproducibility and reliability.

3.4 Detection of Slide Baseline Signal Without Added Antigens

The detection of mycotoxins is based on an indirect c competitive immunoassay format as shown in Fig. 1. The baseline signal is obtained from the antigen microarray when only monoclonal antibody, but no target antigens, is added to the microarray.

1. Incubate the microarray slide separately in three solutions containing only an individual antibody at a high concentration (1 μg/mL).
2. Add a fluorescent secondary antibody (Cy3-labeled anti-mouse IgG) at 10 μg/mL to bind with the bound monoclonal antibodies, and incubate in the dark for 40 min. Measure the fluorescent images of the slide with 543 nm excitation using a microarray scanner and analyze the data with the ProScanArrays Analysis software.
3. Optimize the concentrations of three antigens for antigen printing by varying the antigen concentration to be 200 μg/mL (*see* **Note 1**).
4. To optimize the concentrations of three monoclonal antibodies for competitive immunoassay, print the microarray slides with antigens of the optimal concentration of 200 μg/mL. Incubate the slides separately in three solutions containing only an individual antibody (concentration varies from 20 to 180 ng/mL for anti-AFB1 and anti-OTA, and from 120 to 280 ng/mL for anti-ZEN antibody) for 1 h in humidity box at room temperature (*see* **Note 2**). The results are depicted in Fig. 2.
5. Briefly wash the slides with PBS for 5 min, incubate the slides further in a Cy3-labeled goat anti-mouse IgG solution (10 μg/mL) in the dark for 40 min.
6. Wash the microarray slides with PBS and DI water, successively for three times. Finally, dry the slides at 37 °C and measure the

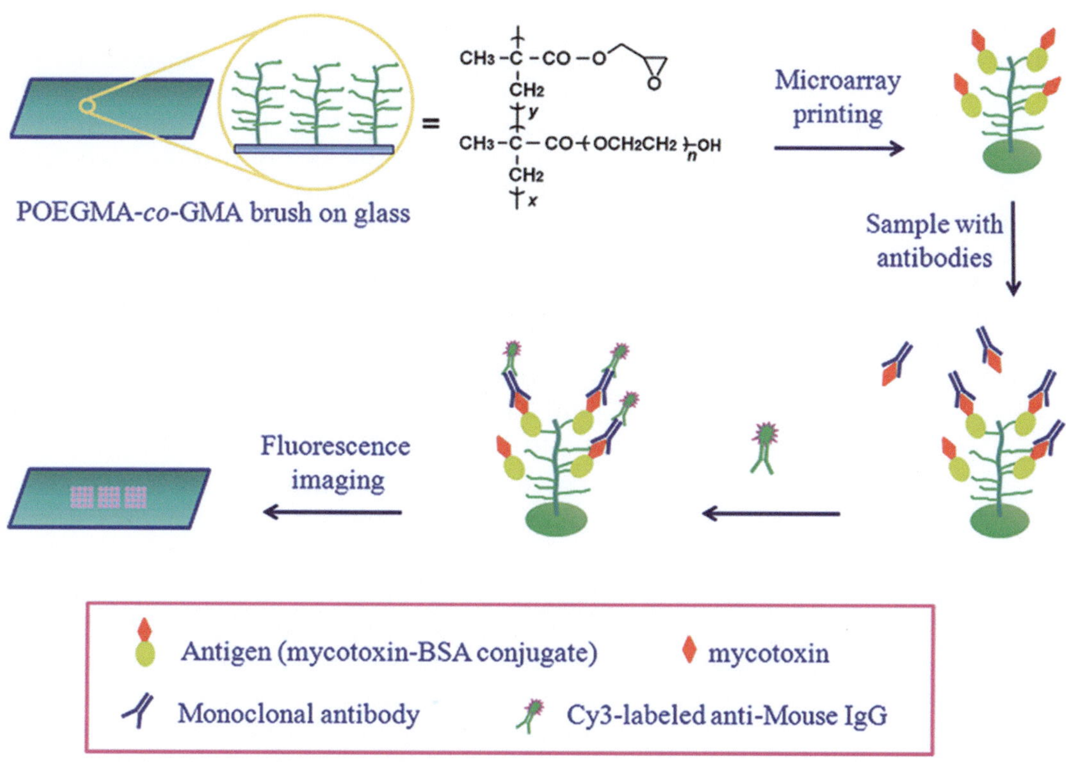

Fig. 1 Schematic illustration of indirect competitive immunoassay on the slide modified with the POEGMA-*co*-GMA brush. For simplicity, only one type of mycotoxins is drawn

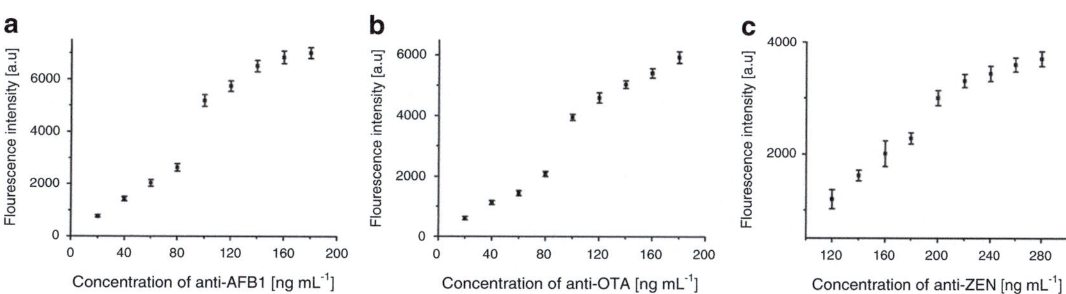

Fig. 2 Fluorescent intensities vs. monoclonal antibody concentrations. The antigen microarray was firstly incubated with monoclonal antibody of different concentration, followed by incubation with 10 μg/mL Cy3-labeled anti-mouse IgG. (**a**) anti-AFB1; (**b**) anti-OTA; (**c**) anti-ZEN. Error bars are standard deviations ($n = 3$) a.u. represents arbitrary units

fluorescence images in order to determine the optimal antibody concentration. The optimal concentrations for the monoclonal antibodies are determined to be 100, 100, and 200 ng/mL for anti-AFB1, anti-OTA, and anti-ZEN, respectively.

3.5 Detection of Mycotoxins

For detection of the target mycotoxin, it will be mixed with the monoclonal antibody and added to the antigen microarray. The target antigen will compete with the corresponding antigen on the microarray to bind with its antibody (Fig. 1). This will result in a weaker fluorescence signal as compared with that obtained in the absence of mycotoxin for quantitative detection.

1. For multiplexed detections of mycotoxins, add a 100 μL solution containing all three monoclonal antibodies and premixed mycotoxins onto the microarray covering the whole surface, and incubate for 1 h. The optimized concentrations of the antibodies are 100, 100, and 200 ng/mL for anti-AFB1, anti-OTA, and anti-ZEN, respectively, and the concentration ranges for mycotoxins are 0.001–10 ng/mL.
2. Briefly wash the slide with PBS. Then incubate it with a Cy3-labeled goat anti-mouse IgG solution (10 μg/mL) for 40 min.
3. Wash the slide with PBS and DI water. Then, scan the slide, and record the fluorescence image (*see* **Note 3**). Perform image analysis and the results are plotted in Fig. 3. The dynamic ranges span for three orders of magnitude and the limits of detection (LOD) are as low as 4, 4, and 3 pg/mL of AFB1, OTA, and ZEN, respectively, as indicated by the circles in Fig. 3 (*see* **Note 4**). The linear range for detection of each antigen is also shown in Fig. 3 (*see* **Note 5**).
4. Finely ground a peanut sample (20 g). Perform an HPLC analysis to confirm the sample contains no mycotoxins.
5. Place 5 g of ground peanut sample in a centrifuge tube (50 mL). To mimic real peanut samples for mycotoxin detection, spike all three mycotoxins with certain amounts to the tube.
6. Add 15 mL of methanol-water (80:20, v/v, 4 % NaCl) to the tube and ultrasonicate it for 10 min to extract the mycotoxins from the spiked peanut sample.
7. After settling for 5 min, remove the supernatant. Dilute 0.2 mL of supernatant with 1.8 mL of PBS.
8. Mix the diluted solution with the three monoclonal antibodies, and detect the mycotoxins with the antigen microarray using the same procedure described above (**steps 1–3**).

4 Notes

1. The fluorescent intensities of the microarray spots increase with the printing antigen concentration until reaching the plateau at 200, 200, and 175 μg/mL for AFB1-BSA, OTA-BSA and

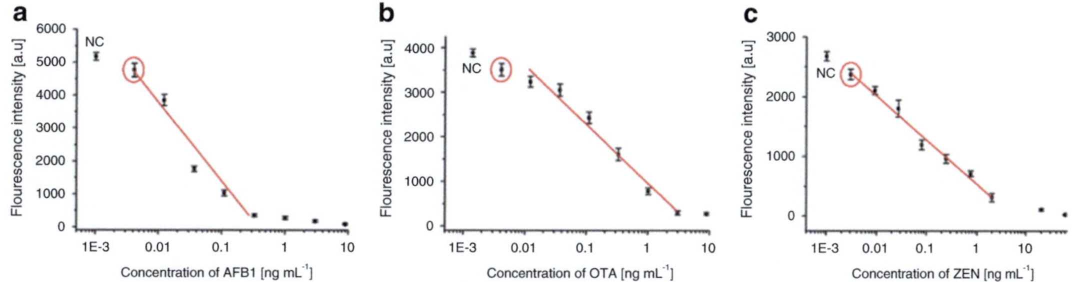

Fig. 3 Fluorescent intensity versus concentration curves for multiplex detection of mycotoxins: AFB1 (**a**), OTA (**b**), and ZEN (**c**). Error bars are standard deviations (n=3). NC is negative control. a.u. represents arbitrary units

ZEN-BSA, respectively. Therefore 200 μg/mL is chosen as an optimal concentration for subsequent antigen printing.

2. In order to achieve low detection limit and wide dynamic range in competitive immunoassay, it is very critical to select the optimal concentrations of the monoclonal antibodies. As shown in Fig. 2, the fluorescent intensity increases with increasing antibody concentration. At a certain concentration range (80–100 ng/mL for anti-AFB1 and anti-OTA, and 180–200 ng/mL for anti-ZEN), the signal exhibits the steepest slope of change, suggesting the highest sensitivity in the competitive immunoassay. Therefore the optimal concentrations for the monoclonal antibodies are determined to be 100, 100, and 200 ng/mL for antibodies of AFB1, OTA, and ZEN, respectively.
3. Multiplexed immunoassays were carried out in a solution containing mycotoxins and their monoclonal antibodies with the optimal concentrations. The representative fluorescent images are shown in Fig. 4. By quantitative analysis of the images, the fluorescent intensities are obtained, and they are plotted against mycotoxin concentrations for three mycotoxins (Fig. 3).
4. The LOD are determined as the values higher than threefold standard deviation of their corresponding negative control (NC) according to the Mb + 3 × Sd definition, where Mb and Sd are the mean value and standard deviation of NC, respectively (n = 3).
5. The linear range of response vs. logarithm of concentration for detection of each antigen is also plotted in Fig. 3, which is further fitted to the equation of $y = a - b \times \log(x)$.

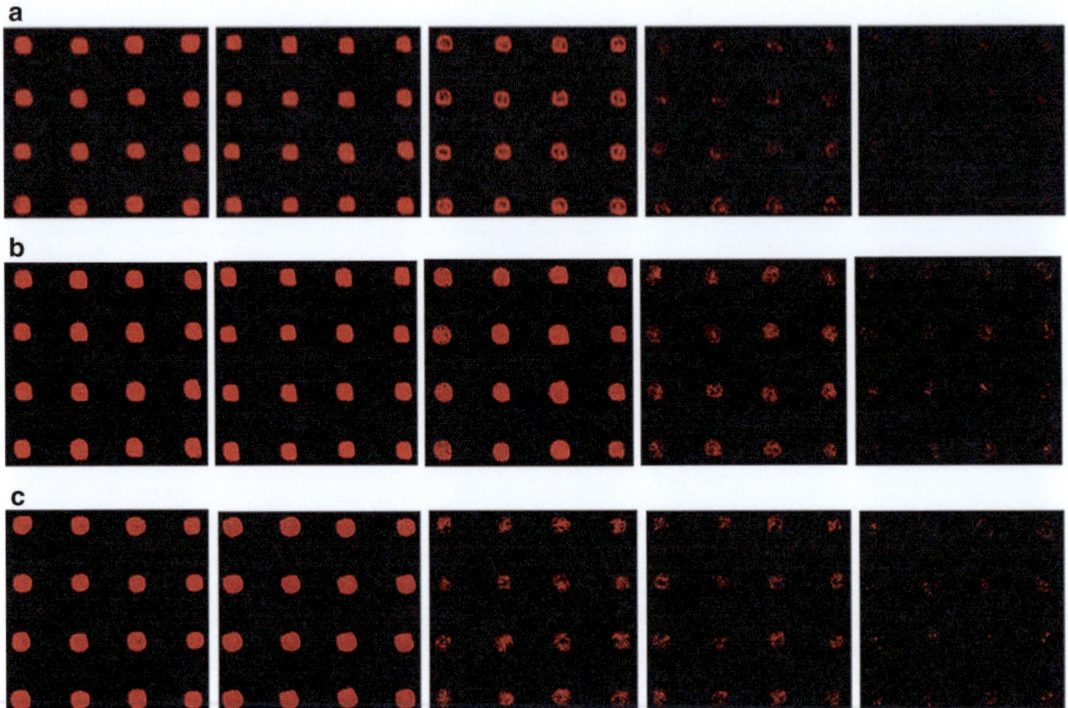

Fig. 4 Representative fluorescent images of microarray detection of mycotoxins. (**a**) 0, 0.012, 0.037, 0.33, 3 ng/mL AFB1; (**b**) 0, 0.012, 0.33, 1, 3 ng/mL OTA; (**c**) 0, 0.003, 0.24, 0.74, 2 ng/mL ZEN (from *left* to *right*)

Acknowledgement

This work was supported by the Project of National Science & Technology Pillar Plan (2012BAB19B09), the Special Fund for Agro-scientific Research in the Public Interest (201203094), and the National Natural Science Foundation of China (21205133, 31101299).

References

1. Wilson DS, Nock S (2003) Recent developments in protein microarray technology. Angew Chem Int Ed 42:494–500
2. Seidel M, Niessner R (2014) Chemiluminescence microarrays in analytical chemistry: a critical review. Anal Bioanal Chem 406:5589–5612
3. Sauceda-Friebe JC, Karsunke XYZ, Vazac S, Biselli S, Niessner R, Knopp D (2011) Regenerable immuno-biochip for screening ochratoxin A in green coffee extract using an automated microarray chip reader with chemiluminescence detection. Anal Chim Acta 689:234–242
4. Wang Y, Liu N, Ning B, Liu M, Lv Z, Sun Z, Peng Y, Chen C, Li J, Gao Z (2012) Simultaneous and rapid detection of six different mycotoxins using an immunochip. Biosensors Bioelectron 34:44–50
5. de Champdore M, Bazzicalupo P, De Napoli L, Montesarchio D, Di Fabio G, Cocozza I,

Parracino A, Rossi M, D'Auria S (2007) A new competitive fluorescence assay for the detection of patulin toxin. Anal Chem 79:751–757

6. Hu W, Liu Y, Lu Z, Li CM (2010) Poly[oligo(ethylene glycol) methacrylate-*co*-glycidyl methacrylate] brush substrate for sensitive surface plasmon resonance imaging protein arrays. Adv Funct Mater 20: 3497–3503
7. Zhang D, Li P, Zhang Q, Zhang W, Huang Y, Ding X, Jiang J (2009) Production of ultrasensitive generic monoclonal antibodies against major aflatoxins using a modified two-step screening procedure. Anal Chim Acta 636:63–69

Chapter 18

Parallel Syntheses of Peptides on Teflon-Patterned Paper Arrays (SyntArrays)

Frédérique Deiss, Yang Yang, and Ratmir Derda

Abstract

Screening of peptides to find the ligands that bind to specific targets is an important step in drug discovery. These high-throughput screens require large number of structural variants of peptides to be synthesized and tested. This chapter describes the generation of arrays of peptides on Teflon-patterned sheets of paper. First, the protocol describes the patterning of paper with a Teflon solution to produce arrays with solvophobic barriers that are able to confine organic solvents. Next, we describe the parallel syntheses of 96 peptides on Teflon-patterned arrays using the SPOT synthesis method.

Key words Peptides, Paper array, Paper patterning, SPOT synthesis, Ligand screening

1 Introduction

Peptide-based therapeutics constitutes a significant fraction of new chemical entities and FDA-approved drugs. Discovery and optimization of peptide ligands that bind to specific targets require testing of a large number of structural variants of the peptides. Library-based methods, such as one-bead-one-compound combinatorial libraries (OBOC) [1] and genetically encoded libraries of peptides and peptides derivatives that are displayed on phage [2, 3], cells [4], or RNA [5], are useful for identification of "hit" peptide sequences from million to billion variants of peptide sequences. On the other hand, the process of hit-to-lead validation requires individual testing of the hit sequences. In this process, the array of peptide-derived ligands is a uniquely convenient tool because it allows parallel synthesis, characterization and testing of relatively large number of hits in uniform conditions. The array can be created by grafting pre-synthesized peptides onto a support at controlled locations [6, 7], or by stepwise assembling the peptides from amino acids directly an a planar support. To accelerate the latter method, Frank and coworkers [8] pioneered the synthesis of peptides as spots on cellulosic membranes (or paper). This solid-phase peptide

Paul C.H. Li et al. (eds.), *Microarray Technology: Methods and Applications*, Methods in Molecular Biology, vol. 1368, DOI 10.1007/978-1-4939-3136-1_18,

synthesis method, dubbed as "SPOT synthesis," was widely adopted in academic and industrial laboratories. It led to the development of commercially available technologies for synthesis of these arrays and to the establishment of service companies providing SPOT arrays on demand. The paper support was shown to be compatible with other classes of combinatorial synthesis to generate small-molecule arrays [9].

Unlike most syntheses on solid supports, which are conducted in stirred or flow-through reactors, in SPOT synthesis, the reagents are deposited onto the porous support to form "spots" of desired sizes. In these conditions, the reactions occur in a static, non-stirred volume of solvent in which the mixing of reagents is slow because it is limited to diffusion only. Moreover, it is problematic to deposit an excess of reagents to drive the reaction to completion, and evaporation of the limited amount of solvent during synthesis can pose a significant problem. To overcome these problems, we developed a method [10] to pattern paper with a solution that consists of a resin of amorphous fluoropolymer (e.g., Teflon solution from DuPont). Patterned Teflon-barriers resist solvent penetration and confine a broad range of organic solvents used in organic syntheses within the barriers. An excess of solvents, when deposited on the solvophilic zones of this Teflon-bound porous support, flows through the support and allows for performing synthesis in a flow-through fashion. The pattern of the porous support that consists of 96 squares (Fig. 1a) follows the exact foot print of a commercial 96-well plate, and this makes the Teflon-patterned array compatible with the use of standard plate-to-plate transfer robotics and plate readers for liquid handling and analysis.

In this chapter, we first describe the steps for patterning the paper support with solvophobic barriers to generate Teflon-patterned arrays with 96 solvophilic zones for the peptide synthesis (Fig. 2). We then provide the details of the method for synthesizing the peptides on paper, which is an adaptation of the protocol published by Hilpert et al. [11]. We describe both the "manual" way, accessible with common equipment found in a chemical/biology laboratory, as well as a "semi-automated" method using a commercially available liquid handling workstation.

2 Materials

2.1 Reagents

1. Teflon solution: 20 % Teflon® AF amorphous fluoropolymer resin (DuPont 400S2-100-1) in solution in methoxyperfluorobutane (Novec HFE-7100 Engineered fluid) (*see* **Note 1**).
2. Sucrose solution (~2.5 mL per array): 1 g/mL of sucrose (or table sugar) in Milli-Q water.
3. Amino acid stock solutions: Determine the need for each 9-fluorenylmethyloxycarbonyl-conjugated amino acid (or

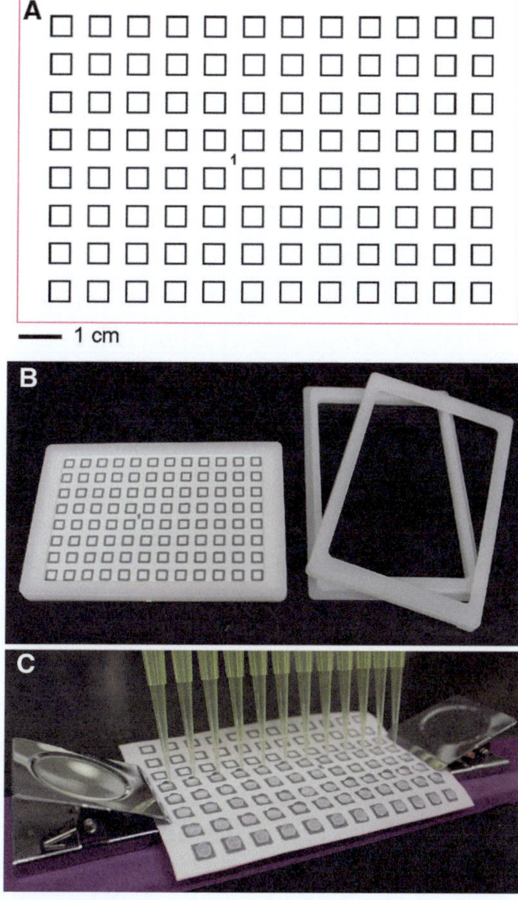

Fig. 1 (**a**) 96-square template for Teflon-patterned array. (**b**) Holder to clamp the paper for convenient handling. (**c**) Alternative holder for the paper made from bulldog clips

Fmoc(amino acid)-OH) for the peptide synthesis using Tables 1, 2, 3 and 4 described in the Subheading 3.3. Prepare a stock solution of each in 0.75 M in *N*-methyl-2-pyrrolidone (NMP). Store the stock solutions in amber glass vials at −20 °C for up to a week (*see* **Note 2**).

4. HOAt stock solution: 1-hydroxy-7-azabenzotriazole (HOAt) at 2.25 M in NMP as indicated in Table 3.
5. DIC stock solution: *N*,*N*′-diisopropylcarbodiimide (DIC) at 1.68 M in NMP as indicated in Table 3. The coupling reagent DIC may be substituted by another one, such as dicyclohexylcarbodiimide (DCC).
6. First β-Ala functionalization solution (used for four arrays, *see* **Note 3**): dissolve 576 mg Fmoc-β-Ala-OH in 9 mL of dimethylformamide (DMF), add 270 μL of 1-methylimidazole and 337.5 μL DIC.

Table 1
Identification of the amino acids to be added at each cycle

	A	B	C	D	E	F	G	H	I	J	K	L	M	N	O	P	Q	R
1	Position		Write peptide sequences in this column	Cycle #														
2	Row	Col		1	2	3	4	5	6	7	8	9	10	11	12	13	14	15
3	1	1-12	TVKHRPDALHPQ	Q	P	H	L	A	D	P	R	H	K	V	T			
4	2	1-12	LTTAPKLPKVTR	R	T	V	K	P	L	K	P	A	T	T	L			
5	3	1-12	GKKQRFRHRNRK	K	R	N	R	H	R	F	R	Q	K	K	G			
6	4	1-12	FHRRIKAGRGDS	S	D	G	R	G	A	K	I	R	R	H	F			
7	5	1-12	PQVTRGDVFTMP	P	M	T	F	V	D	G	R	T	V	Q	P			
8	6	1-12	LTGKNFPMFHRN	N	R	H	F	M	P	F	N	K	G	T	L			
9	7	1-12	MHRMPSFLPTTL	L	T	T	P	L	F	S	P	M	R	H	M			
10	8	1-12	GWQPPARARIG	G	I	R	A	R	A	P	P	Q	W	G				
11																		

Example of an array with eight different peptides, where each row has 12 replicates of the same peptide

Table 2
Maps of the 96-zone array at each coupling cycle (left)

	A	B	C	D	E	F	G	H	I	J	K	L	M	N	O	P	Q	R	S	T	U	V	W	X	Y	Z
1	1	1	1	1	1	1	1	1	1	1	1	1	1			Amino acids count per cycle:										
2	1	Q	Q	Q	Q	Q	Q	Q	Q	Q	Q	Q	Q													
3	2	R	R	R	R	R	R	R	R	R	R	R	R			G	P	A	V	L	I	M	C	F	Y	
4	3	K	K	K	K	K	K	K	K	K	K	K	K			12	12	0	0	12	0	0	0	0	0	
5	4	S	S	S	S	S	S	S	S	S	S	S	S													
6	5	P	P	P	P	P	P	P	P	P	P	P	P			W	H	K	R	Q	N	E	D	S	T	
7	6	N	N	N	N	N	N	N	N	N	N	N	N			0	0	12	12	12	12	0	0	12	0	
8	7	L	L	L	L	L	L	L	L	L	L	L	L													
9	8	G	G	G	G	G	G	G	G	G	G	G	G													
10																										
11	2	1	1	1	1	1	1	1	1	1	1	1	1													
12	1	P	P	P	P	P	P	P	P	P	P	P	P													
13	2	T	T	T	T	T	T	T	T	T	T	T	T			G	P	A	V	L	I	M	C	F	Y	
14	3	R	R	R	R	R	R	R	R	R	R	R	R			0	12	0	0	0	12	12	0	0	0	
15	4	D	D	D	D	D	D	D	D	D	D	D	D													
16	5	M	M	M	M	M	M	M	M	M	M	M	M			W	H	K	R	Q	N	E	D	S	T	
17	6	R	R	R	R	R	R	R	R	R	R	R	R			0	0	0	24	0	0	0	12	0	24	
18	7	T	T	T	T	T	T	T	T	T	T	T	T													
19	8	I	I	I	I	I	I	I	I	I	I	I	I													
20																										

Number of occurrences of each amino acid for each cycle. This example describes the first two cycles for the synthesis of the eight peptides from Table 1

Table 3
Calculation of the quantities of reagents for the coupling of the amino acids

	A	B	C	D	E	F	G	H	I	J	K
1	Cycles	start	last		Volume per spot in µL:		15	Number of coupling:		1	
2		1	7		Number array		4	Dead volume in µL:		50	
3											
4		Amino acids	MW	Vtotal	m_{aa}	HOAT	HOAT sol	DIC	+NMP	DIC in NMP	+ NMP (for aa)
5	Tot.		g/mol	µL	mg	mg	µL of NMP	uL	µL of NMP	µL total	µL
6	48	A	311.3	4800	672	294	960	254	706	960	2880
7	0	C	585.7	0	0	0	0	0	0	0	0
8	9	D	411.5	900	167	55	180	48	132	180	540
9	3	E	425.5	300	57	18	60	16	44	60	180
10	18	F	387.4	1800	314	110	360	95	265	360	1080
11	27	G	297.3	2700	361	165	540	143	397	540	1620
12	51	H	619.7	5100	1422	312	1020	270	750	1020	3060
13	18	I	353.4	1800	286	110	360	95	265	360	1080
14	21	K	468.5	2100	443	129	420	111	309	420	1260
15	72	L	353.4	7200	1145	441	1440	380	1060	1440	4320
16	18	M	371.5	1800	301	110	360	95	265	360	1080
17	18	N	596.7	1800	483	110	360	95	265	360	1080
18	72	P	337.4	7200	1093	441	1440	380	1060	1440	4320
19	54	Q	610.7	5400	1484	331	1080	285	795	1080	3240
20	33	R	648.8	3300	963	202	660	174	486	660	1980
21	78	S	383.4	7800	1346	478	1560	412	1148	1560	4680
22	75	T	397.5	7500	1342	459	1500	396	1104	1500	4500
23	15	V	339.4	1500	229	92	300	79	221	300	900
24	12	W	526.6	1200	284	73	240	63	177	240	720
25	30	Y	459.5	3000	620	184	600	159	441	600	1800
26	672	TOTAL			10844	4116	13440	3551	9889	13440	40320
27					MW g/mol	136.1	MW/d mL/mo	156.6			
28	Suggested stock of HOAT and DIC solutions to prepare in NMP:					4939	16128	4261	11867	16128	

L	M	N	O	P	Q	R	S	T	U	V	W	X	Y	Z	AA	AB
		Cycle #														
	aa	15	14	13	12	11	10	9	8	7	6	5	4	3	2	1
	A	0	0	0	0	0	0	0	0	9	9	9	0	9	12	0
	C	0	0	0	0	0	0	0	0	0	0	0	0	0	0	0
	D	0	0	0	0	0	0	0	0	0	3	0	6	0	0	0
	E	0	0	0	0	0	0	0	0	0	3	0	0	0	0	0
	F	0	0	0	0	0	0	0	0	0	3	0	3	6	3	3
	G	0	0	0	0	0	0	0	0	0	6	3	6	6	3	3
	H	0	0	0	0	0	0	0	0	6	12	12	6	3	3	9
	I	0	0	0	0	0	0	0	0	0	0	3	6	3	6	0
	K	0	0	0	0	0	0	0	0	0	3	3	3	6	3	3
	L	0	0	0	0	0	0	0	0	0	12	6	15	6	6	27
	M	0	0	0	0	0	0	0	0	12	0	3	3	0	0	0
	N	0	0	0	0	0	0	0	0	0	6	0	3	0	9	0
	P	0	0	0	0	0	0	0	0	0	15	18	6	6	12	15
	Q	0	0	0	0	0	0	0	0	15	3	3	6	9	12	6
	R	0	0	0	0	0	0	0	0	0	0	6	0	12	6	9
	S	0	0	0	0	0	0	0	0	27	6	9	15	0	15	6
	T	0	0	0	0	0	0	0	0	12	9	15	15	15	0	9
	V	0	0	0	0	0	0	0	0	6	0	0	0	9	0	0
	W	0	0	0	0	0	0	0	0	0	3	3	0	3	3	0
	Y	0	0	0	0	0	0	0	0	9	3	3	3	3	3	6
	Tot.	0	0	0	0	0	0	0	0	96	96	96	96	96	96	96

This example corresponds to four arrays of 32 peptides shown in Figure 5a and 5c

7. Second β-Ala activated solution (used for four arrays, *see* **Note 3**): Dissolve 1261 mg of Fmoc-β-Ala-OH in 5.4 mL of NMP (or use the 0.75 M stock solution). Fifteen minutes before spotting, add 1.8 mL of HOAt solution (551 mg in 1.8 mL NMP or the 2.25 M stock solution) and 1.8 mL of DIC solution (475 µL of DIC plus 1325 µL NMP or the 1.68 M stock solution). Mix and let react for ~10 min.
8. Capping A solution: 2 % (v/v) of acetic anhydride in DMF (typical volume prepared: 100 mL).
9. Capping B solution (protect from light): 2 % (v/v) of acetic anhydride + 2 % (v/v) of *N*,*N*-diisopropylethylamine (DIPEA) in DMF (typical volume prepared: 50 mL).
10. 20 % Piperidine solution: 20 % piperidine in DMF (typical volume prepared: 200 mL).

Table 4
Maps of the master plates prepared for the coupling of the amino acids for each cycle

	A	B	C	D	E	F	G	H	I	J	K	L	M	N	O
1	Which cycles?		start	last	Add		180	µL /well of stock solution of amino acid							
2			1	7	in each well of the DWP following map(s) below										
3															
4			Cycle		1		Cycle		2						
5			1	2	3	4	5	6	7	8	9	10	11	12	
6	Deep well	a	P	Q	Y	P	Q	L	F	Q			HOAt sol.	DIC sol.	1
7	plate #1	b	L	H	R	P	L	R	S	P					2
8		c	T	L	L	T	K	S	N	Q					3
9		d	T	L	Q	L	I	A	N	P					4
10		e	S	L	G	F	S	W	Q	A			530	530	5
11		f	L	L	P	S	R	I	A	P			µL/well	µL/well	6
12		g	H	L	H	R	S	N	H	A					7
13		h	R	P	Y	K	Y	S	P	G					8
[..]	[..]														
37			Cycle		7		Cycle		0						
38			1	2	3	4	5	6	7	8	9	10	11	12	
39	Deep well	a	S	M	M	S	0	0	0	0			HOAt sol.	DIC sol.	1
40	plate #4	b	T	A	Q	S	0	0	0	0					2
41		c	V	H	S	S	0	0	0	0					3
42		d	Y	Q	A	T	0	0	0	0					4
43		e	Y	T	T	H	0	0	0	0			290	290	5
44		f	S	A	S	M	0	0	0	0			µL/well	µL/well	6
45		g	Y	S	S	Q	0	0	0	0					7
46		h	Q	V	M	Q	0	0	0	0					8
47															

11. Cleavage A solution: Mix 90 % (v/v) trifluoroacetic acid (TFA), 3 % (v/v) triisopropylsilane (TIPS), 2 % (v/v) Milli-Q water, 1 % (w/v) of phenol, and 4 % (v/v) dichloromethane (DCM) (*see* **Note 4**). The typical volume is 60 mL for four arrays. WARNING: TFA and phenol are corrosive and toxic, so they should be handled with care.
12. Cleavage B solution: Mix 50 % (v/v) TFA, 3 % (v/v) TIPS, 2 % (v/v) Milli-Q water, 1 % (w/v) of phenol, 44 % (v/v) DCM (*see* **Note 4**). The typical volume is 60 mL for four arrays. WARNING: TFA and phenol are corrosive and toxic, so they should be handled with care.

13. Bromophenol blue staining solution (optional): 0.003 % (w/v) of bromophenol blue in methanol. Protect the solution from light.

2.2 Apparatus/Instruments

1. Paper: the protocol and quantities are described for Whatman filter paper grade 50, but other cellulosic substrate can be used, although the spotting volume might need some adjustments.
2. Solid ink printer (e.g., Xerox ColorQube 8570 series).
3. Multichannel pipettors or fluid handling workstation (e.g., BioTek Precision XS).
4. Frame holder to clamp the paper (Fig. 1b) or any reliable in-house made system to maintain the paper above the surface during deposition (e.g., clamps; Fig. 1c). Special holders are also available from SyntArray at www.syntarray.com.
5. Common vials, glass bottles, reagent vessels, pipette tips, pipettors used for preparation of the solutions and distribution of reagents and solvents.
6. A vacuum pump or aspirator to aspirate excess reagents from the paper after reaction (Fig. 3).
7. Small plastic bags to store the arrays (e.g., freezer bags, plastic bags).

3 Method

3.1 Patterning of Paper to Generate Solvent-Resistant Arrays

1. Cut the paper into several 6″ × 11″ bands.
2. Open a drawing file (*see* **Note 5**) with the desired pattern (e.g., 96-square, Fig. 1a; three arrays can fit on one band of paper). Print the pattern (without any scaling) on the paper using a solid ink printer.
3. After printing the solid-ink pattern on the paper, place it in the oven at 120 °C for 5 min. This step yields a wax-patterned paper array [12] that has areas resistant to aqueous solutions (Fig. 2a) (*see* **Note 6**).
4. Cut the paper along the red-colored frame (Fig. 1a). If you have a frame holder (Fig. 1b), place the paper in it. Alternatively, cut the printed paper array to the required dimensions and clamp it between two magnetic bulldog clips (Fig. 1c).
5. Distribute 25 μL of the sucrose solution per zone using a multichannel pipette (the sucrose solution protects the non-waxed zones from impregnation with Teflon during the next step). Control the success of the deposition by ensuring the presence of the hemispherical drop on each zone (Fig. 2b).
6. Distribute 3 × 800 μL of the diluted Teflon solution with a P1000 pipettor over the array (Fig. 2c). Ensure that the Teflon

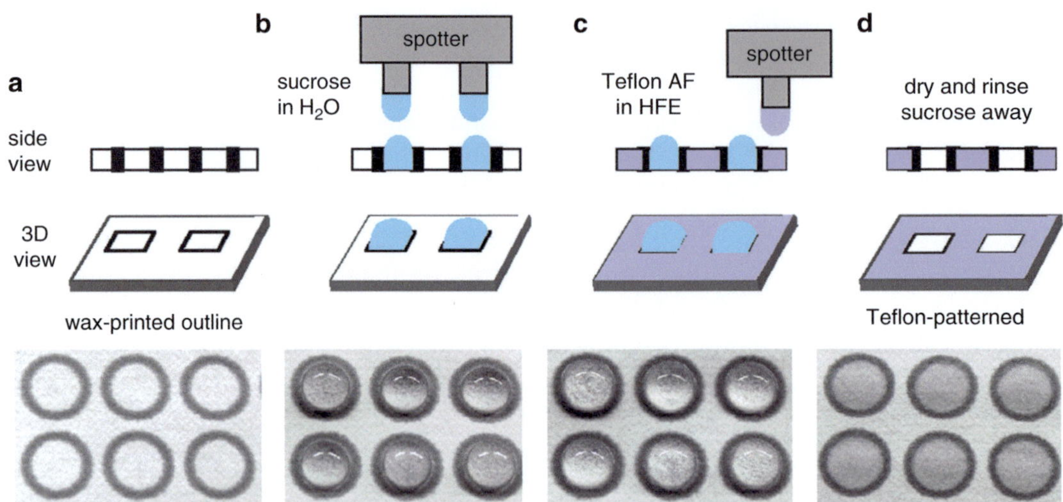

Fig. 2 Step-by-step description of *the* Teflon-patterning process. (**a**) Wax-patterning of the paper, (**b**) spotting of the sucrose protective solution, (**c**) distribution of the Teflon solution made in HFE and (**d**) rinsing of the array after evaporation of the solvent (Reproduced from ref. [10] with permission from Wiley)

solution covers the entire array; all the areas around the zones with the drop of sugar on the paper should appear uniformly translucent. Pay special attention to the area of the paper along the holder because this is where the deposition of Teflon commonly fails (*see* **Note 7**).

7. Evaporate the solvent to allow for the formation of a Teflon film on the paper. The best method is air-drying the paper at room temperature for ~1 h. When working with more than one array, each one should have 2 in. of space above and below it for proper air circulation (*see* **Note 8**).
8. Once the solvent evaporated, the areas of paper patterned with Teflon have a bright white reflective appearance. At this point, remove the top part of the frame holder to dry the borders of paper that were blocked by the holder. Drying is complete when the marks of the borders are invisible. If necessary, the arrays can be left to dry overnight.
9. Immerse the arrays in cold water for 15–20 min. Change the water twice to completely remove the protective sucrose solution (Fig. 2d) (*see* **Note 9**).
10. Check the Teflon patterning for any defects: when the paper is wet, any defect should have a translucent appearance, instead of the bright white reflective appearance indicative of the Teflon-impregnated areas (*see* **Note 10**).
11. Air-dry the arrays or use a heat gun to accelerate the drying.
12. The arrays can be stored in sealed bags at room temperature for several months before use.

Fig. 3 96-nozzle vacuum apparatus that allows simultaneous aspiration of reagents through the paper array from the 96 zones. It was custom-made to fit the dimensions of the SyntArray paper holder (Reproduced from ref. [10] with permission from Wiley)

3.2 Functionalization of the Paper Array with a βAla-βAla Linker

The procedure is described for working with four arrays, since we recommend functionalizing by batches of at least four arrays. If more arrays (or less) are required, scale the reagents accordingly.

1. Mark the top right corners of the Teflon-patterned arrays using a pencil to assist in remembering the orientation of the arrays throughout the synthesis.
2. Place the arrays in a holder using a forceps handling only the edge of the array. Make sure that the array is centered properly in the holder and that there is equal empty space on all four sides of the array.
3. Prepare the first β-Ala functionalization solution by dissolving the pre-weighed Fmoc-β-Ala-OH in DMF. Immediately before spotting, add 1-methylimidazole to the mixture, and mix by vortexing. Add *N*,*N*′-diisopropylcarbodiimide (DIC) and mix by vortexing again.
4. Spot 15 μL of this functionalization solution on each zone on the paper array. Reaction will occur between Fmoc-β-Ala-OH and the cellulose structure of the paper (Fig. 4, reaction a).
5. Stack up to 4 arrays on top of one another in the fume hood. Cover the arrays to prevent evaporation and allow the reaction to proceed in the dark for *3 h*.
6. Once the reaction is completed, place each array on top of a 96-nozzle vacuum apparatus (Fig. 3) and aspirate the reaction solution through each of the 96 zones on the paper using vacuum suction.
7. To wash the zones, use a multichannel pipette to spot on each zone 15 μL of DMF from a vessel reagent or "boat" previously filled with 50 mL of DMF.
8. Aspirate the wash solution through the paper, as previously described in **step 6**.
9. Repeat **steps 7–8** for a total of four washes.

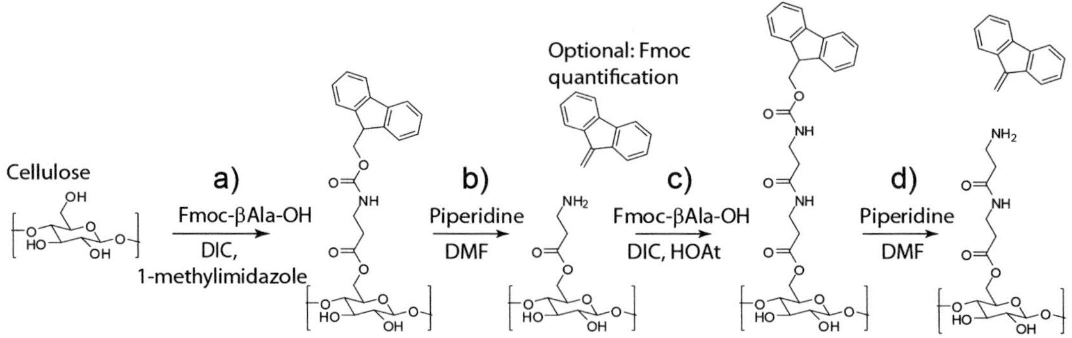

Fig. 4 Functionalization of the paper with a βAla-βAla linker. (**a**) First β-Ala functionalization, (**b**) deprotection of the N-terminus of β-Ala, (**c**) second β-Ala activation, (**d**) deprotection of the N-terminus of βAla-βAla

PP (possible Pause Point): The synthesis can be interrupted between some steps of synthesis. When this happens, we denote it by PP (*see* **Note 11**).

10. To deprotect the N-terminus of the β-alanine bound to the paper, spot 15 μL of solution of piperidine (a mild base) on each zone. For convenience, the piperidine solution as first put in a boat (1.5 mL per array, plus the adequate excess for the dead volume of the boat, for example 3 mL).
11. Let the deprotection reaction (Fig. 4, reaction b) proceed for *5 min*.
12. Aspirate the solution, as previously described in **step 6**.
13. Repeat **steps 10–12** once to complete the deprotection reaction.
14. Repeat **steps 6–9** to wash the arrays.
15. OPTIONAL: To verify the successful β-Ala functionalization of the paper, immerse the array in a container with the bromophenol blue solution (15 mL, poured in the container before to add the paper), and shake the container gently for 1–2 min until a blue color is visible. At this stage, the blue color, indicative of the presence of the amino group from β-Ala, should be uniform in every zone, confirming the successful functionalization of the array. Wash the paper array with 30 mL of methanol for 30 s to remove the blue dye and repeat this washing step four times, until no blue color is released from the paper (*see* **Note 12**).
16. Allow the arrays to air-dry. If the wash solution is aspirated thoroughly, the arrays will dry within 5 min (*see* **Note 13**).

PP = Possible Pause Point (*see* **Note 11**)

17. Prepare the second β-Ala activated solution by dissolving the pre-weighed Fmoc-β-Ala-OH in NMP. Add the HOaT solution, mix by vortexing. Add the DIC solution and mix again by

vortexing. Allow the activation of the amino acid by HOAt and DIC to proceed for *10 min* before spotting.

18. Spot 15 μL of solution on each zone.
19. Allow the peptide coupling reaction (Fig. 4, reaction c) to proceed for *20–30 min*, preferably in the dark, and then aspirate the excess solution through each zone.
20. Prepare Capping solutions A and B or use the premade stock solutions.
21. Spot 15 μL of Capping A solution on each zone.
22. Allow the reaction to proceed for *5 min*, and then aspirate the excess of reagent.
23. Spot 15 μL of Capping B solution on each zone.
24. Allow the reaction to proceed for *10 min*, and then aspirate the excess of reagent.
25. Wash the zones on the paper arrays by following **steps 6–9**.

    PP = Possible Pause Point (*see* **Note 11**)

26. Deprotect the N-terminus of the second β-alanine by following the **steps 10–16** (Fig. 4, reaction d).
27. The arrays modified with the β-Ala-β-Ala linkers can be stored until use.

    PP = Possible Pause Point (*see* **Note 11**)

3.3 Construction of Tables for the Amino Acid Mapping and Automatic Calculations of the Required Reagents for Peptide Synthesis

To facilitate the calculation of the required volumes of amino acids solution, we describe the construction of a set of four tables using EXCEL. These tables yield the maps of the master plates with the required amount of reagents. This section describes in detail how to prepare these tables. Once the tables are constructed, they can be used with very minimal input from the user.

1. Prepare a table to identify the amino acids required at each synthesis cycle. In Table 1, the user add in column C the sequence of the peptide (using the convention: Nterminus-XYZ-Cterminus) for each position indicated in columns A and B. The amino acid required for each cycle to synthesize the peptide (from the C-terminus to the N-terminus) is generated in columns D to R. The letter for each amino acid is extracted from the sequence, i.e., by using formula (1) for cell D3:

$$= \mathrm{IF}\left(\left(\mathrm{LEN}(\$C3) - D\$3 + 1\right) > 0, \mathrm{MID}\left(\$C3, \mathrm{LEN}(\$C3) - D\$3 + 1, 1\right), ""\right) \quad (1)$$

2. Table 2 generates a map for each amino acid to be added at each position of the 8 × 12 array and a total counts for each amino acid per cycle. This second table is generated automatically from the sequences input in column C in Table 1 during

a

	A	B	C	D	E	F	G	H	I	J
1	Position		Write peptide sequences in this column	Cycle #						
2	Row	Col		1	2	3	4	5	6	7
3	1	1-3	STPIQQP	P	Q	Q	I	P	T	S
4	2	1-3	TDSLRLL	L	L	R	L	S	D	T
5	3	1-3	VQPLHKT	T	K	H	L	P	Q	V
6	4	1-3	YHQTTIT	T	I	T	T	Q	H	Y
7	5	1-3	YSIPKSS	S	S	K	P	I	S	Y
8	6	1-3	SPWDARL	L	R	A	D	W	P	S
9	7	1-3	YAAHRSH	H	S	R	H	A	A	Y
10	8	1-3	QALSVYR	R	Y	V	S	L	A	Q
11	1	4-6	MPKYYLQ	Q	L	Y	Y	K	P	M
12	2	4-6	ANTTPRH	H	R	P	T	T	N	A
13	3	4-6	HFRSGSL	L	S	G	S	R	F	H
14	4	4-6	QNTTTAL	L	A	T	T	T	N	Q
15	5	4-6	TKTDTWL	L	W	T	D	T	K	T
16	6	4-6	ALAHRIL	L	I	R	H	A	L	A
17	7	4-6	STPMQNL	L	N	Q	M	P	T	S
18	8	4-6	VLPGRSP	P	S	R	G	P	L	V
19	1	7-9	MHAPPFY	Y	F	P	P	A	H	M
[...]	[...]	[...]	[...]							
34	8	10-12	QAHTVGK	K	G	V	T	H	A	Q

b

	A	B	C	D	E	F	G	H	I	J
1	Position		Write peptide sequences in this column	Cycle #						
2	Row	Col		1	2	3	4	5	6	7
3	1	1;7	GETRAPL	L	P	A	R	T	E	G
4	2	1;7	STASYTR	R	T	Y	S	A	T	S
5	3	1;7	YAGPYQH	H	Q	Y	P	G	A	Y
6	4	1;7	YLTMPTP	P	T	P	M	T	L	Y
7	5	1;7	SPWDARL	L	R	A	D	W	P	S
8	6	1;7	EPLQLKM	M	K	L	Q	L	P	E
9	7	1;7	VIPHVLS	S	L	V	H	P	I	V
10	8	1;7	GRGDS	S	D	G	R	G		
11	1	2;8	SILPYPY	Y	P	Y	P	L	I	S
12	2	2;8	YAAHRSH	H	S	R	H	A	A	Y
13	3	2;8	QALSVYR	R	Y	V	S	L	A	Q
14	4	2;8	HAIYPRH	H	R	P	Y	I	A	H
15	5	2;8	MPKYYLQ	Q	L	Y	Y	K	P	M
16	6	2;8	GVKALST	T	S	L	A	K	V	G
17	7	2;8	ANTTPRH	H	R	P	T	T	N	A
18	8	2;8	GGRDS	S	D	R	G	G		
19	1	3;9	HFRSGSL	L	S	G	S	R	F	H
[...]	[...]	[...]	[...]							
50	8	6;12	GGRDS	S	D	R	G	G		

c

	A	B	C	D	E	F	G	H	I	J	K	L	M	N
1	1	1	1	1	2	2	2	3	3	3	4	4	4	
2	1	P	P	P	Q	Q	Q	Y	Y	Y	P	P	P	
3	2	L	L	L	H	H	H	R	R	R	P	P	P	
4	3	T	T	T	L	L	L	L	L	L	T	T	T	
[...]	[...]	[...]												

d

	A	B	C	D	E	F	G	H	I	J	K	L	M	N
1	1	1	2	3	4	5	6	1	2	3	4	5	6	
2	1	L	Y	L	L	L	L	L	Y	L	L	L	L	
3	2	R	H	L	G	L	W	R	H	L	G	L	W	
4	3	H	R	L	L	L	Y	H	R	L	L	L	Y	
[...]	[...]	[...]												

Fig. 5 Examples of Table 1 for arrays of 32 peptides with grouped replicates (**a**) and arrays of 48 peptides in two block arrays (**b**). Excerpt of Table 2 (first rows) for the example described in a (**c**) and excerpt of Table 2 (first rows) for the example described in b (d)

step 1. A formula generates the map by copying the amino acid of each position from columns D to R of Tables 1 and 2, e.g., formula (2) for cell B2 in Table 2,

$$= \text{INDIRECT}\left(\text{ADDRESS}\left((1 + \$A2 + (B\$1 - 1) * 8), 4 + \$A\$1, 1, 1, \text{"Table1"}\right)\right) \quad (2)$$

where *A1* represents the cycle number, *3+A1* the column number and *(1+$A2+(B$1−1)*8)* the row number in Table 1 from which the amino acid is imported. The term *($B1−1)*8* is required for arrays where different peptides are synthesized in one row as in the examples presented in Figure 5a and b.

The location of the replicates used on the array can easily be changed by modifying the numbers in row 1, columns B–M in Table 2: e.g., the 3 replicates are grouped in Figure 5c whereas the 2 replicates in Figure 5d are separatead to form two blocks displaying the 48 peptides.

For the next synthesis cycles, the only cell that requires modifications in the formula is the cycle number. For example, the formula (3) for cycle 2 (cell \$A\$11) for cell M19 is given by,

$$= \mathrm{INDIRECT}\left(\mathrm{ADDRESS}\left(\left(1+\$A19+\left(B\$1-1\right)*8\right),3+\$A\$11,1,1,"\mathrm{Table1}"\right)\right) \quad (3)$$

This table is particularly useful if the spotting of each activated amino acid solutions is performed manually because Table 2 can be used as a print-out map during the experiment.

A formula is required to calculate the number of occurrences of each amino acid in each cycle, e.g., to calculate for the 12 glycines needed in cycle 1, formula (4) for cell P4 of Table 2 is given by

$$= \mathrm{COUNTIF}\left(\$B\$2:\$M\$9,P3\right) \quad (4)$$

where *\$B\$2:\$M\$9* represents the range of the full map for cycle 1, and *P3* is the cell containing the targeted letter for the amino acid counting (i.e., G).

3. Table 3 calculates the amounts of each amino acid needed for a chosen set of synthesis cycles. Table 3 is based on the example presented in Figure 5a and c. The input of the user is limited to the following six cells: value of the first cycle number (B2) and last cycle number (C2), number of arrays synthesized (G2), volume of solution to be added on each zone (G1, this value can vary depending on the type of paper used), number of repetition of the coupling (J1, only change this value for special cases when the coupling reaction—*see* Subheading 3.4, **step 6**—is performed more than once at each cycle), and the dead volume of the used vessel (J2, typically for deep well plate, this value is 50 μL).

   The "cycle" table on the right-hand side of Table 3 is constructed by repeating formula (4) indicating the letter of the amino acid from column M and adding two conditions (the IF functions) to ensure counting only the considered cycle if it is part of the range of cycles (e.g., 7 as defined in the range of cycles described by cells B2 and C2). For example, for cell AA6, there were 12 A (or alanine), see formula (5) as follows,

$$= \mathrm{IF}\ (\$B\$2 <= \mathrm{IF}\ (AA\$5 <= \$C\$2, \mathrm{COUNTIF}\ ('\mathrm{Table2}'!\$B\$12:\$M\$19,\$M6),0, \mathrm{COUNTIF}('\mathrm{Table2}'!\$B\$12:\$M\$19,\$M6),0) \quad (5)$$

   where *AA\$5* is the considered cycle (i.e., 2), *\$B\$12:\$M\$19* the map of cycle 2 in the second table (values from the table in Figure 5c; cells not visible in the presented excerpt).

4. Column A of Table 3 corresponds to the sum of all the amino acids counted in the "cycle" table described in **step 3**. For example, for cell A6, a value of 48 was obtained from formula (6) as follows

$$= \text{SUM}(\$N6 : \$AB6) \tag{6}$$

5. The total volume of activated amino acid solutions required (*see* Subheading 3.4 **steps 4–6**) are calculated in column D, by combining the user inputs and the total amino acid occurrences from column A. For example, for cell D6, the value of 4800 μL was obtained from formula (7) as follows

$$= \text{A6} * \$G\$1 * \$J\$1 * 250 / 180 * \$G\$2 + (\text{IF}(\text{A6} > 0, \text{A6} / 3,0) * \$J\$2) \tag{7}$$

where the factor *250/180* accounts for the excess volume, and the IF function is used to determine the number of wells of the master plate by dividing the total number of occurrences for each amino acid (A6) by the number of replicates of each peptide per array (here, three replicates are used for the 32 peptides, and so *A6/3* is used), and to count the dead volume (J2) for each well.

6. The amount of amino acids required to achieve a final concentration of 0.45 M in the activated amino acid solutions (amino acid solution mixed with HOAt and DIC) is calculated as follows: total volume (columnD) × molecular weight (columnC) × concentration with corrective factor for the units. The formula (8) for E6 is:

total volume (column D) × molecular weight (column C) ×
concentration with corrective factor for the units. The formula (8) for E6 is given by

$$= (\text{C6} * \text{D6} * 0.45) / 1000 \tag{8}$$

The amounts of the activator HOAt and coupling reagent DIC are calculated similarly with an additional reagent-specific factor (0.5 for HOAt and 0.25 for DIC [11]). The formulae (8′) and (8″) for F6 and H6 are given, respectively, by:

$$= 0.9 * \$F\$27 * 0.5 * \text{D6} / 1000 \tag{8′}$$

$$= 1.35 * \$H\$27 * (\text{D6} / 1000) * 0.25 \tag{8″}$$

where F27 and H27 corresponds respectively to the molecular weight of HOAt (i.e., 135.1 g/mol) and the molecular weight of DIC divided by its density (i.e., 156.6 mL/mol).

7. The volumes for individual reagents are calculated as fractions from the total volume: 3/5 for the amino acid solution

Which cycles ?	start	last		Volume per spot in μL:	15	Number of coupling:	1
	1	7		Number of arrays:	4	Dead volume in μL:	50

Totals	Amino Acid	Vtotal	aa in NMP	HOAt sol.	DIC sol.
		μL	μL	μL	μL
48	A	4800	2880	960	960
0	C	0	0	0	0
9	D	900	540	180	180
3	E	300	180	60	60
18	F	1800	1080	360	360
27	G	2700	1620	540	540
51	H	5100	3060	1020	1020
18	I	1800	1080	360	360
21	K	2100	1260	420	420
72	L	7200	4320	1440	1440
18	M	1800	1080	360	360
18	N	1800	1080	360	360
72	P	7200	4320	1440	1440
54	Q	5400	3240	1080	1080
33	R	3300	1980	660	660
78	S	7800	4680	1560	1560
75	T	7500	4500	1500	1500
15	V	1500	900	300	300
12	W	1200	720	240	240
30	Y	3000	1800	600	600
672			TOTAL	13440	13440

Row	col 1-3	col 4-6	col 7-9	col 10-12
1	STPIQQP	MPKYYLQ	MHAPPFY	STPIQQP
2	TDSLRLL	ANTTPRH	QLMNASR	SHHQKPP
3	VQPLHKT	HFRSGSL	SYHSFNL	SLSLIQT
4	YHQTTIT	QNTTTAL	APRTFNQ	TWYFGPL
5	YSIPKSS	TKTDTWL	TGHSAQG	HSTKVAF
6	SPWDARL	ALAHRIL	SPTGWAP	MPGSLPS
7	YAAHRSH	STPMQNL	SHSLLHH	QEPLTAR
8	QALSVYR	VLPGRSP	MGLQTPY	QAHTVGK

Stock of HOAt sol.:	4939	mg	in 16128	μL of NMP
Stock of DIC sol.:	4261	μL of DIC +	11867	μL of NMP

Fig. 6 Example of a compact version of Table 3. Cells needed for calculations were hidden. Only the required volumes of the amino acids stock solutions, HOAt stock solution, and DIC stock solution are displayed

(column K), 1/5 for HOAt solution (column G), and 1/5 for DIC solution (column J). Since DIC is a liquid, it is the sum of DIC volume (column H) and NMP volume (column I) which makes 1/5 of the total volume.

8. The exact volumes of stock solutions of HOAt and DIC are then calculated in row 26; the row 28 contains the working volumes (i.e., the value in row is 26×1.2, where 1.2 is the excess factor).

9. Table 4 calculates the amount of all the amino acids and reagents required for any particular synthesis cycle(s). A more compact table (Figure 6) can be created by hiding the cells of all the intermediates and focusing on the volumes of the reagents needed for the preparation of the solutions of activated amino acid.

10. Table 4 generates the maps of master plates for each coupling cycle. Typically, multiple deep well plates contain the stock solutions of amino acid at 0.75 M, and the required amount of activator (HOAt) and coupling reagent (DIC) for each cycle,

which will allow for the preparation of the activated amino acid solutions.

No input is required from the user in this table. The range of cycles is imported from Table 3 (cells B2 and C2), and the amino acid map for each cycle from the first table (in this example, we used values of the Figure 5a. For example, the formula (9) for cell C6 is as follows:

$$= \text{INDIRECT}\left(\text{ADDRESS}\left((2+\$O6),3+E\$4,1,1,"\text{Table1}"\right)\right) \tag{9}$$

where the location of the amino acid for each well is defined from its row number (\$O6) and the considered cycle (E\$4). In this example, *(2+\$O6),3+E\$4* corresponds to cell D3 of the table in Figure 5a, and reports a proline (or P).

The cycle numbers, as in cell E4 or I4 of Table 4, are displayed in correlation with the range of cycles selected using formula (10), which for cell I4 is as follows:

$$= \text{IF}\left(\$C\$2+1 <= \$D\$2, \$C\$2+1, 0\right) \tag{10}$$

where the term \$C\$2 +1 is the cycle number (i.e., 2), and the term *+1* needs to be incremented for each additional cycle. For example, for cell E37, the formula (10′) is given as follows:

$$= \text{IF}\left(\$C\$2+6 <= \$D\$2, \$C\$2+6, 0\right) \tag{10'}$$

11. Finally, the volumes for the stock solutions of amino acid (G1) and the two other reagents (same volume for DIC and HOAt, so the same formula is used to display it in the last two columns of each deep well plate, e.g., cells M10 and N10 for the first cycle) are calculated from data of Table 3 with the following formulae respectively:

$$= (\text{'Table4'!}\$G\$1 * \text{'Table4'!}\$J\$1 * 250/180 * \text{'Table4'!}\$L\$1 * 3 + \text{'Table4'!}\$L\$2) * 3/5 \tag{11}$$

$$= IF(\$D\$2 \quad \$C\$2 >= 0, \text{COUNTIF}(\$C4:\$J4,">0") * \$G\$1/3 * 4 + \text{'Table3'!}\$L\$2, 0) \tag{12}$$

In formula (11), all the references to 'Table 3' were previously described in **step 3**. The ratio 250/180 accounts for an excess of liquid and the multiplier 3 represents the number of replicates to be spotted per array. The IF function in formula (12) counts the number of cycles in this particular deep well plate. In the example of Table 4, plate # 1 contains solutions for two cycles; whereas plate # 4 contains solutions for one cycle. The ratio 1/3 corresponds to the volume of reagent relative to the calculated volume of amino acid solution in cell G1 and the multiplier 4 represents the number of amino acid solutions to which the reagents need to be added.

3.4 Syntheses of the Peptides Through Successive Addition of the Amino Acids

1. Calculate the amounts of amino acids and coupling reagents needed. Generate the maps of the master plates using Tables 1, 2, 3, and 4, as described in Subheading 3.3.
2. Prepare the master plates (deep well plate) by filling each well with the required amounts of the stock solutions of amino acids, HOAt and DIC following the map of Table 4 (*see* **Note 14**).
3. Dry the paper arrays completely prior to the coupling of the amino acids.
4. Using an 8-channel pipettor, transfer the pre-calculated volumes of stock solutions (HOAt and DIC) to the wells of the master plate for the required cycle. In the specific example of Table 4: 60 μL of 2.25 M HOAt and 60 μL of 1.68 M DIC are added to the well containing 180 μL of amino acid solutions. Mix the solutions well by pipetting up and down multiple times.
5. Allow the activation of the amino acid to proceed in the master plate for *10 min*.
6. Spot 15 μL of the activated amino acid solutions from the master plate onto the paper array(s) following the map of Table 2 using an 8-channel pipettor and allow the reaction to proceed for *30 min*, preferably in the dark (*see* **Note 15**).
7. Aspirate the excess solution, as described in **step 6** of Subheading 3.2 (*see* **Note 16**).
8. Cap the unreacted sites, as described in **steps 20–25** of Subheading 3.2.

   PP = Possible Pause Point (*see* **Note 11**)

9. Deprotect the N-terminus of the newly coupled amino acid, as indicated in **steps 10–16** of Subheading 3.2 (*see* **Note 17**).

   During the last washing step, the master plate that contains the solutions of amino acids for the next cycle can be warmed up to room temperature.

10. Ensure that the arrays are completely dry, and then proceed to the next synthesis cycle by repeating the **steps 4–9** until all amino acids are coupled.

3.5 Cleavage of the Side-Chain Protecting Group

1. Thaw phenol at room temperature for 15 min before weighing it.
2. Prepare the Cleavage A and Cleavage B solutions; the reagents must be added following the order given in Subheading 2.1. Slowly swirl the solution after each addition of reagents during the preparation.
3. Place each array in an individual container (e.g., small polypropylene box or glass dish). Make sure the array face upwards using the pencil mark on the top right corner of the array. Add ~15 mL of Cleavage A solution to the container and submerge the array in it. If air bubbles are trapped under the array, lift the

corner of the array with a forceps and submerge the array again until all bubbles disappear.

4. Close each container and allow the cleavage reaction to proceed for *30 min* in the dark.
5. Open the container(s) and pour the excess Cleavage A solution into a designated waste disposal container (e.g., "corrosive"). Then, wash the arrays with 4 × 15 mL of DCM. (*see* **Note 18**).
6. Repeat **steps 3** and **4** with 15 mL of Cleavage B solution, but allow the reaction to proceed for *3 h instead of 30 min*.
7. Wash the arrays with 4 × 15 mL of DCM and 2 × 15 mL of ethanol. Air-dry the arrays in a fume hood.
8. The arrays can be used for biological assays immediately or they can be stored at −80 °C until use.

3.6 Semi-automated Synthesis on Patterned-Paper Microarray

In this section, we described the adaptation of protocols detailed in Subheadings 3.1 and 3.2 to achieve an automated generation of paper arrays of peptides with a liquid-handling workstation (BioTek Precision XS). Examples of the programs for the workstation can be found in the supporting information of Deiss et al. [10]. We anticipate the user to be familiar with the general operation of the Biotek instrument as well as basic programming of the instrument and the associated terminology.

3.6.1 Automated Deposition of Teflon

1. In the program, define the paper as a new receiving vessel, based on a 96-well plate format: add a vertical offset along the *z*-axis to define as the bottom of the vessel the height of the paper in the holder on the platform of the liquid-handling workstation.
2. Adjust the specification files of the dispensing system to allow spotting of the solutions from a 5-mm height above the paper.
3. Adjust the dispensing rate to 1 (the slowest) for the distribution of the viscous sucrose solution.
4. Prepare a patterning program [10]. Aspirate with the 8-channel dispenser 8 × 100 μL of sucrose solution from a boat and dispense four times 8 × 25 μL in the first four columns. Loop three times to ensure deposition of sucrose solution on the 96 zones of the arrays. Dispense 20 μL of Teflon solution using the single-channel bulk dispenser between each zone using a horizontal offset; loop the dispensing 88 times. Dispense 25 μL of Teflon solution on the left and right side of the arrays using the same dispenser 8 times for each side. Repeat the program four times for the four positions of the paper arrays on the workstation.
5. Follow **steps 1–4** of Subheading 3.1 to prepare the paper array. Place four holders with wax-patterned paper arrays on the workstation matching the position of the receiving vessels. A

stacker loaded with multiple holders can also be used for the automated deposition of Teflon on a larger number of paper arrays.

6. Fill a boat on the workstation with 50 mL of the sucrose solution.
7. For the distribution of the Teflon solution, a bulk dispenser is recommended as it allows the solution to be dispensed from a closed bottle, which prevents evaporation of the solvent.

   Initialize the single-channel dispenser, use the "prime" function to rinse the tubing with 2000 μL of HFE-7100 (i.e., methoxyperfluorobutane). Then, prime the tubing with 1500 μL of Teflon solution.
8. Load the Teflon patterning program prepared at **step 4** and start the run.
9. When the program has proceeded for the four arrays, the arrays can be put aside and a new batch of four arrays can be prepared by simply refilling the boat with the sucrose solution and starting the run again without unloading/reloading the program.
10. Process the Teflon-patterned paper arrays, as described in **steps 7–11** of Subheading 3.1.
11. When finished with the full batch of arrays, clean the single-channel dispenser. This is achieved by purging 2000 μL back from the Teflon solution bottle, and priming it with 4000 μL of HFE-7100 to rinse the residual Teflon in the dispenser. Prime with 3 × 4000 μL of ddH_2O, then use the "purge" function to empty the tubing by purging a fictive liquid volume of 2000 μL prior to shutdown.

3.6.2 Automated Functionalization of the Paper Array with a βAla-βAla Linker

1. Follow the functionalization protocol described in Subheading 3.2 (**steps 1** and **3**). At **step 4**, use the liquid-handling workstation to spot 15 μL of the first β-Ala functionalization solution. Use a script similar to the one used for the deposition of the sucrose solution, as described in **step 4** of Subheading 3.6. Aspirate 8 × 90 μL from a boat containing 9 mL of first β-Ala functionalization solution, and dispense six times 8 × 15 μL of solution on the array. Loop twice for the entire array. Repeat four times for the four paper vessels on the platform.
2. For the Fmoc-deprotection detailed in **steps 10–16** of Subheading 3.2, use one program for both depositions of piperidine solution. Aspirate 8 × 90 μL from a boat containing 9 mL of piperidine, dispense six times 8 × 15 μL of solution on the array. Loop twice for the entire array. Repeat four times for the four paper vessels on the platform. Add a 5-min timer followed by a "pause" using the function "stand-by until user resume" in "replenish of supply" to allow the time for the user

to transfer the plate to the 96-nozzle aspirator. Then, repeat the same program for the second spotting of piperidine.

3. The four washing steps are combined in one program. Aspirate 8 × 90 μL from a boat containing 50 mL of DMF, dispense six times 8 × 15 μL of solution on the array. Loop twice for the entire array. Repeat four times for the four paper vessels on the platform. Add a "pause" using the function "stand-by until user resume" in "replenish of supply" to allow the time for the user to transfer the plate to the 96-nozzle aspirator, and then place the holders back on the platform. Repeat the program four times.
4. For the spotting of the second β-Ala activated solution, as described in **step 18** of Subheading 3.2, use the program made for **step 1** but fill the boat with the second βAla activated solution.
5. For the capping step described in Subheading 3.2 **steps 20–25**, prepare a program similar to the Fmoc deprotection in **step 13**. Aspirate 8 × 90 μL from a boat containing 9 mL of Capping A solution, and dispense six times 8 × 15 μL of the solution on the array. Loop twice for the entire array. Repeat four times for the four paper vessels on the platform. Add a 5-min timer followed by a pause using the function "stand-by until user resume" in "replenish of supply" to allow the time for the user to transfer the plate to the 96-nozzle aspirator. Repeat the program and change the location of the boat. Fill the second boat with 9 mL of the Capping B solution. Change the timer from 5 to 10 min.

All the steps of washing, Fmoc-deprotection, and capping can be performed by running the programs written in **step 13**. The programs required for the coupling cycle of the amino acids solution are created for each cycle, using programs similar to those described in the previous section. Briefly, there are three key steps: (1) the automated mixing of the DIC, HOAt and amino acid solutions, (2) a 10-min incubation and, (3) the spotting of the solutions onto the arrays (8 × 15 μL at a time). Examples of programs are available *elsewhere* [10].

3.7 Cleavage of the Peptides from the Paper for Characterization

1. Using a 1-hole punch with a diameter of 3.1-mm (1/8 in.) to punch out the areas of paper that contain the peptides. The area should be smaller than the size of one square (4 mm × 4 mm) of the paper array.
2. Place the pieces of punched paper in 2-mL glass vials.
3. Place all the open vials in a glass desiccator and evacuate the air from the desiccator (e.g., using vacuum or aspirator).
4. The peptides will be cleaved off the paper by ammonia. Generate ammonia gas by mixing 20 g of sodium hydroxide

(NaOH), 20 g of ammonium chloride (NH_4Cl) and 10 mL of water in a 500-mL Erlenmeyer flask in a fume hood. Shake the flask to start the gas flow. Use a tubing to connect the flask to the desiccator and fill it with ammonia gas. Allow the cleavage reaction to proceed overnight. WARNING: Ammonia gas is corrosive and irritating; this step must be performed in a fume hood.

5. After the dry aminolysis, open the desiccator and release the ammonia gas for 30 min in the fume hood.
6. Add 100–200 μL of water to each vial, mix by vortexing and incubate for 1 h to desorb the peptides from the paper.
7. Transfer the solutions into new vials and perform the peptide analysis (e.g., characterization by LC-MS).

3.8 Ethanol Sterilization of Paper for Biological Assays

If you require sterile arrays for biological assays such as ligand binding assays, follow this protocol prior to the assay.

1. Submerge the paper arrays in 95 % ethanol for 30 min.
2. Discard ethanol and air-dry the paper in a bio-hood for 1 h.
3. Rinse the arrays by submerging them in sterile basal media (e.g., DMEM) for 5 min. Repeat the rinsing step three times.
4. Rinse twice with the media required for your biological assay.

4 Notes

1. The commercial solution of Teflon AF is very viscous. To prepare a dilute solution of Teflon, pipette the commercial solution slowly and transfer it into a bottle already containing methoxyperfluorobutane (or HFE-7100). Mix by pipetting up and down multiple times until the solution does not appear to be viscous anymore.
2. Warm up the bottles containing amino acids to room temperature for 15–30 min before opening the bottles. The Fmoc-protected derivatives of arginine, cysteine, asparagine, glutamine and histidine require extensive vortexing and time to dissolve.
3. The BioTek Precision XS workstation can handle four arrays at a time. It will take ~1.5 mL of solution per array. The reagent vessel—50 mL polypropylene “boat”—has a dead volume of 3 mL.
4. The Cleavage solutions A and B must be prepared fresh before the step of cleavage of the protecting groups, as described in the Subheading 3.5. The Cleavage B solution must be protected from light.
5. The Teflon solution does not impregnate well the cellulosic fibers of the paper coated with the solid ink (or wax). Therefore,

the use of a stroke of 0.8 pt for Whatman paper No. 50 (or 1 pt for thicker paper like Whatman paper grade 1) maximizes the area accessible to the Teflon solution for the formation of the solvophobic barriers.

6. Do not stack the paper arrays in the oven to ensure proper distribution of the heat and to avoid transfer of wax from one sheet of paper to another.
7. After deposition of the sucrose solution, minimize the delay before depositing the Teflon solution, because the sucrose solution can drip through the paper then allows a "puddle" of Teflon to form on top of the flattened drop of sucrose.
8. Drying of the paper arrays can be accelerated by placing them in an oven at 60 °C for 20 min.
9. For arrays that are dried overnight, the sucrose solution will harden and a long washing time is required to remove it.
10. The defects observed on the paper after the wash can be marked with a pencil. This mark will be used to facilitate the subsequent "repair" of the defects with Teflon solution.
11. At PP (Possible Pause Point), ensure that the paper arrays are dry. If needed, wash the arrays twice with methanol (or ethanol) and dry them by a natural air flow in a fume hood or by using *cold* air from an air gun/hair dryer. Place the arrays in a resealable plastic bag, and store them at −20 °C for a short time or at −80 °C for longer time. The arrays have to be warmed up to room temperature for 30 min prior to resuming the synthesis.
12. The presence of the bromophenol blue dye on the paper does not affect the subsequent peptide synthesis.
13. To remove the excess DMF from the paper arrays and accelerate their drying, spray the arrays with ethanol and aspirate it.
14. Master plates can be prepared ahead of time and stored for up to 1 week at −20 °C. If a master plate is used on the same day, it can be stored at 4 °C.
15. If the optional step for bromophenol blue staining test (*see* **step 15** of the Subheading 3.2) was performed, at this **step 6** of Subheading 3.4, the zones should have a yellow-green color.
16. Glutamine and arginine can precipitate during the coupling reaction, and thus the arrays need to be aspirated for a longer time. If after 2 min, the excess solution is still not completely removed by the vacuum, add DMF on the problematic zones and continue to aspirate for another minute.
17. After Fmoc-deprotection the zones should become blue-colored. The color intensity depends on the nature of the amino acid, and thus the staining gives only qualitative infor-

mation. If the color fades after a few coupling cycles, repeat **step 15** of Subheading 3.2.

18. In the presence of Cleavage A and B solutions, the paper arrays are fragile and might get damaged by a strong flow. Use caution when slowly pouring DCM onto the arrays.

References

1. Lam KS et al (1991) A new type of synthetic peptide library for identifying ligand-binding activity. Nature 354(6348):82–84
2. Kehoe JW, Kay BK (2005) Filamentous phage display in the new millennium. Chem Rev 105(11):4056–4072
3. Koivunen E, Wang B, Ruoslahti E (1995) Phage libraries displaying cyclic peptides with different ring sizes: ligand specificities of the RGD-directed integrins. Nat Biotechnol 13(3): 265–270
4. Boder ET, Wittrup KD (1997) Yeast surface display for screening combinatorial polypeptide libraries. Nat Biotechnol 15(6): 553–557
5. Roberts RW, Szostak JW (1997) RNA-peptide fusions for the in vitro selection of peptides and proteins. Proc Natl Acad Sci U S A 94(23):12297–12302
6. Pauloehrl T et al (2013) Spatially controlled surface immobilization of nonmodified peptides. Angew Chem-Int Ed 52(37): 9714–9718
7. Pirrung MC (1997) Spatially addressable combinatorial libraries. Chem Rev 97(2):473–488
8. Frank R (1992) Spot-synthesis: an easy technique for the positionally addressable, parallel chemical synthesis on a membrane support. Tetrahedron 48(42):9217–9232
9. Blackwell HE (2006) Hitting the SPOT: small-molecule macroarrays advance combinatorial synthesis. Curr Opin Chem Biol 10(3):203–212
10. Deiss F et al (2014) Flow-through synthesis on teflon-patterned paper to produce peptide arrays for cell-based assays. Angew Chem-Int Ed 53(25):6374–6377
11. Hilpert K, Winkler DFH, Hancock REW (2007) Peptide arrays on cellulose support: SPOT synthesis, a time and cost efficient method for synthesis of large numbers of peptides in a parallel and addressable fashion. Nat Protoc 2(6):1333–1349
12. Martinez AW et al (2007) Patterned paper as a platform for inexpensive, low-volume, portable bioassays. Angew Chem Int Ed 46(8): 1318–1320

Chapter 19

Cell Microarrays for Biomedical Applications

Mario Rothbauer, Verena Charwat, and Peter Ertl

Abstract

In this chapter the state of the art of live cell microarrays for high-throughput biological assays are reviewed. The fabrication of novel microarrays with respect to material science and cell patterning methods is included. A main focus of the chapter is on various aspects of the application of cell microarrays by providing selected examples in research fields such as biomaterials, stem cell biology and neuroscience. Additionally, the importance of microfluidic technologies for high-throughput on-chip live-cell microarrays is highlighted for single-cell and multi-cell assays as well as for 3D tissue constructs.

Key words Cell microarray, Microfluidic, Integrated microdevices, Single cell, Organoids, Spheroids

1 Introduction

The implementation of micromachining and micro-scale technologies for biomedical applications enables the advanced in vitro cell analysis using cellular microarrays, microfluidic systems, and micro-scale diagnostics. The greatest benefit of miniaturized cell analysis systems is the ability to provide quantitative data in real time with high reliability and sensitivity, which are key parameters for any cell-based assay. An additional advantage of cell-based microarrays is their inherent high-throughput capability, which allows for large-scale screening of single cells, multi-cell populations, and spheroids. The interest in cell microarrays is also reflected in a rapid increase in the number of publications over a period of 10 years. Figure 1 provides an overview of number of manuscripts published between 2000 and 2013 based on article search in the ScienceDirect database using keywords such as microarray, cell microarrays, and 3D and single-cell microarrays.

Miniaturization of cell assays using cell microarrays increases not only throughput but also significantly reduces the consumption of reagents and the requirement of cellular material, which are key criteria when using clinical grade reagents and primary cell cultures. The proven cost reduction of cell-based microarrays

Paul C.H. Li et al. (eds.), *Microarray Technology: Methods and Applications*, Methods in Molecular Biology, vol. 1368, DOI 10.1007/978-1-4939-3136-1_19,

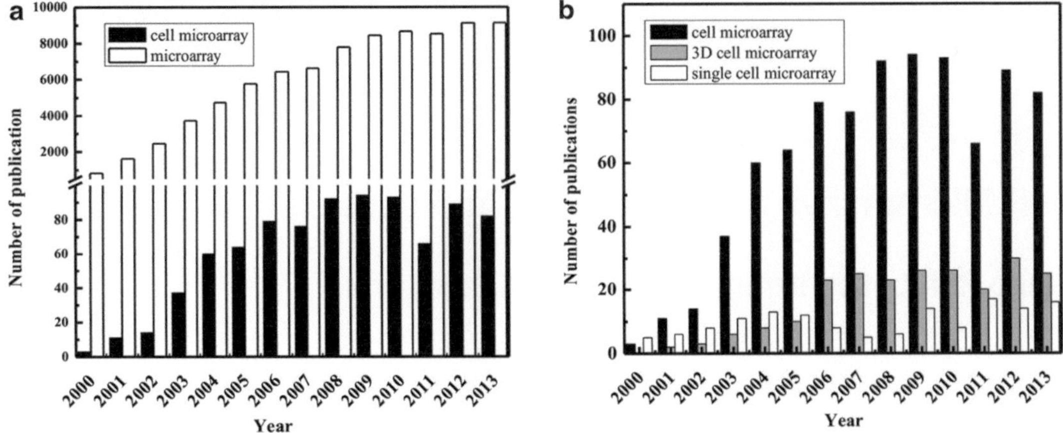

Fig. 1 (**a**) Histogram illustrating the number of publications containing the keywords "microarrays" and "live cell microarrays" over a 13-year period. (**b**) Histogram illustrating the number of publications related to "single cell," "3D/ spheroid," and "cell microarrays" between 2000 and 2013

makes them a highly attractive tool for a wide range of applications in pharmacology, toxicology and stem cell research [1]. Consequently, cell microarrays have been explored for pharmacological applications to determine gene expression, cell-to-surface interaction, extracellular matrix (ECM) production, cell migration and proliferation [2]. Additionally a variety of cell microarrays have been used to study alterations of intracellular/extracellular biochemistry, cell morphology, motility and adhesion, survival/apoptosis, and proliferative properties.

Generally, there are two strategies for the fabrication of microarrays for cell analysis and they involve either direct or indirect cell patterning approaches. The indirect method involves the placing of cells on top of pre-modified surfaces that allow for cell attachment. The most important application of indirect patterning is the so-called "reverse transfection," developed by the Sabbatini group in 2001, where small spots of vector constructs are printed on a slide, and a cell layer is then cultivated on the slide [3]. Functional assays are subsequently performed to identify effects of gene and protein overexpression or knockdown. Indirect cell patterning requires proper surface chemistry and applied functionalization procedures become a determining factor in the successful fabrication of cell microarrays. On the other hand, the direct cell pattering approach is mainly based on integrated geometric features within the microdevice including channels, grooves, wells, and pillars that allow for efficient cell capture. A recent technological advancement of cell-based microarrays involves their combination with microfluidics to enable nutrient supply and waste removal for optimum cell culture conditions. Microfluidics is considered to be a technology that allows for the precise manipulation and control of very small fluid volumes down to pL scale. The main advantages of integrating microfluidic channels to cell-based microarrays is the ability to regulate and transport fluids, soluble factors, drug candidates, and bioactive substances at specific solution concentrations and gradients. The application of microfluidics has already shown to create new opportunities for the spatial and temporal control of cell proliferation and stimuli [4]. For instance, microfluidic cell assays have been used for conducting fast screening experiments, evaluating drug-related toxicity, and elucidating optimal cell culture conditions. The current trend towards the integration of sensory systems into cell-based microsystems leads to the creation of fully automated, highly integrated multifunctional Lab-on-a-Chip (LOC) systems for biomedical, biomaterial, and pharmaceutical research [5–9].

2 Live-Cell Microarrays

Understanding the impact of bioactive substances on cell cultures is a fundamental aspect of many biomedical research fields ranging from cell biology studies to drug testing and development of optimized cell cultivation strategies. Originating from the DNA microarray technology, patterns of small molecules and peptides were initially used to assess their suitability as cell adhesion promoters in large-scale screening efforts. For instance, screening of cell adhesion-promoting proteins was investigated by Ito et al. in 2005 using photochemistry for microarray patterning [10].

In another early study, different cell types were seeded on a peptide array to analyze cellular activities including cell adhesion and functional phosphorylation in response to different stimuli [11]. In a similar study, a large range of peptides was screened for their potential of binding Jurkat cells on top of predefined patterns [12]. There were other functional screening studies which investigated immune responses of antigen-specific T-cells mediated by major histocompatibility complex (MHC) using the peptide-MHC arrays [13, 14]. In addition to peptides, another substance of great interest for cell interaction studies is the glycans, which are known to play important cell biological roles including cell–cell signaling and immune responses. As an example, selective binding of $CD4^+$ T cells to carbohydrates was demonstrated by microarray screening [15]. Additionally, cell–membrane interactions have been studied in the microarray format where a large number of cell adhesion molecules were characterized in the presence of different membrane compositions [16, 17]. Furthermore, cell microarrays have also been used for functional analysis of polymers for application in cell culture handling. In 2004, Anderson et al. screened a polymer library for transfection efficiency in cancer cells with the goal to find a nonviral DNA vector for gene therapy in cancer treatment

[18]. Another interesting microarray screening for polymers was aimed at identifying relationships between surface chemical structure and related protein adsorption), and the effect of polymer on cell adhesion [19].

In this respect, cell microarrays demonstrated their usefulness to screen for cellular response to synthetic and natural biomaterials such as synthetic polymers and ECM-derived adhesion promoters, respectively [20–22]. The focus of these studies is based on surface chemistry and surface topography as well as biological interactions with surface-patterned biomolecules. For instance, heterogenic polymer-based microarrays can be used to identify new biomaterials that support adsorption of ECM-derived adhesion promoters, thus promoting cell adhesion [23]. Another approach implemented switchable thermo-responsive surface microarrays to support cell adhesion with consecutive nonenzymatic sample release upon temperature decrease [24, 25]. An alternative application of cell microarrays is the screening of antifouling surfaces that are able to resist bacterial adhesion, which is important to prevent biofilm formation at the surfaces of biomaterials [26–28]. Furthermore, microarrays containing different topographic patterns have been used to elucidate the interplay between surface topography and cellular behavior, and to find improved biomaterials for cell culture applications [29, 30].

The use of antibody microarrays as a platform for high-throughput screening of immune cells in blood to detect specific surface markers on immune cells constitutes a promising approach. Here, single cells and whole cell populations can readily be captured on the patterned antibody regions and functionalized microstructures because of specific cell-to-surface interactions [31, 32]. One prominent example involves the use of antibody arrays for phenotyping of blood cells to identify subpopulations of $CD19^+$ B lymphocytes, $CD16^+$ neutrophils, $CD36^+$ monocytes as well as $CD4^+/CD8^+$ T cells [33–35]. In a similar manner, antibody arrays have been applied for serotyping prokaryotic cells [36, 37]. Overall speaking, the antibody-based microarray technology can advance state-of-the-art monitoring of disease and pathological conditions (HIV, leukemia, circulating tumor cells, etc.) by increasing the sample throughput as well as the throughput of antigens to be tested.

In recent years, live-cell microarrays have mainly been employed for parallelization and high-throughput analyses in the field of cell biology, tissue engineering and tissue regeneration to gain a deeper understanding of dynamic cell response. For instance, cell-based microarrays have been applied for investigation of cell proliferation and morphology changes [38], protein expression [39], transfection of cell cultures; imaging of (single) cells and tissue constructs [40], as well as for cell–cell, cell–surface, and cell–substance interaction studies. We especially want to emphasize the importance of cell microarray technology in the field of neuroscience, stem cell research as well as cell-to-surface interaction studies. Figure 2

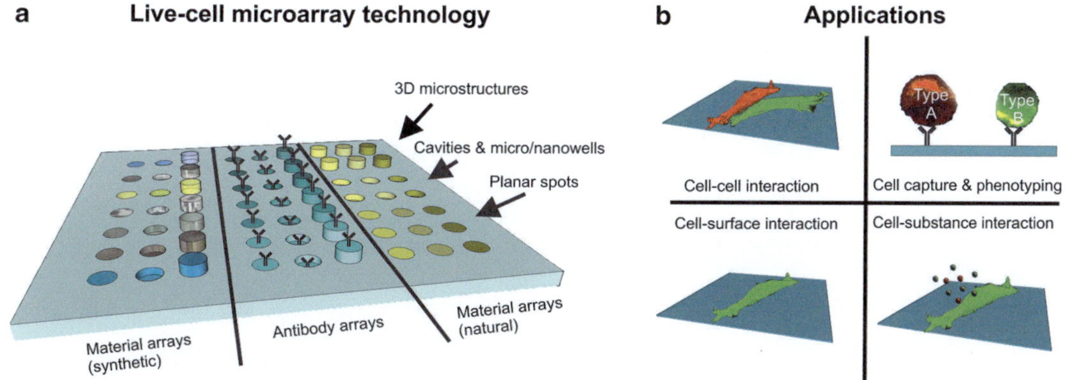

Fig. 2 Live-cell microarray technology. (**a**) Microarrays based on planar patterns, wells and 3D microstructures, (**b**) various applications of live-cell microarray for cell analysis

shows a schematic overview of life-cell microarray technology based on planar spots, cavities, and microwells as well as 3D microstructures for cell analysis applications.

With the advent of cell-based personalized therapies, cell microarrays have been increasingly applied in stem cell research for the analysis of the fate of cellular processes and of biomaterial interaction, and for the screening of cell-specific stimuli [41]. As for the fate of cellular processes, where the spatiotemporal control of the cellular microenvironment plays a major role, microarrays have been applied to control cell morphology as well as migration, differentiation, proliferation and the health status of stem cells [42–44]. As an example, the responses of human embryonic stem cells (hES) to biomaterials were investigated using microarrays revealing the impact of individual compositions of biomaterials on stem cell [20, 21]. Furthermore, ECM microarrays have been employed for high-throughput analysis of environmental factors that guide stem cell function and fate. To account for the natural 3D microenvironment of cocultures that is present in native tissues, spheroid microarrays [45] have been recently developed to improve the differentiation efficiency of multipotent mesenchymal stem cells (MSCs). In this study, the osteogenic and adipogenic differentiation efficiency as well as phenotype maintenance (*see* Fig. 3) was significantly increased by introduction of 100 μm-sized cell spheroid microarrays.

Another prominent application of live-cell microarrays is in the field of neuroscience, where the analysis of single neurons as well as large neuronal populations is of highest interest for understanding brain development including the onset and progression of degenerative diseases. Here, live-cell microarrays are able to overcome limitations of conventional analysis technologies that mainly record neuronal data based on the activity of cellular clusters, thus

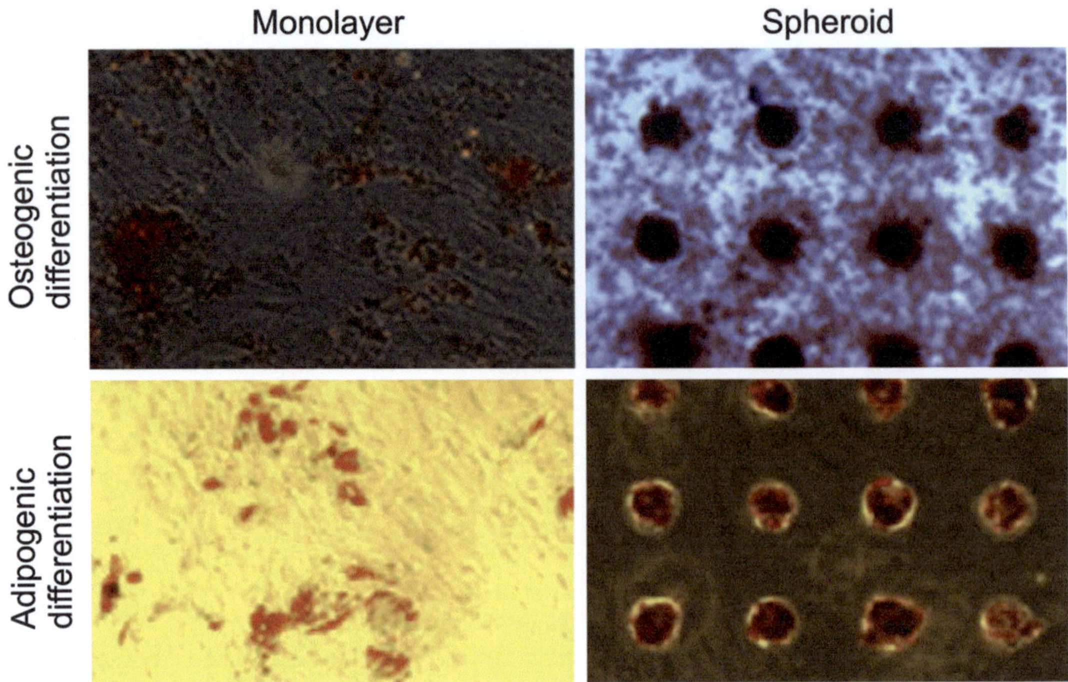

Fig. 3 Comparison between monolayer culture and spheroid in the live-cell microarray technology on the capacity of mesenchymal stem cells (MSCs) to differentiate in osteogenic and adipogenic lineage (adapted from ref. 45, with permission from Elsevier)

only providing information on subpopulations of neurons [46]. It is important to note that spatially resolved analysis of neuronal populations have highlighted that local as well as global cell densities play a crucial role in neural network activity [47, 48]. Consequently, cell patterning approaches combined with multi-electrode arrays (MEA) containing 4,096 single electrodes have shown to retain the key properties of random neuronal networks such as transmission, short-term plasticity as well as bulk network activity [49]. In a similar manner, MEAs with an even higher spot density of 11,011 microelectrodes has been used to visualize network topography and action potential propagation of neurons upon stimulation and record dynamic responses with single-cell resolution [50]. Figure 4 shows examples of microarray layouts for the investigation of neural networks.

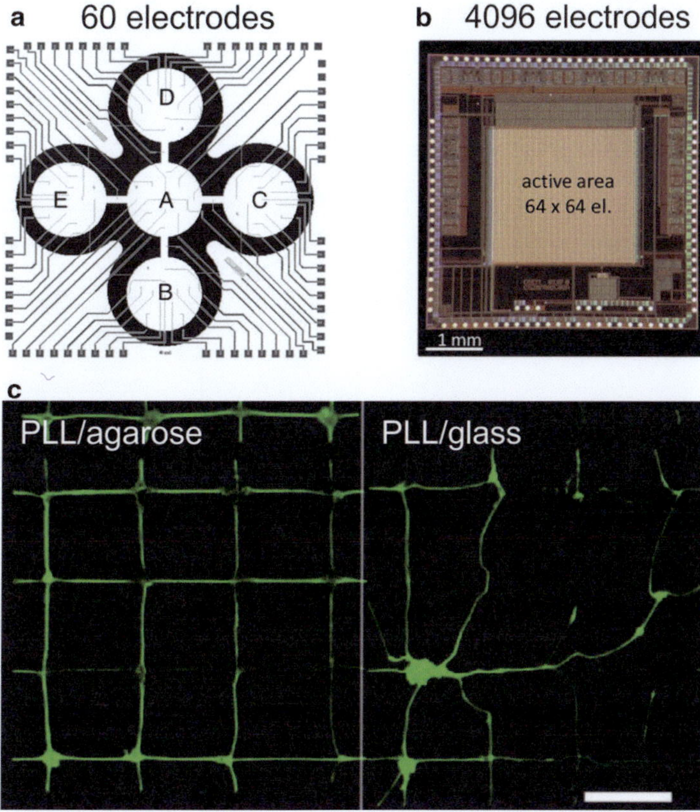

Fig. 4 Live-cell microarray for applications in neuroscience. (**a**) Microdevice with 60 electrodes for the analysis of multiple cellular populations (*A–E*) (adapted from ref. 46 with permission from Elsevier). (**b**) Multi-electrode array MEA biosensor featuring 4096 electrodes for high-resolution measurements of neuronal networks. (**c**) Micro-contact printed PLL/agarose surface patterns for establishment of a symmetric network of primary neurons. Scale bar, 200 μm

3 Microfluidic Live-Cell Microarrays

As outlined in the previous section, cell-based microarrays have enabled deeper insights into important cell biological aspects such as cell-to-cell and cell-to-surface interactions, and have mainly been applied for toxicological screenings of pharmaceutical compounds and nanomaterials. Recent advancements of live-cell microarrays included the integration of microfluidic channels that allow for further miniaturization and automation of cell-based assays. Developed in the early 1980s, microfabrication and MEMS technology were revolutionized by Whitesides and colleagues in the late 1990s with the introduction of soft lithography, which made microfluidic technology available to a broad scientific community [51]. The main advantage of microfluidics for cell analysis

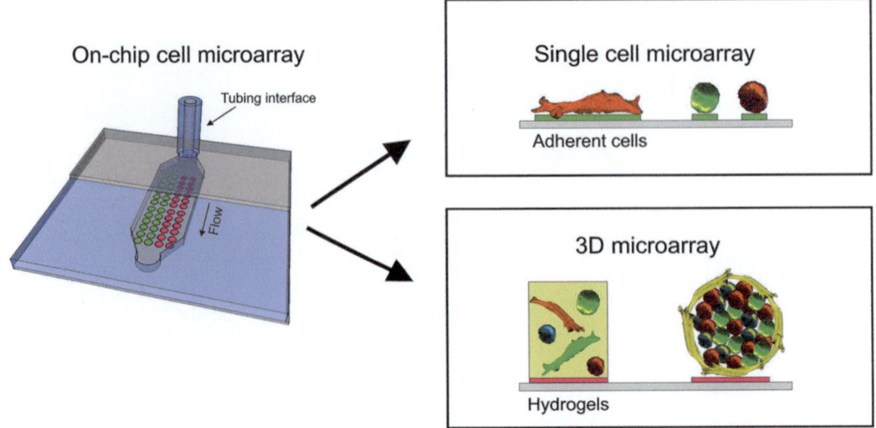

Fig. 5 Schematic overview of microfluidic single-cell and 3D microarray technologies

is the incorporation of automated fluid handling routines, which allows for reproducible cell culture conditions. Some of the many benefits of microfluidics for cell culture handling include control over surface chemistry, topography, and geometry as well as the precise transport of fluids and soluble factors (growth factors etc.). Therefore, the increasing effort on combining microfluidics with various cell patterning methods has led to the development of next-generation live-cell microarrays over the last decade [52–55]. For instance, one of the major issues addressed by microfluidic live-cell microarrays is the inherent heterogeneity of cell populations by developing high-throughput single-cell microarrays that recapitulate the biological situation, thus providing in vitro data that are more relevant to in vivo situations [56].

Based on the relevance of microfluidic live-cell microarrays for industrial and clinical applications, the various advancements of these microarrays for single-cell and three-dimensional assays (*see* Fig. 5) are described in more detail in the following two sections.

3.1 Microfluidic Single-Cell Microarrays

Practical applications of microfluidic single-cell arrays include tumor biology, stem cell biology, antibiotic resistance screening, as well as single-cell immune-typing. As indicated above, microfluidic single-cell microarrays are ideally suited to assess the heterogeneity within a cell population by analyzing the responses of a large number of individual cells to provide information on subpopulation distribution, cellular activities, and the ratio of responding and non-responding cells. It is important to highlight that the intrinsic cellular heterogeneity of single circulating tumor cells (CTCs) determines the metastatic potential, thus further highlighting the importance of single-cell analysis approaches. The main emphasis in tumor biology, therefore, is concerned with understanding the formation and growth of primary tumors, local tumor cell invasion, migration and extravasation, and finally, tumor cell metastasis.

However, one of the major limitations of assessing for example the metastatic potential of CTCs is the inherent difficulty of isolating 10–100 rare cells in blood within high background of normal blood cells (10^9 to 10^{10} cells per mL). Microfluidic single-cell arrays can facilitate the study of CTCs by providing the diagnostic tools capable of isolating and analyzing CTCs using surface marker-based and marker-free methods. While surface marker-based methods predominantly employ magnetic beads for cell capture [57, 58], marker-free microfluidic isolation methods use pillars and flow focusing approaches [59–61]. A prominent example of single-cell analysis using microfluidic single-cell microarrays is the application of genotyping and mechano-typing, also called "deformability cytometry," which has been established for identification of malignant and benign cells [62–64]. Another example of a microfluidic single-cell microarray integrates an array of PDMS-based cell-capture pockets (*see* Fig. 6a) that can be used to detect tumor proliferation and apoptosis following the administration of anticancer agents [65]. More recently, single-cell gene profiling has been used to identify different populations of CTCs thus highlighting that cellular heterogeneity is a major factor in cell-based assays [66]. Furthermore, it could be shown that expression profiles of CTCs diverged distinctly from those of well-established cancer cell lines, thus questioning the suitability of conventional in vitro models for drug discovery and cancer therapy research.

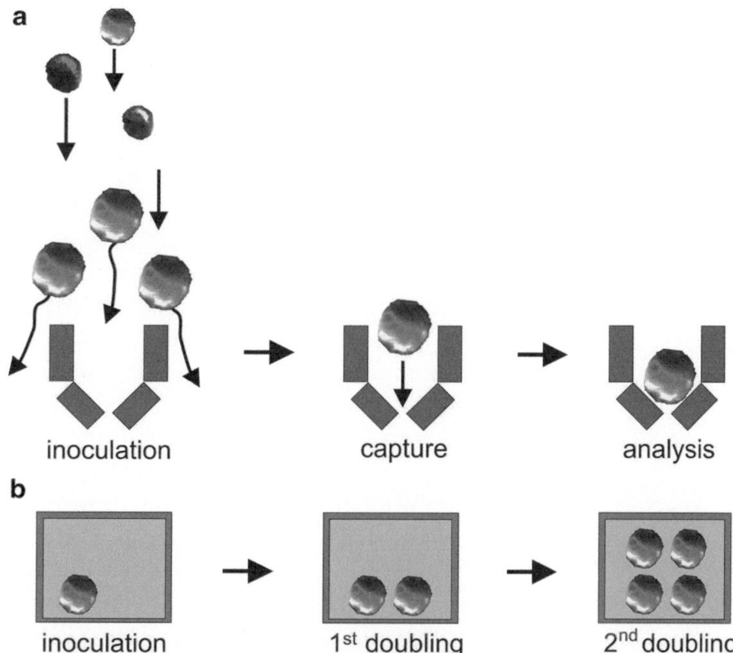

Fig. 6 Microfluidic cell microarray technologies for single-cell analysis. Schematic representation of single-cell microfluidic devices for (**a**) cell capture and analysis, and (**b**) proliferation studies at the single-cell level

Similarly microfluidic single-cell microfluidic microarrays with integrated cell-capture pockets have been applied for analysis of signaling dynamics of hematopoietic stem cells as well as their cell division, time-resolved viability, and cell migration and motility analysis [67]. Another approach uses micro-arrayed 4.1 nL nano-pockets (*see* Fig. 6b) for investigation of rare hematopoietic stem cells, where proliferation studies of single cells have been conducted [68]. Based on these technological advances, single-cell microfluidic approaches can potentially be applied for bacterial pathogenesis research. For instance, to investigate how antibiotic resistance can arise in bacteria, the microfluidic technology has been employed that provided evidence on the establishment of resistant populations from one single bacterium [69]. In another microfluidic approach, the impact of paracrine signaling on clotting capabilities of blood was investigated using single perfused pockets inside microchannels [70].

Alternative approaches for single-cell immuno-typing are based on single-cell nanowell arrays. In one application, T cells were captured by gravity sedimentation within the nanowells and subsequently stimulated. Cell analysis was accomplished using ELISA and immunofluorescence staining to provide biological information with single-cell resolution [71, 72]. Similarly, microarrayed nanowells were applied for on-chip analysis of the secretome of $CD4^+$ T cells [73]. In another approach implemented by electrophysiological sensor recordings on single cells with high spatiotemporal resolution, the neuronal network activity was investigated [74]. Moreover, single live-cell microarrays in combination with an array of cantilevers have been established to measure the mass of cancer cells, thus revealing information on individual cells under different physiological conditions in a noninvasive manner [75].

3.2 Microfluidic 3D Cell Microarrays

While microfluidic single-cell assays provide information on heterogeneity within a cell population, microfluidic 3D cell microarrays are used to further investigate cell-to-cell interaction and communication of heterotypic cultures including the impact of cellular secretome of individual cells on bulk cell cultures. Additional advantages of employing microfluidic 3D cell microarrays for cell culture applications include the improved in vivo-like microenvironment where cells are surrounded by ECM, are in direct contact with each other (either in a homotypic or heterotypic way) and are in combination with controlled nutrient supply and waste removal. Various studies have shown that 3D culture techniques based on aggregates, spheroids and tissue scaffolds or hydrogels help cells to retain their natural functionality. For instance, it has been shown that hepatocytes can exclusively maintain their physiologically relevant phenotype within a 3D context [76]. Additionally, mesenchymal stem cells (MSC)-derived hepatocytes exhibited key functions including urea synthesis and metabolite clearance only when cultured within a three-dimensional cellular microenvironment [77]. Moreover, the 3D spheroid systems have been applied extensively for immune-activation, as well as for engineering of native tissues including cartilage, lung, liver, kidney, gut, bone, brain, pancreatic, and cardiac organoids in vitro [78–86]. Furthermore, various spheroid cultures have been presented as viable and novel tools to establish vascular structures and to investigate network formation in vascularization research [87, 88]. More recently, there are developments that incorporate microstructures in microfabricated systems to increase the variety of functional spheroid geometries (*see* Fig. 7), and these microstructures include stripes, triangles, and star-shaped objects [89, 90]. Based on these advances, the microfluidic 3D cell microarrays represent a valuable tool for high-throughput screening applications such as improved drug screening and tissue engineering.

Similar to the microfluidic systems containing integrated pockets, on-chip U-shaped microstructure arrays (*see* Fig. 8a) have been employed as an effective method for the generation of multicellular spheroids (MCS) [91]. It is demonstrated that in situ fabrication can replace an expensive cleanroom setup for creating the PEG-based microstructures (pockets) within microchannels. The epithelial HepG2 tumor cell spheroids do respond to doxorubicin

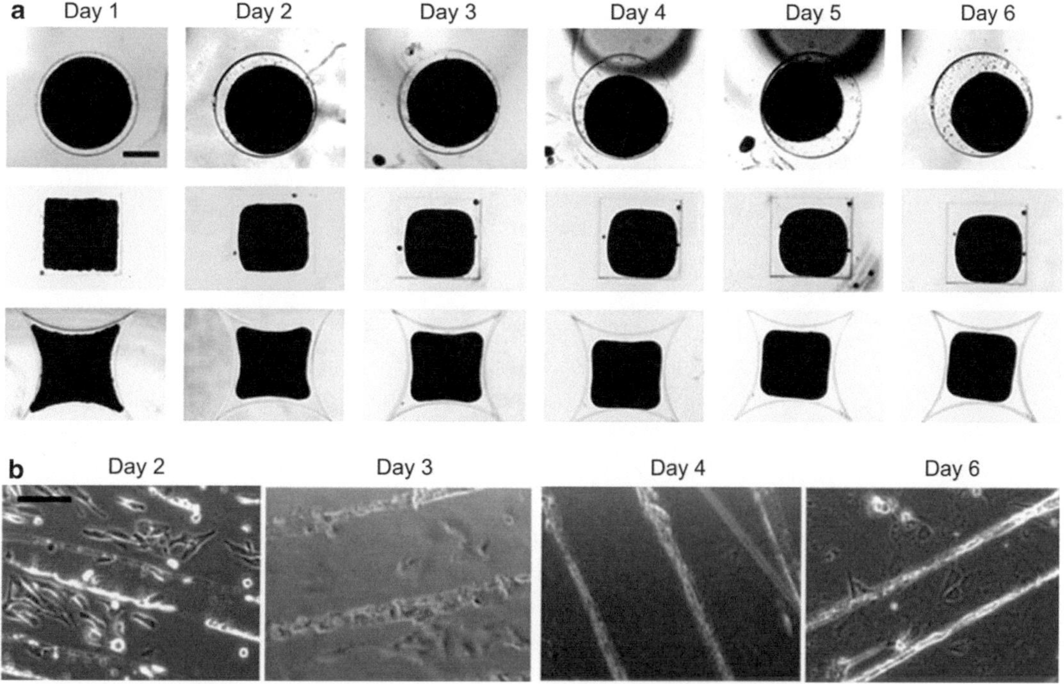

Fig. 7 Organoid geometries on chip. (**a**) Millimeter-scaled 3D biomaterial-free tissue constructs with the *star-like*, *round*, and *square* geometry. Scale bar, 500 μm. (**b**) Formation of cardiac organoids with the *stripe* geometry. Scale bar, 100 μm

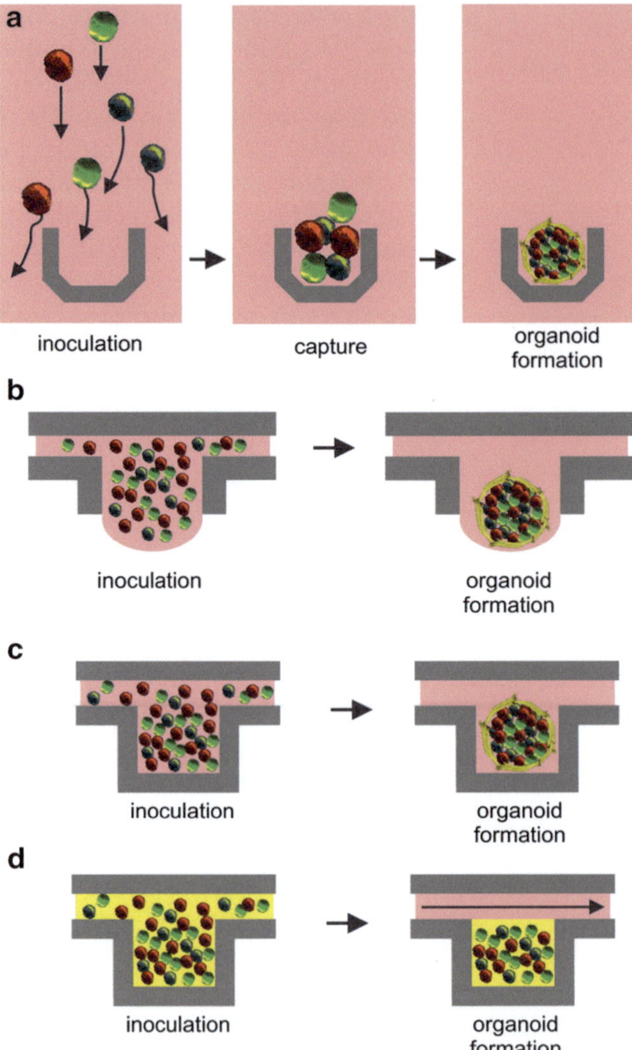

Fig. 8 Microfluidic approaches to establish 3D cell cultures. (**a**) Microchannel with integrated U-shaped pockets for capture of cell populations and consecutive organoid formation. (**b**) Microfluidic devices based on formation of organoids using hanging drop techniques. (**c**) Microfluidic chamber arrays with cell capture pockets for organoid formation within microchannels. (**d**) Microfluidic devices based on cell-laden hydrogels for 3D organoid construct formation

treatment, and this response could effectively demonstrate a 3D liver tissue construct is superior to the conventional 2D cultures. In a similar study, an increased chemotherapeutic resistance has been reported for spheroids generated from cells obtained from patients with terminal epithelial ovarian carcinoma, which was related to an enhanced expression of kallikrein-related peptidases in the spheroid cell culture [92]. Approaches based on the well-known hanging drop technique have also been developed for microfluidic live-cell microarray (*see* Fig. 8b). For instance, parallel formation of spheroids of different cell types was achieved on the hanging drop for consecutive in-line bioactivation and pharmaceutical compound evaluation assays [93]. PDMS-silicon hybrid devices containing integrated pyramid-like micro-cavity arrays were used for short-term MCF-7 breast cancer and long-term HepG2 liver spheroid culture analysis [94]. Using this microfluidic 3D cell microarray, cell viability, albumin secretion, and respiratory activity were recorded in a high-throughput manner. Another study reported a three-layer PDMS/PC membrane microfluidic system featuring integrated cell capture chambers (*see* Fig. 8c) capable of forming prostate cancer co-culture spheroids to recapitulate the growth behavior of PC-3 cancer cells within a bone metastatic prostate cancer microenvironment [95]. Results of study showed that spheroid culture of $CD133^{+}$ PC-3 cells remained in the quiescent state and as an undifferentiated phenotype, thus preserving the relevant surface markers of cancer stem cells (CSCs). These cells are believed to play a major role in metastasis and may become a promising avenue for anticancer therapy. Other microfluidic devices as shown in Fig. 8d have utilized 3D cocultures that were embedded within a basement membrane hydrogel, rather than organoid structures [96]. Such systems have been used for chemotherapeutic drug testing using a three-dimensional hydrogel-based MCF-7 breast cancer spheroid model [97].

4 Conclusion and Future Prospects

This chapter reviews live-cell microarrays as a versatile platform for high-throughput cell analysis, where cells are exposed to a range of stimuli in a highly parallelized manner. The ability to obtain a large amount of data from a single experiment using live-cell microarray technology represents an ideal approach to gain deeper insights into cellular phenotypes, which is of special relevance in the context of system biology, disease modeling and personalized medicine. In general, two major fields of applications can be defined, which can be associated with (a) screening of small substance (chemical) and genomic libraries and (b) evaluation of cell–microenvironment interactions. Consequently, live-cell microarrays have proven to be useful for a wide range of biomedical applications including investigation of cell signaling in healthy and diseased systems, cell–cell and cell–matrix interaction studies, drug screening, cell sorting, and cell phenotype characterization.

In the light of an increasing demand of robust, reliable, and reproducible cytotoxicity screening assays for disease modeling, pharmaceutical compound testing and cell-based therapies, cell microarrays are expected to play a major role in future biomedical science. The necessity for advanced in vitro cell analysis systems has therefore provided the opportunity to develop automated cell culture systems capable of monitoring single cells, multi cell populations and spheroids. Here, the combination of microfluidics with microarray technology allows for further miniaturization, automation and large volume testing even using complex biological systems. For instance, microfluidic 3D cell microarrays containing live tissue analogues can be envisioned for high-throughput drug screening with in vivo relevance. This latest trend of combining microarrays, microfluidics and 3D cell culture technology includes the reliable establishment of multi-organs-on-a-chip and human-on-a-chip systems that mimics the complex interplay of multiple organs in one single device. One prominent multi-layer multi-organ-chip system for long-term cultivation of liver and skin organoids has been applied for long-term monitoring of cellular metabolic activity including glucose consumption, lactate dehydrogenase (LDH) and lactate production in the presence of a direct long-term exposure to fluid flow [98]. In addition, the impact of troglitazole (an antidiabetic drug) on the metabolic activity was investigated over a 6-day exposure period. However, practical application of organ-on-a-chip technology for clinical testing requires to address the limitations of various components associated with systems integration such as micropumps, microheaters, microdegassers, and microsensor arrays, automation, and miniaturization. Figure 9 highlights key aspects of the need to further develop microfluidic cell microarray technology for high-throughput and high-content testing of living cell cultures.

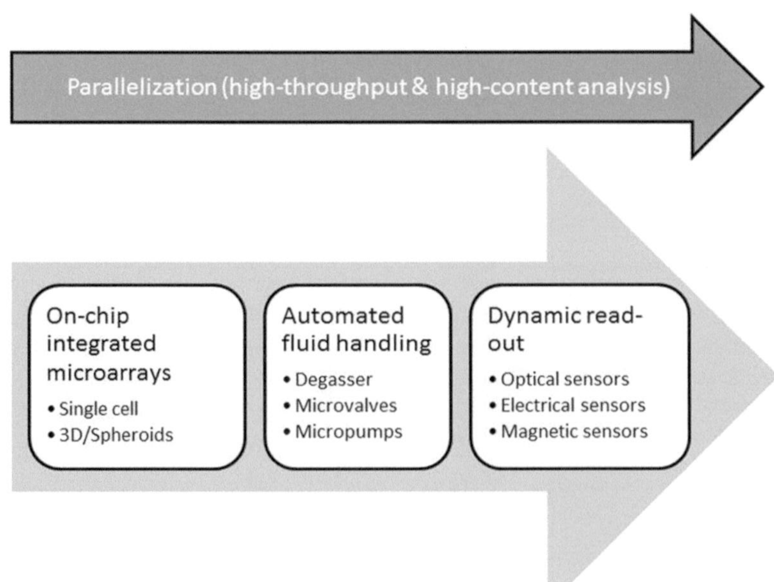

Fig. 9 Requirements for live-cell microarray technology with respect to high-content and high-throughput screening

References

1. Fernandes TG, Diogo MM, Clark DS, Dordick JS, Cabral JMS (2009) High-throughput cellular microarray platforms: applications in drug discovery, toxicology and stem cell research. Trends Biotechnol 27:342–349
2. Angres B (2005) Cell microarrays. Expert Rev Mol Diagn 5:769–779
3. Ziauddin J, Sabatini DM (2001) Microarrays of cells expressing defined cDNAs. Nature 411:107–110
4. West J, Becker M, Tombrink S, Manz A (2008) Micro total analysis systems: latest achievements. Anal Chem 80:4403–4419
5. Dittrich PS, Manz A (2006) Lab-on-a-chip: microfluidics in drug discovery. Nat Rev Drug Discov 5:210–218
6. Richter L, Charwat V, Jungreuthmayer C, Bellutti F, Brueckl H, Ertl P (2011) Monitoring cellular stress responses to nanoparticles using a lab-on-a-chip. Lab Chip 11:2551–2560
7. Charwat V, Rothbauer M, Tedde SF, Hayden O, Bosch JJ, Muellner P, Hainberger R, Ertl P (2013) Monitoring dynamic interactions of tumor cells with tissue and immune cells in a lab-on-a-chip. Anal Chem 85:11471–11478
8. Novak R, Wartmann D, Mathies RA, Dostálek J, Ertl P (2015) Microfluidic platform for multiplexed cell sampling and time-resolved spr-based cytokine sensing. In: Lacković I, Vasic D (eds.) 6th European conference of the international federation for medical and biological engineering—MBEC 2014, 7–11 September 2014, Dubrovnik, Croatia, Springer. pp. 785–788
9. Ertl P, Sticker D, Charwat V, Kasper C, Lepperdinger G (2014) Lab-on-a-chip technologies for stem cell analysis. Trends Biotechnol 32:245–253
10. Ito Y, Nogawa M, Takeda M, Shibuya T (2005) Photo-reactive polyvinylalcohol for photo-immobilized microarray. Biomaterials 26:211–216
11. Falsey JR, Renil M, Park S, Li S, Lam KS (2001) Peptide and small molecule microarray for high throughput cell adhesion and functional assays. Bioconjug Chem 12:346–353
12. Xu QC, Miyamoto S, Lam KS (2004) A novel approach to chemical microarray using ketone-modified macromolecular scaffolds: application in micro cell-adhesion assay. Mol Divers 8:301–310
13. Soen Y, Chen DS, Kraft DL, Davis MM, Brown PO (2003) Detection and characterization of cellular immune responses using peptide-MHC microarrays. PLoS Biol 1:E65
14. Stone JD, Demkowicz WE Jr, Stern LJ (2005) HLA-restricted epitope identification and detection of functional T cell responses by using MHC-peptide and costimulatory microarrays. Proc Natl Acad Sci U S A 102:3744–3749
15. Nimrichter L, Gargir A, Gortler M, Altstock RT, Shtevi A, Weisshaus O, Fire E, Dotan N, Schnaar RL (2004) Intact cell adhesion to glycan microarrays. Glycobiology 14:197–203
16. Hovis JS, Boxer SG (2001) Patterning and composition arrays of supported lipid bilayers by microcontact printing. Langmuir 17:3400–3405
17. Groves JT, Mahal LK, Bertozzi CR (2001) Control of cell adhesion and growth with micropatterned supported lipid membranes. Langmuir 17:5129–5133
18. Anderson DG, Peng W, Akinc A, Hossain N, Kohn A, Padera R, Langer R, Sawicki JA (2004) A polymer library approach to suicide gene therapy for cancer. Proc Natl Acad Sci U S A 101:16028–16033
19. Yang J, Mei Y, Hook AL, Taylor M, Urquhart AJ, Bogatyrev SR, Langer R, Anderson DG, Davies MC, Alexander MR (2010) Polymer surface functionalities that control human embryoid body cell adhesion revealed by high throughput surface characterization of combinatorial material microarrays. Biomaterials 31:8827–8838
20. Anderson DG, Levenberg S, Langer R (2004) Nanoliter-scale synthesis of arrayed biomaterials and application to human embryonic stem cells. Nat Biotechnol 22:863–866
21. Flaim CJ, Chien S, Bhatia SN (2005) An extracellular matrix microarray for probing cellular differentiation. Nat Methods 2:119–125
22. Ankam S, Teo BKK, Kukumberg M, Yim EKF (2013) High throughput screening to investigate the interaction of stem cells with their extracellular microenvironment. Organogenesis 9:128–142
23. Mei Y, Saha K, Bogatyrev SR, Yang J, Hook AL, Kalcioglu ZI, Cho SW, Mitalipova M, Pyzocha N, Rojas F, Van Vliet KJ, Davies MC, Alexander MR, Langer R, Jaenisch R, Anderson DG (2010) Combinatorial development of biomaterials for clonal growth of human pluripotent stem cells. Nat Mater 9:768–778
24. Zhang R, Mjoseng HK, Hoeve MA, Bauer NG, Pells S, Besseling R, Velugotla S, Tourniaire G, Kishen REB, Tsenkina Y, Armit C, Duffy CRE, Helfen M, Edenhofer F, de Sousa PA, Bradley M (2013) A thermoresponsive and chemically defined hydrogel for long-

term culture of human embryonic stem cells. Nat Commun 4(1335):1–10

25. Cheng XH, Wang YB, Hanein Y, Bohringer KF, Ratner BD (2004) Novel cell patterning using microheater-controlled thermoresponsive plasma films. J Biomed Mater Res A 70A:159–168
26. Hook AL, Chang CY, Yang J, Atkinson S, Langer R, Anderson DG, Davies MC, Williams P, Alexander MR (2013) Discovery of novel materials with broad resistance to bacterial attachment using combinatorial polymer microarrays. Adv Mater 25:2542–2547
27. Anglin E, Davey R, Herrid M, Hope S, Kurkuri M, Pasic P, Hor M, Fenech M, Thissen H, Voelcker NH (2010) Cell microarrays for the screening of factors that allow the enrichment of bovine testicular cells. Cytometry Part A: the journal of the International Society for Analytical Cytology 77:881–889
28. Pernagallo S, Wu M, Gallagher MP, Bradley M (2011) Colonising new frontiers-microarrays reveal biofilm modulating polymers. J Mater Chem 21:96–101
29. Unadkat HV, Hulsman M, Cornelissen K, Papenburg BJ, Truckenmuller RK, Carpenter AE, Wessling M, Post GF, Uetz M, Reinders MJ, Stamatialis D, van Blitterswijk CA, de Boer J (2011) An algorithm-based topographical biomaterials library to instruct cell fate. Proc Natl Acad Sci U S A 108:16565–16570
30. Moe AA, Suryana M, Marcy G, Lim SK, Ankam S, Goh JZ, Jin J, Teo BK, Law JB, Low HY, Goh EL, Sheetz MP, Yim EK (2012) Microarray with micro- and nano-topographies enables identification of the optimal topography for directing the differentiation of primary murine neural progenitor cells. Small 8:3050–3061
31. Ohnaga T, Shimada Y, Moriyama M, Kishi H, Obata T, Takata K, Okumura T, Nagata T, Muraguchi A, Tsukada K (2013) Polymeric microfluidic devices exhibiting sufficient capture of cancer cell line for isolation of circulating tumor cells. Biomed Microdevices 15:611–616
32. Li N, Ho CM (2008) Photolithographic patterning of organosilane monolayer for generating large area two-dimensional B lymphocyte arrays. Lab Chip 8:2105–2112
33. Ellmark P, Hogerkorp CM, Ek S, Belov L, Berglund M, Rosenquist R, Christopherson RI, Borrebaeck CAK (2008) Phenotypic protein profiling of different B cell sub-populations using antibody CD-microarrays. Cancer Lett 265:98–106
34. Sekine K, Revzin A, Tompkins RG, Toner M (2006) Panning of multiple subsets of leukocytes on antibody-decorated poly(ethylene) glycol-coated glass slides. J Immunol Methods 313:96–109
35. Wu JQ, Wang B, Belov L, Chrisp J, Learmont J, Dyer WB, Zaunders J, Cunningham AL, Dwyer DE, Saksena NK (2007) Antibody microarray analysis of cell surface antigens on CD4+ and CD8+ T cells from HIV+ individuals correlates with disease stages. Retrovirology 4:83
36. Gao J, Liu C, Liu D, Wang Z, Dong S (2010) Antibody microarray-based strategies for detection of bacteria by lectin-conjugated gold nanoparticle probes. Talanta 81:1816–1820
37. Marimon JM, Monasterio A, Ercibengoa M, Pascual J, Prieto I, Simon L, Perez-Trallero E (2010) Antibody microarray typing, a novel technique for Streptococcus pneumoniae serotyping. J Microbiol Methods 80:274–280
38. Niu W, Narayanaswamy R, Scouras A, Hart GT, Davies J, Ellington AD, Iyer VR, Marcotte EM (2006) Systematic profiling of cellular phenotypes with spotted cell microarrays reveals new mating pheromone response genes. Faseb J 20:A928–A928
39. Bochner BR, Gadzinski P, Panomitros E (2001) Phenotype MicroArrays for high-throughput phenotypic testing and assay of gene function. Genome Res 11:1246–1255
40. Schwenk JM, Stoll D, Templin MF, Joos TO (2002) Cell microarrays: an emerging technology for the characterization of antibodies. Biotechniques 33:S54–S61
41. Chin VI, Taupin P, Sanga S, Scheel J, Gage FH, Bhatia SN (2004) Microfabricated platform for studying stem cell fates. Biotechnol Bioeng 88:399–415
42. Woodruff K, Fidalgo LM, Gobaa S, Lutolf MP, Maerkl SJ (2013) Live mammalian cell arrays. Nat Methods 10:550–552
43. Hong BJ, Sunkara V, Park JW (2005) DNA microarrays on nanoscale-controlled surface. Nucleic Acids Res 33:e106
44. Sunkara V, Hong BJ, Park JW (2007) Sensitivity enhancement of DNA microarray on nanoscale controlled surface by using a streptavidin-fluorophore conjugate. Biosens Bioelectron 22:1532–1537
45. Wang WJ, Itaka K, Ohba S, Nishiyama N, Chung UI, Yamasaki Y, Kataoka K (2009) 3D spheroid culture system on micropatterned substrates for improved differentiation efficiency of multipotent mesenchymal stem cells. Biomaterials 30:2705–2715
46. Berdondini L, Chippalone M, van der Wal PD, Imfeld K, de Rooij NF, Koudelka-Hep M, Tedesco M, Martinoia S, van Pelt J, Le Masson G, Garenne A (2006) A microelectrode array

(MEA) integrated with clustering structures for investigating in vitro neurodynamics in confined interconnected sub-populations of neurons. Sensor Actuat B-Chem 114:530–541

47. Maccione A, Gandolfo M, Tedesco M, Nieus T, Imfeld K, Martinoia S, Berdondini L (2010) Experimental investigation on spontaneously active hippocampal cultures recorded by means of high-density MEAs: analysis of the spatial resolution effects. Frontiers in Neuroengineering 3(4):1–12
48. Maccione A, Garofalo M, Nieus T, Tedesco M, Berdondini L, Martinoia S (2012) Multiscale functional connectivity estimation on low-density neuronal cultures recorded by high-density CMOS Micro Electrode Arrays. J Neurosci Methods 207:161–171
49. Marconi E, Nieus T, Maccione A, Valente P, Simi A, Messa M, Dante S, Baldelli P, Berdondini L, Benfenati F (2012) Emergent functional properties of neuronal networks with controlled topology. PLoS One 7:e34648
50. Bakkum DJ, Frey U, Radivojevic M, Russell TL, Muller J, Fiscella M, Takahashi H, Hierlemann A (2013) Tracking axonal action potential propagation on a high-density microelectrode array across hundreds of sites. Nat Commun 4(2181):1–12
51. Xia YN, Whitesides GM (1998) Soft lithography. Angew Chem Int Ed 37:551–575
52. Khademhosseini A, Yeh J, Jon S, Eng G, Suh KY, Burdick JA, Langer R (2004) Molded polyethylene glycol microstructures for capturing cells within microfluidic channels. Lab Chip 4:425–430
53. Chiu DT, Jeon NL, Huang S, Kane RS, Wargo CJ, Choi IS, Ingber DE, Whitesides GM (2000) Patterned deposition of cells and proteins onto surfaces by using three-dimensional microfluidic systems. Proc Natl Acad Sci U S A 97:2408–2413
54. Khademhosseini A, Suh KY, Jon S, Eng G, Yeh J, Chen GJ, Langer R (2004) A soft lithographic approach to fabricate patterned microfluidic channels. Anal Chem 76:3675–3681
55. Whitesides GM, Ostuni E, Takayama S, Jiang X, Ingber DE (2001) Soft lithography in biology and biochemistry. Annu Rev Biomed Eng 3:335–373
56. Narsinh KH, Sun N, Sanchez-Freire V, Lee AS, Almeida P, Hu SJ, Jan T, Wilson KD, Leong D, Rosenberg J, Yao M, Robbins RC, Wu JC (2011) Single cell transcriptional profiling reveals heterogeneity of human induced pluripotent stem cells. J Clin Invest 121:1217–1221
57. Kang JH, Krause S, Tobin H, Mammoto A, Kanapathipillai M, Ingber DE (2012) A combined micromagnetic-microfluidic device for rapid capture and culture of rare circulating tumor cells. Lab Chip 12:2175–2181
58. Nagrath S, Sequist LV, Maheswaran S, Bell DW, Irimia D, Ulkus L, Smith MR, Kwak EL, Digumarthy S, Muzikansky A, Ryan P, Balis UJ, Tompkins RG, Haber DA, Toner M (2007) Isolation of rare circulating tumour cells in cancer patients by microchip technology. Nature 450:1235–1239
59. Karabacak NM, Spuhler PS, Fachin F, Lim EJ, Pai V, Ozkumur E, Martel JM, Kojic N, Smith K, Chen PI, Yang J, Hwang H, Morgan B, Trautwein J, Barber TA, Stott SL, Maheswaran S, Kapur R, Haber DA, Toner M (2014) Microfluidic, marker-free isolation of circulating tumor cells from blood samples. Nat Protoc 9:694–710
60. Hur SC, Mach AJ, Di Carlo D (2011) High-throughput size-based rare cell enrichment using microscale vortices. Biomicrofluidics 5(022206):1–10
61. Mach AJ, Kim JH, Arshi A, Hur SC, Di Carlo D (2011) Automated cellular sample preparation using a Centrifuge-on-a-Chip. Lab Chip 11:2827–2834
62. Dalerba P, Kalisky T, Sahoo D, Rajendran PS, Rothenberg ME, Leyrat AA, Sim S, Okamoto J, Johnston DM, Qian DL, Zabala M, Bueno J, Neff NF, Wang JB, Shelton AA, Visser B, Hisamori S, Shimono Y, van de Wetering M, Clevers H, Clarke MF, Quake SR (2011) Single-cell dissection of transcriptional heterogeneity in human colon tumors. Nat Biotechnol 29:1120–1127
63. Gossett DR, Tse HTK, Lee SA, Ying Y, Lindgren AG, Yang OO, Rao JY, Clark AT, Di Carlo D (2012) Hydrodynamic stretching of single cells for large population mechanical phenotyping. Proc Natl Acad Sci U S A 109:7630–7635
64. Tse HTK, Gossett DR, Moon YS, Masaeli M, Sohsman M, Ying Y, Mislick K, Adams RP, Rao JY, Di Carlo D (2013) Quantitative diagnosis of malignant pleural effusions by single-cell mechanophenotyping. Sci Transl Med 5(212):212ra163, 1–9
65. Wlodkowic D, Faley S, Zagnoni M, Wikswo JP, Cooper JM (2009) Microfluidic single-cell array cytometry for the analysis of tumor apoptosis. Anal Chem 81:5517–5523
66. Powell AA, Talasaz AH, Zhang H, Coram MA, Reddy A, Deng G, Telli ML, Advani RH, Carlson RW, Mollick JA, Sheth S, Kurian AW, Ford JM, Stockdale FE, Quake SR, Pease RF, Mindrinos MN, Bhanot G, Dairkee SH, Davis RW, Jeffrey SS (2012) Single cell profiling of circulating tumor cells: transcriptional hetero-

geneity and diversity from breast cancer cell lines. PLoS One 7:e33788

67. Faley SL, Copland M, Wlodkowic D, Kolch W, Seale KT, Wikswo JP, Cooper JM (2009) Microfluidic single cell arrays to interrogate signalling dynamics of individual, patient-derived hematopoietic stem cells. Lab Chip 9:2659–2664
68. Lecault V, VanInsberghe M, Sekulovic S, Knapp DJHF, Wohrer S, Bowden W, Viel F, McLaughlin T, Jarandehei A, Miller M, Falconnet D, White AK, Kent DG, Copley MR, Taghipour F, Eaves CJ, Humphries RK, Piret JM, Hansen CL (2011) High-throughput analysis of single hematopoietic stem cell proliferation in microfluidic cell culture arrays. Nat Methods 8:581–586
69. Boedicker JQ, Li L, Kline TR, Ismagilov RF (2008) Detecting bacteria and determining their susceptibility to antibiotics by stochastic confinement in nanoliter droplets using plug-based microfluidics. Lab Chip 8:1265–1272
70. Kastrup CJ, Boedicker JQ, Pomerantsev AP, Moayeri M, Bian Y, Pompano RR, Kline TR, Sylvestre P, Shen F, Leppla SH, Tang WJ, Ismagilov RF (2008) Spatial localization of bacteria controls coagulation of human blood by 'quorum acting'. Nat Chem Biol 4:742–750
71. Han Q, Bagheri N, Bradshaw EM, Hafler DA, Lauffenburger DA, Love JC (2012) Polyfunctional responses by human T cells result from sequential release of cytokines. Proc Natl Acad Sci U S A 109:1607–1612
72. Yamanaka YJ, Szeto GL, Gierahn TM, Forcier TL, Benedict KF, Brefo MSN, Lauffenburger DA, Irvine DJ, Love JC (2012) Cellular barcodes for efficiently profiling single-cell secretory responses by microengraving. Anal Chem 84:10531–10536
73. Jin A, Ozawa T, Tajiri K, Obata T, Kondo S, Kinoshita K, Kadowaki S, Takahashi K, Sugiyama T, Kishi H, Muraguchi A (2009) A rapid and efficient single-cell manipulation method for screening antigen-specific antibody-secreting cells from human peripheral blood. Nat Med 15:1088–1092
74. Berdondini L, Imfeld K, Maccione A, Tedesco M, Neukom S, Koudelka-Hep M, Martinoia S (2009) Active pixel sensor array for high spatio-temporal resolution electrophysiological recordings from single cell to large scale neuronal networks. Lab Chip 9:2644–2651
75. Park K, Jang J, Irimia D, Sturgis J, Lee J, Robinson JP, Toner M, Bashir R (2008) 'Living cantilever arrays' for characterization of mass of single live cells in fluids. Lab Chip 8: 1034–1041
76. Bierwolf J, Lutgehetmann M, Feng K, Erbes J, Deichmann S, Toronyi E, Stieglitz C, Nashan B, Ma PX, Pollok JM (2011) Primary rat hepatocyte culture on 3D nanofibrous polymer scaffolds for toxicology and pharmaceutical research. Biotechnol Bioeng 108:141–150
77. Li J, Tao R, Wu W, Cao HC, Xin JJ, Li J, Guo J, Jiang LY, Gao CY, Demetriou AA, Farkas DL, Li LJ (2010) 3D PLGA scaffolds improve differentiation and function of bone marrow mesenchymal stem cell-derived hepatocytes. Stem Cells Dev 19:1427–1436
78. Salmenpera P, Kankuri E, Bizik J, Siren V, Virtanen I, Takahashi S, Leiss M, Fassler R, Vaheri A (2008) Formation and activation of fibroblast spheroids depend on fibronectin-integrin interaction. Exp Cell Res 314:3444–3452
79. Bao BA, Lai CP, Naus CC, Morgan JR (2012) Pannexin1 drives multicellular aggregate compaction via a signaling cascade that remodels the actin cytoskeleton. J Biol Chem 287:8407–8416
80. Jukes JM, Both SK, Leusink A, Sterk LMT, Van Blitterswijk CA, De Boer J (2008) Endochondral bone tissue engineering using embryonic stem cells. Proc Natl Acad Sci U S A 105:6840–6845
81. Lumelsky N, Blondel O, Laeng P, Velasco I, Ravin R, McKay R (2001) Differentiation of embryonic stem cells to insulin-secreting structures similar to pancreatic islets. Science 292:1389–1394
82. Kehat I, Kenyagin-Karsenti D, Snir M, Segev H, Amit M, Gepstein A, Livne E, Binah O, Itskovitz-Eldor J, Gepstein L (2001) Human embryonic stem cells can differentiate into myocytes with structural and functional properties of cardiomyocytes. J Clin Invest 108:407–414
83. Carraro A, Hsu WM, Kulig KM, Cheung WS, Miller ML, Weinberg EJ, Swart EF, Kaazempur-Mofrad M, Borenstein JT, Vacanti JP, Neville C (2008) In vitro analysis of a hepatic device with intrinsic microvascular-based channels. Biomed Microdevices 10:795–805
84. Jang KJ, Cho HS, Kang do H, Bae WG, Kwon TH, Suy KY (2011) Fluid-shear-stress-induced translocation of aquaporin-2 and reorganization of actin cytoskeleton in renal tubular epithelial cells. Integrative biology : quantitative biosciences from nano to macro 3:134–141
85. Kimura H, Yamamoto T, Sakai H, Sakai Y, Fujii T (2008) An integrated microfluidic system for long-term perfusion culture and on-line monitoring of intestinal tissue models. Lab Chip 8:741–746

86. Park J, Koito H, Li J, Han A (2009) Microfluidic compartmentalized co-culture platform for CNS axon myelination research. Biomed Microdevices 11:1145–1153
87. Gu A, Shively JE (2011) Angiopoietins-1 and -2 play opposing roles in endothelial sprouting of embryoid bodies in 3D culture and their receptor Tie-2 associates with the cell-cell adhesion molecule PECAM1. Exp Cell Res 317:2171–2182
88. Rouwkema J, de Boer J, Van Blitterswijk CA (2006) Endothelial cells assemble into a 3-dimensional prevascular network in a bone tissue engineering construct. Tissue Eng 12:2685–2693
89. Rivron NC, Vrij EJ, Rouwkema J, Le Gac S, van den Berg A, Truckenmuller RK, van Blitterswijk CA (2012) Tissue deformation spatially modulates VEGF signaling and angiogenesis. Proc Natl Acad Sci U S A 109:6886–6891
90. Khademhosseini A, Eng G, Yeh J, Kucharczyk PA, Langer R, Vunjak-Novakovic G, Radisic M (2007) Microfluidic patterning for fabrication of contractile cardiac organoids. Biomed Microdevices 9:149–157
91. Fu CY, Tseng SY, Yang SM, Hsu L, Liu CH, Chang HY (2014) A microfluidic chip with a U-shaped microstructure array for multicellular spheroid formation, culturing and analysis. Biofabrication 6(1):015009. doi:10.1088/1758-5082/6/1/015009
92. Dong Y, Tan OL, Loessner D, Stephens C, Walpole C, Boyle GM, Parsons PG, Clements JA (2010) Kallikrein-related peptidase 7 promotes multicellular aggregation via the alpha(5) beta(1) integrin pathway and paclitaxel chemoresistance in serous epithelial ovarian carcinoma. Cancer Res 70:2624–2633
93. Frey O, Misun PM, Fluri DA, Hengstler JG, Hierlemann A (2014) Reconfigurable microfluidic hanging drop network for multi-tissue interaction and analysis. Nat Commun 5:4250. doi:10.1038/ncomms5250
94. Torisawa YS, Takagi A, Nashimoto Y, Yasukawa T, Shiku H, Matsue T (2007) A multicellular spheroid array to and viability realize spheroid formation, culture, assay on a chip. Biomaterials 28:559–566
95. Hsiao AY, Torisawa YS, Tung YC, Sud S, Taichman RS, Pienta KJ, Takayama S (2009) Microfluidic system for formation of PC-3 prostate cancer co-culture spheroids. Biomaterials 30:3020–3027
96. Xu Z, Gao Y, Hao Y, Li E, Wang Y, Zhang J, Wang W, Gao Z, Wang Q (2013) Application of a microfluidic chip-based 3D co-culture to test drug sensitivity for individualized treatment of lung cancer. Biomaterials 34:4109–4117
97. Torisawa YS, Shiku H, Yasukawa T, Nishizawa M, Matsue T (2005) Multi-channel 3-D cell culture device integrated on a silicon chip for anticancer drug sensitivity test. Biomaterials 26:2165–2172
98. Wagner I, Materne EM, Brincker S, Sussbier U, Fradrich C, Busek M, Sonntag F, Sakharov DA, Trushkin EV, Tonevitsky AG, Lauster R, Marx U (2013) A dynamic multi-organ-chip for long-term cultivation and substance testing proven by 3D human liver and skin tissue coculture. Lab Chip 13:3538–3547

Index

Paul C.H. Li et al. (eds.), *Microarray Technology: Methods and Applications*, Methods in Molecular Biology, vol. 1368,
DOI 10.1007/978-1-4939-3136-1,

O

P

Q

R

S

T

FSC
www.fsc.org
MIX
Papier aus verantwortungsvollen Quellen
Paper from responsible sources
FSC® C105338

If you have any concerns about our products,
you can contact us on
ProductSafety@springernature.com

In case Publisher is established outside the EU,
the EU authorized representative is:
Springer Nature Customer Service Center GmbH
Europaplatz 3, 69115 Heidelberg, Germany

Printed by Libri Plureos GmbH
in Hamburg, Germany